D0927963

Statistical Control

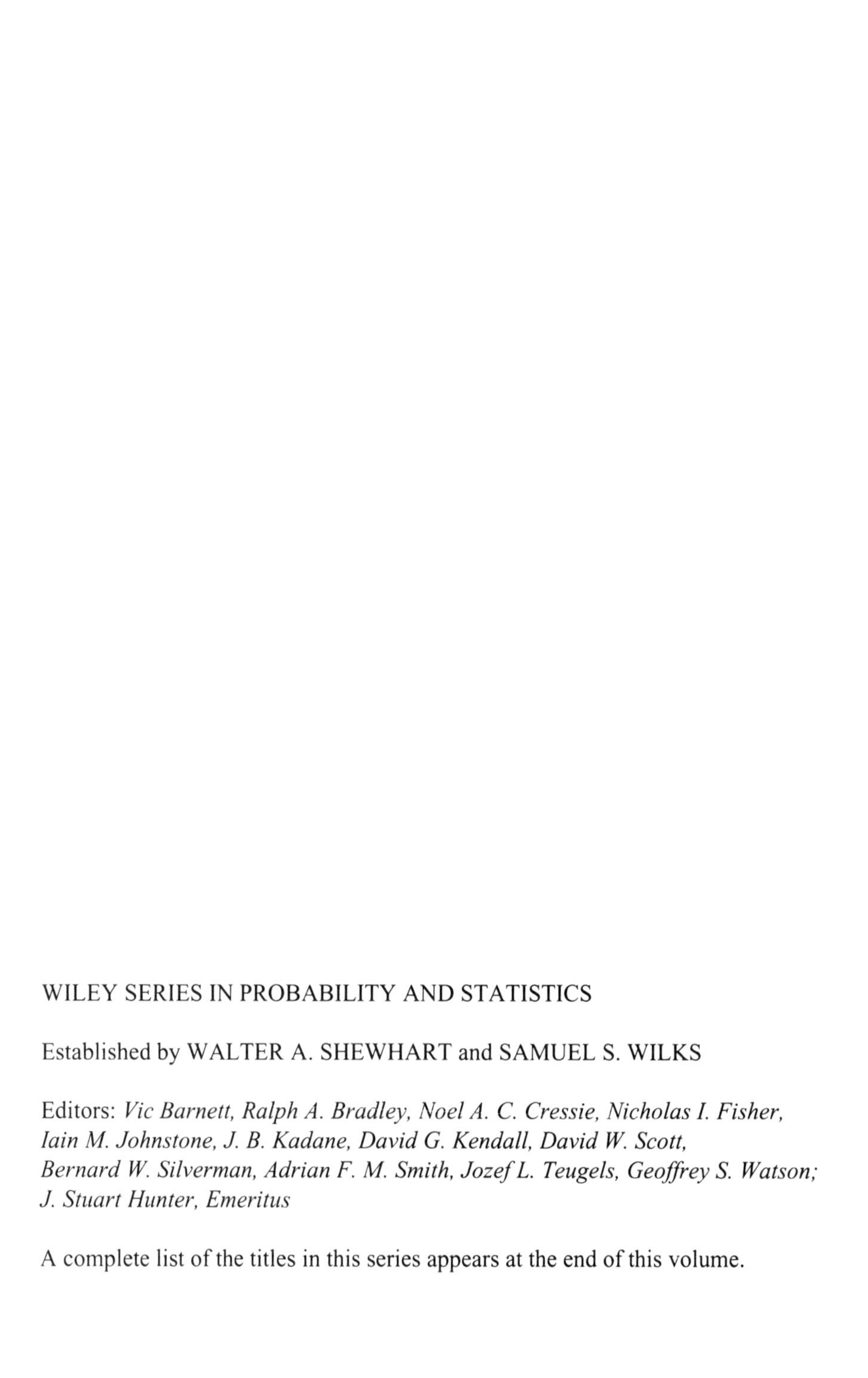

WILEY SERIES IN PROBABILITY AND STATISTICS

Established by WALTER A. SHEWHART and SAMUEL S. WILKS

A complete list of the titles in this series appears at the end of this volume.

Statistical Control

By Monitoring and Feedback Adjustment

GEORGE BOX

University of Wisconsin
Madison, Wisconsin

ALBERTO LUCEÑO

University of Cantabria
Santander, Spain

A Wiley-Interscience Publication
JOHN WILEY & SONS, INC.
New York • Chichester • Weinheim • Brisbane • Singapore • Toronto

This book is printed on acid-free paper. ♾

Published simultaneously in Canada.

Library of Congress Cataloging in Publication Data:

Box, George

Statistical control by monitoring and feedback adjustment/George Box, Alberto Luceño.

p. cm. — (Wiley series in probability and statistics. Applied probability and statistics)

"A Wiley-Interscience publication."

Includes bibliographical references and index.

ISBN 0-471-19046-2 (cloth : alk. paper)

1. Process control—Statistical methods. 2. Feedback control systems. I. Luceño, Alberto. II. Title. III. Series.

TS156.8.B678 1997

629.8′3—dc21 97-459

Printed in the United States of America

10 9 8 7 6 5 4 3 2 1

To Claire and Marian

Contents

Preface

Efficient process control is a key element in the maintenance and improvement of quality and productivity. This book is about two complementary statistical approaches to achieve such control. These are *process monitoring* and *process adjustment.*

Process monitoring, using standard quality control techniques, provides an ongoing check on the stability of the process and points to problems whose elimination can reduce variation and permanently improve the system.

Process adjustment uses feedback control to compensate those sources of drifting variation that cannot be eliminated in this way.

Clearly the two approaches are complementary and considerable advantage is to be gained by augmenting the more commonly used quality control techniques with feedback methods. In this book, in addition to discussing process monitoring, we describe techniques for feedback adjustment specifically designed for application in the statistical process control (SPC) environment.

In our discussion of feedback adjustment we first describe methods of discrete *proportional-integral* control that can be applied with adjustment charts no more difficult to use than Shewhart monitoring charts. Also we show how such schemes can be designed to avoid the need for excessively large adjustments. In the SPC environment, however, these adaptations from engineering process control (EPC) can only take us so far. They are inappropriate when substantial cost is associated with making an adjustment. To meet this need, feedback schemes of a quite different kind are introduced which minimize the overall cost of control. These are put into effect using bounded adjustment charts that indicate at which point a process adjustment should be made and the appropriate size of that adjustment. For those processes for which the cost of sampling and of testing the product is not negligible, we show how the sampling rate may be chosen to take account of this additional cost.

Later we return to a deeper consideration of the central problem of process monitoring—that of efficient detection of specific signals in appropriately modeled noise. Cuscore statistics are introduced to achieve this. In particular, these statistics clarify the different conditions which favor the use of Shewhart charts, exponentially weighted charts, and Cusum charts. Finally we consider an important question which arises in the complementary use of monitoring and adjustment, that of, how to conduct efficient monitoring at the same time that the process is being adjusted by feedback control.

Because we believe that the broadening of SPC to include feedback adjustment can result in major improvements in industrial efficiency we have made a special effort to make the contents of this book available to the widest possible audience. Little or no previous knowledge of the subject is assumed and we have introduced, as the necessity arises, just enough technical background for the understanding of the techniques under discussion. Considerable simplification has been possible because of recent research that shows the surprising degree of robustness (insensitivity to assumption) of the feedback methods we discuss. Matter that can be omitted at first reading is indicated by small print and more technical material is given in appendices. The use of calculus and operational methods is avoided (sometimes at the expense of brevity), but, for interested readers, references are given to more theoretical treatments.

Earlier drafts of some of this material were begun by one of the authors while a visiting fellow at the Institute for Advanced Study in the Behavioral Sciences at Stanford, a visit partially supported by a grant from the Guggenheim Foundation. Both authors are particularly grateful to the National Science Foundation (Grant DMI—9414765), and to the Dirección General de Investigación Científica y Técnica of Spain (Grant PB95-0583) for making possible the research on which much of the material in the later chapters is based. Also, thanks are due to The University of Cantabria, Santander, Spain, the Center For Quality and Productivity, College of Engineering University of Wisconsin, Madison, Wisconsin, and to the IBM Corporation for providing support during the writing of this book.

We are grateful to many colleagues with whom we have worked and with whom these problems have been discussed over many years, particularly to George A. Barnard, Tim Kramer, Jose Ramírez, Bovas Abraham, J. S. Hunter, J. F. McGregor, Søren Bisgaard, Daniel Peña, and Albert Pratt. For invaluable help in the preparation of the manuscript we are indebted to Judy Pagel, Deanna Knickerbocker, Virginia Heaton, Charles Romenesko, Suzanne Kestner, Murat Kulahci, and Ernesto Barrios.

Especially we wish to thank Claire Box for many years of editorial and computational support and for almost daily contributions to the realization of this book.

Madison, Wisconsin
Santander, Spain

GEORGE BOX
ALBERTO LUCEÑO

CHAPTER 1

Control in a Nonstationary World

"Now here, *you see it takes all the running you can do, to keep in the same place."*

Through the Looking Glass, LEWIS CARROLL

1.1 INTRODUCTION

Control is a continuous endeavor to keep measures of quality as nearly as possible equal to their target values for indefinite periods of time. This is not easily achieved.

If you have a house you know that you must work hard to keep it habitable—the tiles on the roof, the paint on the walls, the washing machine, the refrigerator, the television, and so forth all need attention from time to time. A car, a friendship, and our own bodies must similarly be continually nurtured or they will not remain in shape very long. The same is true of industrial processes. As Deming (1986) says, "no process, except in artificial demonstrations by use of random numbers, is steady and unwavering." If left to themselves, machines do not stay adjusted, components wear out, and managers and operators forget, miscommunicate, and change jobs. Thus a stable stationary state is an unnatural one and its achievement requires a hard and continuous fight. It is hard because it requires us to try to undo the effects of the second law of thermodynamics. This law says that, *left to itself,* the entropy (or disorganization) of any system can never decrease. While it is not possible *totally* to defeat this inexorable law by control, two different techniques that can help to tame it are *process monitoring* and *process adjustment.* Depending on circumstances, either (or, much more likely, both) will be needed.

By *process monitoring,* we mean the use of, for example, Shewhart charts, EWMA (exponentially weighted moving average) charts, and Cusum (cumulative sum) charts, which can continually check the desired stable state of the system. The use of such charts can then lead to the elimination of assignable causes pointed to by discrepant behavior. Process monitoring parallels statistical *hypothesis testing*. Its properties are described in terms of probabilities such as the probability of a point falling outside the action limits of a Shewhart chart. It is part of statistical process control (SPC).

By *process adjustment* we will mean the use of feedback control to maintain the process as close as possible to some desired target value. It employs statistical *estimation* rather than hypothesis testing and its success is measured by, for example, the output standard deviation and the *cost* of control. It is often regarded as the domain of the control engineer, and is then called engineering process control (EPC), and it is usually thought of as something that must be put into effect with automatic equipment.

Thus if you ask a statistical quality control practitioner and a control engineer what they mean when they speak of process control, you are likely to receive very different answers. On the one hand, the quality control practitioner will most likely talk about the uses of control charts for process *monitoring*. On the other hand, the control engineer will talk about such things as feedback control, proportional integral controllers, and so forth for process *adjustment*. SPC and EPC originated in different industries—the *parts* industry and the *process* industry. The control objectives in these two industries were often very different. The parts industry was attempting to reproduce individual items as accurately as possible—for example, to manufacture steel rods with diameters having *smallest possible variation* about a fixed target value T. The process industries were concerned with yields of product, percentage conversion of chemicals, and measures of purity. They were attempting to obtain the *highest possible mean values* for these measures with smallest variation. Moreover, while in the parts industry properties of feed materials such as steel sheet could be reasonably well controlled, in the process industry important external variables such as ambient temperature and the properties of natural feedstocks were often uncontrollable and had to be compensated by manipulating some related variable, usually[1] by a system of feedback control. Again in the parts industry the *cost of adjustment* was frequently substantial, involving the stopping of the machine or the replacement of a tool, and the *frequency of monitoring* the process might also be an appreciable cost factor. In contrast, the only noncapital cost involved in automatic process adjustment was often, but not invariably, the cost of being off target.

[1] Feedforward and other forms of control are sometimes used but are not discussed in this book.

More recently, the sharply drawn lines dividing the parts industry and the process industry have begun to disappear. One reason is that some processes like the manufacturing of computer chips are hybrids, having certain aspects of the parts industry and others of the chemical industry. Another reason is that conglomerate companies, in which both kinds of manufacture occur, are now much more common. A third reason is that because of the "quality revolution" a greater awareness of the importance of control has led each industry to experiment with the control technology of the other.

The difference between these two approaches has in the past sometimes led to "turf wars" and acrimony between the groups practicing SPC and EPC. This is unfortunate because both are important with long and distinguished records of practical achievement. Serious inefficiency can occur when these tools are not used together and appropriately coordinated. Thus, in the past, automatic feedback control has sometimes been applied without first removing major process "bugs" that could have been detected and eliminated with SPC methods and would have allowed more effective feedback control. On the other hand, Shewhart charts, which are intended to monitor the process, have sometimes been used by practitioners of statistical process control for *adjustment* of the process (i.e., to perform feedback control). When used for this purpose, these charts can be very inefficient. It is now generally accepted that activities of monitoring and adjustment should not be carried out separately, but should be performed simultaneously. It is by the complementary use of these methods that fully efficient control can be achieved with least cost.

Some of these issues were addressed earlier by Box and Jenkins (1962, 1963, 1970) and by Box, Jenkins, and MacGregor (1974), and many more authors have been concerned with various aspects of the resulting problems. We are indebted in particular to Adams (1988); Adams, Lowry, and Woodall (1992); Adams and Woodall (1989); Alwan and Roberts (1988); Aström (1970); Aström and Wittenmark (1984, 1989); Bagshaw and Johnson (1975); Barnard (1959); Bather (1963); Bergh and MacGregor (1987); Bergman and Klefsjö (1994); Bohlin (1971); Brook and Evans (1972); Crowder (1986, 1987, 1992); Deming (1986); Faltin, Hahn, Tucker, and Vander Wiel (1993); Gan (1991); Grant and Leavenworth (1972); Hahn, Faltin, Tucker, Richards, and Vander Wiel (1988); Harris (1988); Harris and MacGregor (1987); Hunter (1986); Hromi (1996); Imai (1986); Ishikawa (1985, 1989, 1990); Jensen and Vardeman (1993); Johnson and Bagshaw (1974); Juran (1988); Kotz and Johnson (1993); Kramer (1989); Lorenzen and Vance (1986); Lucas (1982); Lucas and Crosier (1982); Lucas and Saccucci (1990); MacGregor (1972, 1976, 1987, 1988); MacGregor and Wong (1980); Montgomery (1997); Montgomery, Keats, Runger, and

Messina (1994); Montgomery and Mastrangelo (1991); Page (1954, 1957,1961); Pitt (1994); Prett and Garcia (1988); Ramirez (1989); Roberts (1959); Ryan (1989); Shewhart (1931); Taguchi (1981, 1986); Taguchi, Elsayed, and Hsiang (1989); Tucker, Faltin, and Vander Wiel (1993); Vander Wiel (1996); Vander Wiel and Tucker (1989); Vander Wiel, Tucker, Faltin, and Doganaksoy (1992); Wadsworth, Stephens, and Godfrey (1986); Wardell, Moskowitz, and Plante (1994); Wheeler and Chambers (1992); Woodall (1983, 1984); Woodall and Adams (1993); and Yashchin (1993).

In this book our purpose is to describe as simply as possible some available techniques for monitoring and adjustment, to underline the different contexts, objectives, and assumptions of their operation and to consider how best they may jointly be used to produce efficient control.

As an introduction to some of the issues involved, we consider next some types of variation we may encounter and some fundamentals concerning the building of models.

1.2 TYPES OF VARIATION

Suppose you sample the product from a production line at equal time intervals and you measure some characteristic of interest. If you plot your data as a *run chart* of the kind shown in Figure 1.1, you will see that your measurements are subject to variation. In order to develop and apply appropriate methods of statistical control you will need to be able to describe and understand different kinds of variation. (We shall sometimes also refer to variation as *the noise* or *the disturbance*.)

In subsequent chapters we will discuss more thoroughly the characterization of variation, but here for preliminary illustration we consider three basic types. Figures 1.1a and 1.1b show time series exhibiting *stationary* variation. Such stationary series vary in a stable manner about a fixed mean. Figure 1.1c shows a *nonstationary* time series that wanders about with no fixed mean.

The series in Figure 1.1a is a *white noise* series. In such a series, the deviations from the mean are *statistically independent;* that is, they are such as might be obtained by drawing *at random* from a large aggregate of possible deviations. (It is supposed that this aggregate is described by a frequency distribution having zero mean and often assumed to be approximately a normal distribution.) A key property of a white noise series is that the *order* in which the data occur tells us nothing about the series. One consequence of this is that past values of the series are of no help in predicting future values.

The series in Figure 1.1b is also stationary but it represents *autocorrelated noise* in which a given deviation from the mean is not statistically

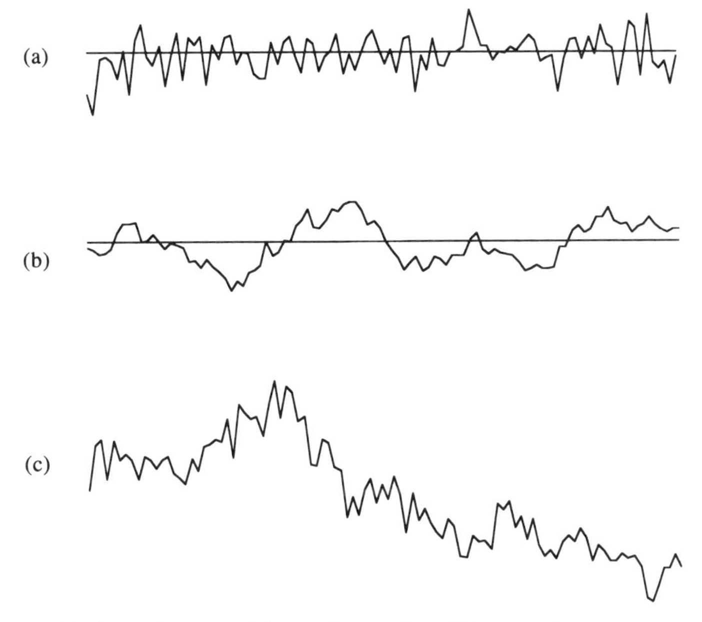

Figure 1.1 (a) A stationary white noise series; (b) a stationary autocorrelated series; (c) a nonstationary series.

independent of adjacent deviations. Statistical *dependence* implies that the probability of one particular deviation depends on the magnitude of others. For example, in Figure 1.1b, positive deviations tend to follow—and to be followed by—positive deviations, and negative deviations by negative deviations.

Figure 1.1c illustrates nonstationary variation. Series of this kind frequently occur in economics and business. We need to consider them here because many processes must be stabilized by using feedback control, some forms of which are described in later chapters. Such control may be necessary when the process is affected by factors that cannot be standardized—such as ambient temperature or the characteristics of natural feedstocks such as oil, wood pulp, or sewage. Now to design an efficient feedback control scheme we need to know something about the *disturbance* we are trying to control—that is, the sequence of deviations that *would* have occurred if *no control* had been applied. Such a disturbance can frequently be represented by a nonstationary series, such as that shown in Figure 1.1c. Usually an operating process affected by such a disturbance will have a control scheme of some sort already in place. Consequently, the output series that we actually see will not be the uncontrolled series but rather the series obtained after the control has been applied. If the control is effective, the controlled series will *not* look like Figure 1.1c but will be a stationary series such as are shown in Figures 1.1a and 1.1b.

In later chapters you will encounter various examples of series, many of which are like those in Figure 1.1. Their properties and the models that describe them will be discussed in greater detail as they are needed. Also, in Chapter 12 you will find a review that provides an elementary account of time series models and of modeling. You may wish to refer to this from time to time.

1.3 MODELS

A topic that is implicit in all that follows is the complementary roles of empiricism and theory. There is a school of thought that believes that all statistical procedures should be derived from clearly stated mathematical models. A different group argues that what is important are empirical procedures justified by the fact that they work. A key consideration in reconciling these ideas concerns the *robustness* of the procedure that is employed—that is, its insensitivity to likely violations of the assumptions on which it is based. It has been said that "all models are wrong but some models are useful." In other words, any model is at best a useful fiction—there never was, or ever will be, an exactly normal distribution or an exact linear relationship. Nevertheless, enormous progress has been made by entertaining such fictions and using them as approximations. Because people realize that models are necessarily approximate, they are often concerned with how *close* the model is to perfection. The implications are that if the model is "nearly right" then the procedure based upon it will be nearly right, and, correspondingly, that if the model is "badly wrong" then so will be the derived procedure. Both these ideas are faulty because they do not take account of robustness (or the lack of it).

To understand the concept of robustness, consider the following example. Suppose that you set up a quality control procedure designed to signal the occurrence of a discrepant observation. The purpose is to improve quality by tracking down a possible fault that has produced the discrepancy and so eliminate it once and for all. Now the tracking down of faults—the search for what Shewhart called *assignable causes*—can be expensive so that you will want your procedure to give as few false alarms as possible. (By a false alarm we mean a signaled discrepancy that is merely due to chance and so can lead to a wild goose chase.) So perhaps you choose a procedure that, *if the assumptions on which it is based are true,* gives false alarms only once in 500 times, that is, with probability $p = 0.002$. One important assumption, let's call it A, on which this calculation rests might be that, when the process is operating properly, the deviations from target are statistically independent; in particular, the correlation between successive deviations from target is zero. Now suppose you knew that in practice this assumption A might not be exactly true and that there could easily be a correlation of, say, 0.3 between successive observations.[2] A question you would naturally ask is how much effect will such a violation of assumption A have on the probability of false alarms? Now there are a number of quality procedures based on various kinds of charts that might be used for looking for a discrepant observation; let's call two such procedures X and Y. Suppose it was true that if you used procedure X the effect of this degree of correlation would change the probability of a false alarm only slightly, say, from one in 500 to one in 400—from 0.0020 to 0.0025. This is not a very big change and will make little difference to any action that you might take. Thus you could say that procedure X was robust to small correlations in the data. But suppose that for procedure Y the same discrepancy produced a change in the probability from one in 500 to one in 5—from 0.002 to 0.200. Then you would conclude that procedure Y was very nonrobust (very sensitive) to small correlations. If, as is in fact quite likely, such correlation might occur, you would certainly not want to use procedure Y to decide when to look for an assignable cause.

As a further example, readers familiar with F statistics will know that two of their uses are (1) to compare two mean squares in the analysis of variance, and (2) to compare the variances of two independent samples of data (as you might need to do, for example, in the comparison of two methods of chemical analysis).

Both tests assume among other things that the data are normally distributed, but it can be shown that the effect of certain likely types of nonnormality on the significance level of the test is greater for application (2) than for application (1). That is, F statistics are more robust to nonnormality when they are used in the analysis of variance than when they are used to compare two independent variances.

More generally, suppose some outcome of interest depends on a particular assumption. This outcome need not be a probability level; it could be a manufacturing cost that you have calculated, or the slope of a fitted straight line that has been estimated. Also, the assumption could be about anything—that the gravitational constant is 32.2 ft/s^2 or

[2] That is, three-tenths of the variation, as measured by the variance, was common to the current observation and the one that preceded it.

that the moon is to a sufficient approximation spherical. In any case the effect of a departure from assumption depends on *two* things:

1. The size of the departure.
2. The sensitivity of the procedure to such a departure.

The general mathematical argument, illustrated in terms of the above example, goes like this:

Suppose an outcome P of some procedure depends (among other things) on a certain assumption A (e.g., the stated probability of false alarm depends on the truth of the assumption that the correlation between successive observations is zero). Let some measure of the departure from this assumption (e.g., the size of the correlation) be denoted by Δ_A and let the consequential effect on the outcome (e.g., the change in the probability of false alarm) be denoted by Δ_P. Then approximately

$$\Delta_P = \Delta_A \frac{dP}{dA}.$$

Thus, as we said, the effect on the procedure of the departure from assumption rests on *two* entities:

1. The size of the departure Δ_A.
2. The rate of change dP/dA with respect to A, which measures the robustness (sensitivity) of the procedure to this assumption.

Thus two different procedures, P_1 and P_2, can differ greatly in sensitivity to identical assumptions, and there is nothing in the derivation from "clearly stated mathematical models" that tells us whether the derived procedure will be robust or not. This is the reason that "sensitivity analyses" are often made, for example, by design engineers. Such investigations could equally well be called robustness studies.

The requirement of robustness may be easy to satisfy if you know what kinds of departures from assumption you expect and fear. But how to get such knowledge can be perplexing. For example, a few years ago a particular type of commercial aircraft was involved in a series of mysterious crashes. Most of these occurred during stormy weather conditions, which, however, would not have been dangerous for most aircraft. One of the authors was told that the problem had arisen from an oversight in the design of the wings. The wings of an aircraft are, of course, deliberately structured to be appropriately flexible, for like trees in a high wind, their safety lies in their ability to bend. Extensive computer simulations had allowed for many types of contortions

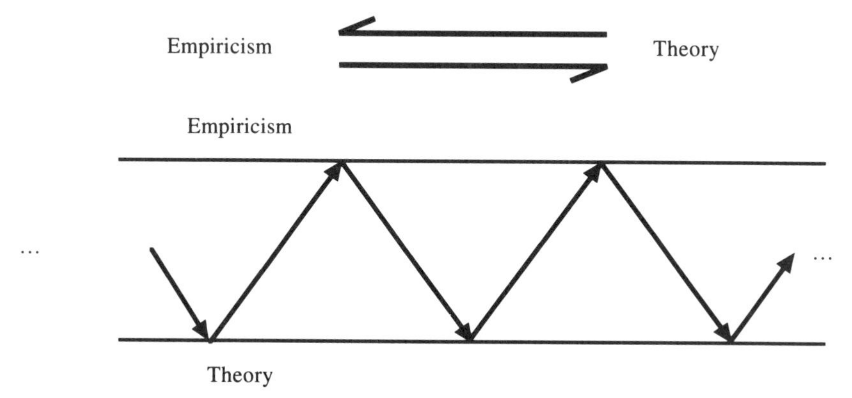

Figure 1.2 The iteration between empiricism and theory.

that the wings might have to undergo, but there was no term in the model that represented their reaction to one particular type of twisting force. Omission of a term in a model is equivalent to assuming that its effect is zero. Unhappily, it turned out that the resulting design was not robust to twisting that could occur under particular meteorological conditions and this led to disaster.[3]

So since all models are wrong, it is very important to know what to worry about; or, to put it in another way, what models are likely to produce procedures that work in practice (where exact assumptions are never true). Obviously, a good approach is to consider the properties of procedures that *do* seem to work in practice. In fact, to obtain a useful procedure one needs *both empiricism and theory*. But—more than that—one needs continuous iteration between them, as is illustrated in Figure 1.2.

That some empirical procedure seems to work well is immediately valuable to the practitioner. But this fact can be of much wider consequence if we ask what would need to be theoretically "true" *for it to work*. The importance of this further step is that the theoretical model we are led to is often useful to solve problems quite different from that in which it originated. Furthermore, such iteration can lead to the development of theory itself and this, in turn, to further practical results.

For example, in the 1950s operational research workers needed to obtain reasonable forecasts for time series of commercial inventories and sales. The mathematical models that were studied at that time were mostly for *stationary* time series that varied about a fixed mean. These were of little help for inventory and sales series, which were not of that kind. So these workers developed an empirical method of their own.

[3] Eventually the problem was solved by structural alterations to the wings.

They used a weighted average of past values of the series to forecast future values. It seemed reasonable that a forecast should place most emphasis (weight) on the last available observation, less on the previous one, and so on; so they arranged that the weights fell off geometrically (exponentially). This resulted in a forecast obtained by "exponential smoothing" of past observations. See, for example, Holt (1957). These empirical forecasts came to be called *exponentially weighted moving averages* (EWMAs) and they had, and still have, very wide application. Later, Muth (1960) identified a particular *nonstationary* mathematical time series model for which the EWMA forecast was optimal, in the sense that it supplied a forecast for which the mean of the squared forecast errors was smallest. This model is called an *integrated moving average* (IMA) time series model. Now in an apparently quite different field—the process industries—a form of feedback control called proportional integral derivative (PID) control is widely used. It turns out (e.g., see[4] Box and Jenkins, 1970, 1976; Box, Jenkins, and Reinsel, 1994) that the same IMA time series model that justified exponential smoothing can also justify PID control. In recent years the IMA model has been found to approximately represent a wide variety of time series arising in many different fields of inquiry. Further development of this type of model to include the so-called ARIMA models and their multivariate counterparts have found wide application in forecasting, in the modeling of dynamic systems, in control theory, and in intervention analysis. Recently (Box and Kramer, 1992), a theoretical underpinning for the IMA model, further discussed in Chapter 5, was developed that partially explains its ubiquity as a useful approximation for many nonstationary disturbances.

1.4 EXPLANATORY MODELS AND WORKING MODELS

Explanatory Models

In this book we discuss control methods for discrete equispaced data. If you believe such observations made on some system can be explained

[4] In this book we make frequent reference to the book *Time Series Analysis, Forecasting and Control*, which has appeared in three editions. The first two editions (1970 and 1976) by Box and Jenkins were published by Holden Day, the third edition (1994), by Box, Jenkins, and Reinsel, was published by Prentice Hall. To avoid tedious repetition we use the letters BJR to refer to this material unless a more specific reference is needed.

by a particular model, then, for an observation y_t made at time t, you can write

$$y_t = m_t + e_t \quad \text{or} \quad e_t = y_t - m_t,$$

where m_t is the value given by the *model* at time t.

In this equation, e_t is simply the part of the observation y_t that is not explained by the model and is called the *residual*. If you want to check an explanatory model using data $y_1, y_2, \ldots, y_n$, then you should calculate values $m_1, m_2, \ldots, m_n$ given by your model and carefully examine the set of residuals $e_1, e_2, \ldots, e_n$ that you obtain. If these residuals contain any significant and explicable pattern, this could point to the need to modify the explanatory model. For example, you might find that the residuals e_t were related to the humidity h_t that occurred at time t when the observation was made; so the residuals plotted against humidity would yield some kind of relationship. Or you might find that each residual was related to the previous one; that is, when the residual e_2 was plotted against e_1, e_3 against e_2, e_4 against e_3, and so on, you found that they were correlated. In both of these cases you would suspect that the model did not explain all that was going on. In the first case you might change the model to include a term that allowed for the effect of humidity; in the second case you might include a term in the model for y_t that allowed for its dependence on the previous value y_{t-1}.

If at some point you found that the residuals calculated from a suitably modified model seemed to be totally informationless, looking like a sequence of random deviations, you might say that, so far as you could tell, the model explained the data.

This iterative process of model checking and model building may be likened to that which might be used in testing the efficiency of a process for extracting gold from ore. To see if you really had got all the gold out, you might test the residual filtrate coming out of the extraction process. If this contained gold, then you would conclude that the extraction process was imperfect and you might try to improve it. Similarly, if residuals contained information, then the current explanatory model would be less than perfect and would need to be modified.

You will notice that in this process of model building you are trying to find out what is *really going on*. Exercises of this kind are extremely important for learning and discovery but the models that they produce are not necessarily good *working* models.

Working Models

We have said that all models are wrong but some are useful. Also, consideration of robustness tells us that we need models that work reasonably well even when, as inevitably happens, technically they are wrong. For instance, perhaps the most frequently used statistical tests for the comparison of means, such as the analysis of variance and the *t* tests, are derived on the assumption that the data are normally distributed. Moderate departures from the assumption of normality usually do not seriously affect the behavior of these tests[5]; so in most real circumstances although the underlying distribution will certainly never be exactly normal this distribution can often provide a good working model.

A story that further illustrates the point concerns the famous Danish physicist Niels Bohr. When during the second world war the first atomic bomb was to be tested, a number of eminent scientists were invited to witness this event. No one was quite sure whether the bomb would actually work and, the evening before the trial, a friend who was passing was surprised to see Bohr nailing a horseshoe over the door of his army hut. The friend remarked "Surely Dr. Bohr, you don't believe that that horseshoe will affect the outcome of the test, do you?" Bohr replied "No, but I understand it works whether you believe in it or not."

The basis on which a working model could be defended is not necessarily that it is true but that it is often (but not invariably) useful.

A Working Model for Process Monitoring

In Chapters 2 and 3 we consider process monitoring using Shewhart and other control charts. Such charts are based on the concept, due to Shewhart and to Deming, of what may be called the *common cause plus special cause* model. Essentially the idea is that when a process is operating in a "state of control" it will vary in a stable manner about a fixed mean and this variation is said to be due to *common* causes. Occasionally, some upset will occur producing aberrant behavior, which when detected by a control chart can initiate a search for its cause. Such an upset is described by Deming as a *special* cause. The model is valuable because it frequently approximates the true situation well enough to be effective in eliminating special causes and so steadily produces process improvement.

[5] Although lack of independence between observations can seriously affect them.

The Shewhart–Deming statistical model may be written

$$y_t = \mu + e_t,$$

in which y_t, the observed value of the quality characteristic at time t, is represented by a fixed mean μ (Greek *mu*) plus a deviation from the mean e_t, often called the "error". Such a model might be written more fully as

$$y_t = \mu(x_1, x_2, \ldots, x_n) + e_t(x_{n+1}, x_{n+2}, \ldots). \qquad (1.1)$$

In this expression $x_1, x_2, \ldots, x_n$ are variables that we *already know* affect the mean and that we have been careful to "tie down" so as to keep μ constant. But Equation (1.1) implies that e_t cannot be dismissed as a "random variable" dropped by angels from a celestial table of random numbers. It depends on $x_{n+1}, x_{n+2}, \ldots$ representing the myriad number of "lurking" variables not so far identified.

Thus the common causes are represented by, $e_t(x_{n+1}, x_{n+2}, \ldots)$, and the discovery of a special cause would correspond to the identification of one of these lurking variables, say, x_{n+1}, enabling us to remove it from the error bracket and to put it in the mean bracket as is indicated in the diagram below:

$$y_t = \mu(x_1, x_2, \ldots, x_n) + e_t(x_{n+1}, x_{n+2}, \ldots).$$

Note that this implies that one person's common cause could be another's special cause. Some years ago, when it was first realized that many Japanese products were designed to specifications previously believed impossible, it was evident that some of the Japanese special causes that they had already eliminated were still common causes to their competition!

State of Control

In the above we said that the term "state of control" meant that the output from the process was varying in a stable manner about a fixed mean. In this book, we broadly interpret the term to mean that the process is approximately stationary with output data resembling the series shown in Figures 1.1a and 1.1b. It might, at first, be thought that autocorrelation as displayed in Figure 1.1b should necessarily point to an assignable cause, which should be eliminated. However, although the cause of the autocorrelation may be known and hence assignable, this does not mean that it can or should be eliminated. For example, autocorrelation frequently occurs as a result of mixing of the product upstream. This mixing may, of course, be an essential step in reducing variation.

For clarity in what follows, we refer to a process stably producing autocorrelated data, like that in Figure 1.1b, as in an *autocorrelated state of control.* The unvarnished term *state of control* will be used to refer to a process producing uncorrelated data like that shown in Figure 1.1a.

Autocorrelation should be distinguished from systematic periodic cycling caused, for example, by improperly adjusted machinery. Such cycling may well point to a source of variation that ought to be eliminated. The apparent "cycles" in highly autocorrelated series, which occur because of the correlation of adjacent observations, are distinguishable from systematic cycles because they have no fixed period.

In the monitoring charts discussed in Chapters 2 and 3 it will usually be supposed that, to an adequate approximation, autocorrelation is absent or may be ignored. The problem of efficiently monitoring autocorrelated processes is discussed in Chapters 10 and 11. Also, the possibility of using feedback control to reduce variation for systems in a state of autocorrelated control is discussed in Chapter 6.

Working Models for Process Adjustment

The common cause–special cause model for process monitoring, although never exact, is widely useful. The models we employ for process adjustment discussed in the later chapters of this book are also of this kind—never exact but widely useful. In more formal justification of the feedback methods we employ it is usually supposed that the process disturbance that we need to compensate might look like Figure 1.1c and can be approximately represented by the nonstationary IMA model referred to earlier. Also, that, when there is process inertia, it can be represented by a first-order dynamic system, such as discussed in Chapter 4, possibly with one or more added periods of delay. As we later demonstrate, the resulting procedures seem to be extremely robust to wide departures from these models and usually work well in practice. They are appropriate for adoption by practitioners since, while retaining good efficiency, they are also easy to understand and to apply. We have in mind that the true efficiency of a procedure might be defined as its theoretical efficiency when the assumptions are "exactly satisfied," multiplied by a robustness factor measuring the loss of efficiency in the real circumstance in which it is used, multiplied by the probability that anyone will actually use it.

1.5 WHAT'S IN THIS BOOK?

In Chapters 2 and 3, the common cause plus special cause model is employed in a discussion of process monitoring and the use of Shewhart, exponential, and Cusum charts. Also, some attention is given to situations where models and procedures may need to be modified.

In Chapter 4, in preparation for a discussion of feedback control, a simple dynamic model is introduced that can allow for common types of process inertia. This also leads to a deeper understanding of exponential smoothing, and its application to forecasting.

In Chapter 5, building on these ideas, some time series models are developed, in particular the integrated moving average (IMA) model, which frequently can approximate nonstationary disturbances that need to be controlled by feedback adjustment.

In Chapter 6 we begin to discuss a form of feedback control in which the adjustment is proportional to the last deviation from the target. This idea is of great antiquity (see Mayr, 1970), having been used to control water clocks in ancient Egypt; to ensure that windmills point into the wind; and by James Watt to control his steam engine. Furthermore, it is the discrete analog of integral control, which is the most important of the three adjustment modes in the engineer's proportional plus integral plus derivative (PID) control. In view of this history, it is hardly surprising that this form of control is extremely robust.

Chapter 7 is about the control of processes containing inertia where the response to adjustment is not immediate. It turns out that an appropriate form of feedback adjustment may then correspond to the engineer's proportional plus integral (PI) control in which the adjustment depends on the *last two* deviations from target instead of the last one.[6] It might be thought that the "best" form of control of this kind would be that which minimized the standard deviation of the output quality characteristic. Somewhat surprisingly, this is not the case. Theoretical schemes that minimize the output standard deviation can be virtually useless because they require excessive adjustment. In this chapter, therefore, we discuss PI control in which adjustment is constrained. Rather remarkably, it turns out that for very small increases in the output standard deviation, very large reductions in the needed manipulation can be produced.

[6] It is easily shown that for discrete PID control the adjustment depends on the last *three* errors but this degree of elaboration seems unnecessary for the applications we consider. (See also Harris, MacGregor, and Wright, 1982.)

The control schemes discussed in Chapters 6 and 7 require repeated adjustment to be made at each opportunity (e.g., after each observation). There are many occasions when such a procedure is not burdensome. The operator can easily be trained to read off and make the appropriate adjustment at the time he/she takes the data. This can be done by using a feedback control chart or a suitably programmed process computer. These procedures are, however, not appropriate when additional cost is incurred in making an adjustment. In Chapter 8, therefore, we consider the problem of feedback control where a cost is incurred each time an adjustment is made. Feedback schemes then call for adjustment only when the need is signaled by a *bounded adjustment* chart.

Chapter 9 carries the idea further and considers the additional possibility that there is a cost of *getting* an observation. This raises the issue of how frequently observations should be taken.

Chapter 10 returns to the discussion of process monitoring. The general problem of process monitoring is here considered as that of looking for a specific kind of signal buried in a specific kind of noise. The general concept of efficient score statistics due to R. A. Fisher leads at once to Cuscore charts (Box and Ramírez, 1992), appropriate to any situation of this kind. In particular, the needed assumptions that would make Shewhart charts, EWMA charts, and Cusum charts fully efficient are considered and elaborated.

In Chapter 11 we consider how the results from earlier chapters can be brought together to produce appropriately coordinated adjustment and monitoring.

In Chapter 12, we present an elementary review of time series models and their identification and estimation useful for reference.

The final section headed Conclusions is a brief summary of important issues discussed in the book and their more general implications for the process of model building.

Answers to exercises and problems and a collection of the time series used for illustration are included at the end of the book.

CHAPTER 2

Control Charts for Frequencies and Proportions

"Find out the cause of this effect. Or rather say, the cause of this defect. For this effect defective comes by cause."

Hamlet, WILLIAM SHAKESPEARE

Good quality usually requires that we reproduce the same thing consistently. If we are manufacturing a particular part for an automobile, we would like the dimensions for each item to be as nearly constant as possible. It would be nice if this could be achieved, once and for all, by carefully adjusting the machine and letting it run. Unfortunately, this would rarely, if ever, result in the production of uniform product. As we said in Chapter 1, extraordinary precautions are needed in practice to ensure that a dimension does not change or drift away from its target value. Again, consider some routine task in a hospital, such as the taking of blood pressure; having found the best way to do this, we would like it to be done that way consistently, but experience shows that very careful planning is needed to ensure that this happens.

Because we live in a nonstationary world—a world in which external factors never stay still—the idea of stationarity, that is, of a stable world in which, without our intervention, things stay put over time, is a purely conceptual one. The concept of stationarity is, however, useful as a background against which the real nonstationary world can be judged. For example, the manufacture of parts is an operation involving machines and people. The parts of a machine are not fixed entities. They are wearing out, changing their dimensions, and losing their adjustment. The behavior of the people who run the machines is not fixed either. A single operator forgets things over time and alters what he/she does. When a number of operators are involved, the opportunities for involuntary changes because of failures to communicate are further multiplied. Thus, if left to itself, any process or system will drift away from its initial state. So the first thing we must understand is that stationarity, and hence uniformity of everything depending on it, is an *unnatural* state that requires a great deal of effort to achieve.

In this book we employ the terms *mean, mean value,* and *true mean* for the long-term mean of a hypothetical stable system and reserve the term *average* or *sample average* for the average of a small sample of data. Now quality *control* is about keeping things *constant* with the mean on target but quality *improvement* is about *changing* things for the better. These aims, however, are not inconsistent. One way to see this is to consider again a statistical model for a stable process written

$$y_t = \mu + e_t. \tag{2.1}$$

The observed value of the quality characteristic at time t is y_t and this is represented as a fixed mean μ (Greek *mu*) and a deviation from the mean e_t. As we saw in Chapter 1, this model might also be written

$$y_t = \mu(x_1, x_2, \ldots, x_n) + e_t(x_{n+1}, x_{n+2}, \ldots), \tag{2.2}$$

where $x_1, x_2, \ldots, x_n$ are variables we know about that affect the mean and that we have been careful to "tie down" and the "error" e_t depends on $x_{n+1}, x_{n+2}, \ldots$ representing the myriad number of "lurking" variables we have not so far identified. Suppose now that by thinking about the process, by observing it, or perhaps even by experimenting with it, the identity of one of these lurking variables (say, x_{n+1}) is discovered. Then we can transfer it from the unknown bracket to the known bracket:

$$y_t = \mu(x_1, x_2, \ldots, x_n, x_{n+1}) + e_t(x_{n+2}, \ldots). \tag{2.3}$$

This can have two results, both good. First, variation will be reduced because there is one fewer unknown component in the error bracket that is "flapping in the wind." Second, now that we know that x_{n+1} affects y_t, we can set x_{n+1} at its best level and so improve overall performance. For example, if y_t is a measure of purity, it may be possible to choose the value of this newly recognized variable x_{n+1} so as to increase the mean purity μ.

An important process model proposed by Shewhart came out of his extensive experience in the producing of standard parts for the telephone industry. It is supposed that it is possible to get the manufactured process into a "state of control," so that most of the time its behavior is approximately represented by Equation (2.1) with the mean μ equal to the target value and, usually, with the error e_t assumed to vary randomly and independently about μ. It is further supposed that, occasionally, the process is affected by the influence of some previously unknown factor whose presence is announced by abnormal behavior recorded on a control chart. This may lead to an "assignable cause" being discovered and so eliminated in the manner illustrated by Equations (2.2) and (2.3). It is important to remember that, like all models, Shewhart's model is an approximation to the truth. However, many years of experience have shown that, in the parts industries at least, it is frequently an extremely useful model. In what follows, we illustrate its use and later discuss some cautions and precautions that need to be considered.

Another way in which quality control and improvement fit together has been described by Juran (1988) and by Imai (1986). It is practically certain that, given appropriate training and empowerment, quality teams can discover better ways to do things. Unfortunately, experience shows that if nothing is done to prevent it, the improved quality, won with such difficulty, may begin to disappear. Over a period of time, carefully adjusted machines will become untuned and people may become less careful in their work. In the natural course of events, therefore, quality will erode in the way illustrated in Figure 2.1. What we must do is to lock in improvements, so that at each stage the new higher level of quality is maintained, as in Figure 2.2.

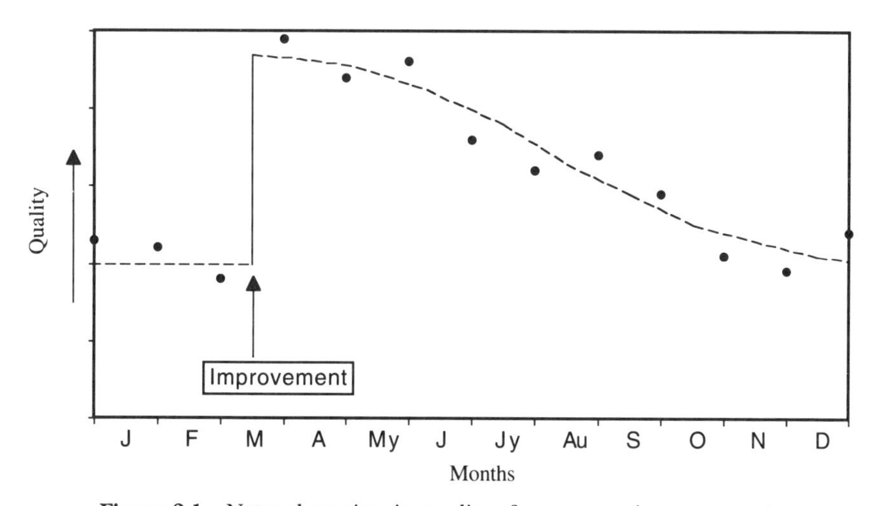

Figure 2.1 Natural erosion in quality after process improvement.

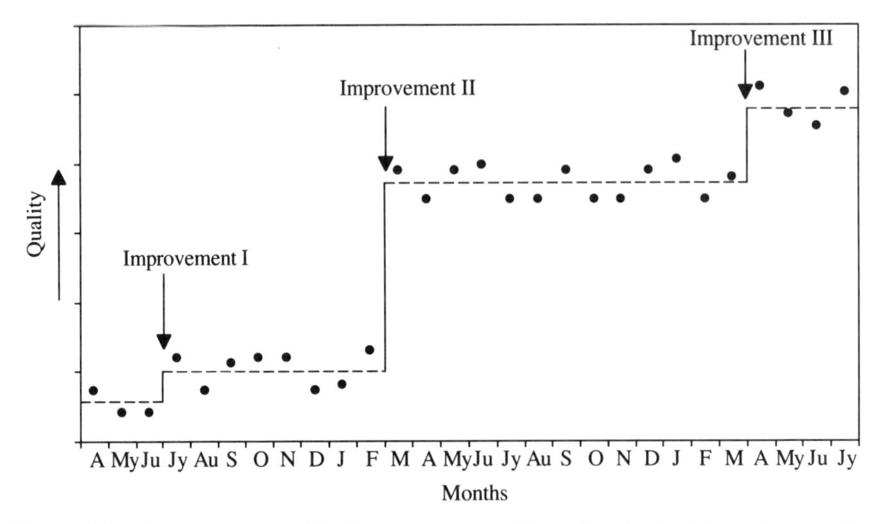

Figure 2.2 Continuous quality improvement with gains locked in using quality control.

2.1 QUALITY CONTROL OF FREQUENCIES: THE c-CHART

An important tool for locking in quality at each stage is the Shewhart quality control chart. There are different kinds of Shewhart charts appropriate to different circumstances. For simplicity, we begin by considering a control chart called a c-chart. This is a chart appropriate for data that occur as *counts,* that is, as *frequencies of occurrence.* The ideas are best illustrated and explained by an example.

A company has an employee Health Clinic where cuts, bruises and other minor injuries sustained at work are treated. It is in the interest of everyone that injuries are minimized, so a rigorous safety program is maintained. To monitor the situation, a count is kept of the number of patients attending the clinic each week. It is found that over an extended period of time the mean number is 20.6. Over a period extending from early May to late November the data were as follows:

Week	1	2	3	4	5	6	7	8	9	10	11	12	13
Patients	28	30	18	21	18	19	21	25	17	23	28	14	20

Week	14	15	16	17	18	19	20	21	22	23	24	25	26
Patients	25	22	12	23	18	16	24	22	16	27	20	28	28

They are plotted in the run chart of Figure 2.3. What are we to conclude from looking at this chart? Was there, for example, an unusually high attendence at the end of May? Was there an upward trend

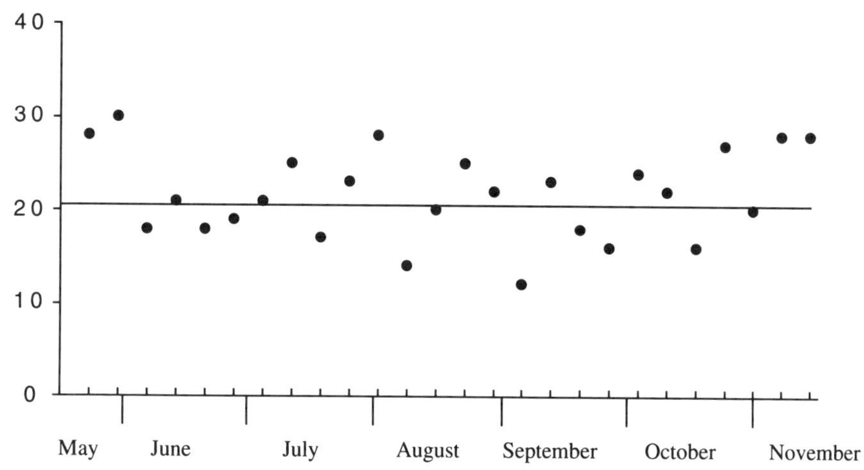

Figure 2.3 A run chart of weekly totals of new minor injuries at an employee clinic.

beginning in the middle of September? If so, should we be looking for assignable causes? For things to change? For people to commend or to blame?

The question of *what to worry about and what not to worry about* is of fundamental importance to all of us in the conduct of our daily lives. But it is particularly so for managers. On the one hand, to worry about everything is unnerving. It is also counterproductive, for it can result in continual tinkering with a correctly operating system in response to imagined phantoms in the data. Such tinkering can increase rather than decrease variability. It can also lead to the praising or blaming of people for things for which they are not responsible, resulting in apathy and resentment. On the other hand, if important changes *are* occurring, we need to find out why. If we can do this, we can correct those things that are making matters worse and preserve those that are making matters better, and so learn how to improve the process.

Returning to the example of the health clinic, obviously, the number of new injuries will not be constant from week to week. The problem is to know how much variation to expect if the causal system remains the same, that is, if the system is in a *state of control.* Remarkably enough, if certain assumptions are true, the question can have a precise answer. Imagine the working week divided up into small intervals of, say, one minute. Then, on the assumption[1] that there is a *small* chance of an

[1] More exactly, two additional assumptions are required, namely, (1) that no more than one injury is likely to occur in one minute and (2) that the numbers of injuries occurring in two nonoverlapping—but not necessarily small—periods of time are statistically independent.

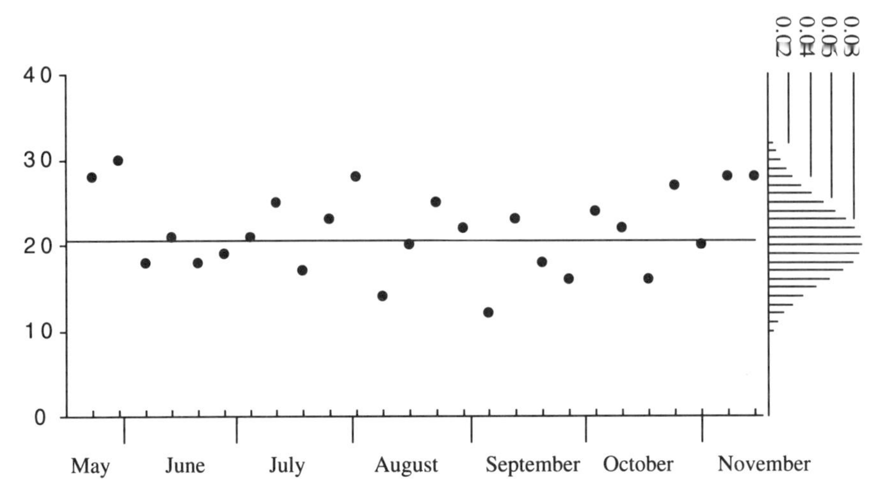

Figure 2.4 Run chart with a Poisson distribution with mean 20.6 as a reference distribution.

injury occurring in any given minute, the number of injuries occurring each week would follow a frequency distribution of exactly known mathematical form called the *Poisson* distribution. We will first suppose that the assumptions are true and later discuss circumstances where they might not be true and what might be done to set matters right. Rather surprisingly, the Poisson distribution can be calculated simply by knowing the *mean* frequency of events (the mean number of injuries per week), which must be constant when the system is in a state of control. Knowing only this number, it is possible to calculate the probability that in a given week there will be no injuries, one injury, two injuries, and so on. The mathematical formula, from which these calculations can be made, will be found in Appendix 2A. Suppose that, from study of records over an extended period of time, the mean number of injuries per week was found to be $\mu = 20.6$. In Figure 2.4 we have calculated the Poisson probability distribution using this value 20.6 for the mean frequency. This distribution is drawn in the right-hand margin of the figure.

The distribution[2] tells us, for example, that the probability of exactly 20 accidents occurring in a given week is just under 0.09 and the

[2] Probabilities for frequencies of accident according to the Poisson distribution with mean 20.6 are as follows:

Accidents	9	10	11	12	13	14	15	16	17	18	19	20
Probability	0.002	0.004	0.008	0.014	0.022	0.032	0.044	0.057	0.069	0.079	0.085	0.088

Accidents	21	22	23	24	25	26	27	28	29	30	31	32
Probability	0.086	0.081	0.072	0.062	0.051	0.041	0.031	0.023	0.016	0.011	0.007	0.005

probability of 15 accidents is just over 0.04. The distribution is that of a stable purely random sequence of events (in this case, injuries) with mean frequency 20.6. It thus provides a yardstick or *reference distribution* against which to judge the numbers that actually occur on a chart. If these look like a random sequence from this reference distribution, then there is no reason to believe that the system is not in a *state of control.* If they do not look like a random sequence of drawings from this reference distribution, we have reason to suppose that something else is going on and it is worth looking for an assignable cause.

What do we mean by a random sequence of drawings from the reference distribution? Imagine a hat full of tickets with numbers on them corresponding to weekly injuries. Thus one ticket may have a 20 on it, another ticket a 15, and so on. Suppose we arrange that the relative frequency of different kinds of tickets in the hat exactly corresponds with our reference distribution. Thus, for example, there would be about twice as many tickets with the number 20 on them as there would be for the number 15. Then a random sequence would be produced by blindly drawing from the hat of thoroughly stirred up tickets and plotting the results in sequence. Quantities that behave like random drawings from a reference distribution are called *random variables.* Inspection of Figure 2.4 suggests that the plotted injuries might well be a random sequence of drawings from the reference distribution and, consequently, there is no reason to seek special causes for what is happening.

In practice, the whole reference distribution is not usually shown on a quality control chart. Instead "tram lines" are drawn denoting what are often called *action limits* and *warning limits.* The solid lines shown in Figure 2.5 are action limits chosen so that, when the process is in a state of control, points will go outside these limits only very rarely. Specifically, only about once in 400 times. About once in 800 times, a point will be above the upper limit, and once in 800 times, a point will be below the lower limit. The dotted lines shown on the figure are the warning limits. When the process is in a state of control, points will fall outside the warning limits about once in 20 times. Once in 40 times they will be above the upper dotted line, and once in 40 times they will be below the lower dotted line.

For reasons we explain shortly, the warning lines are usually called the two sigma (2σ) limits and the action lines the three sigma (3σ) limits.

Note that, in Figure 2.5, no points fall outside the action limits. Two points, out of a total of 26, fall almost on the warning limits, but this is not surprising particularly since these are not close together and once in 20 times we expect a point to fall beyond these limits by chance.

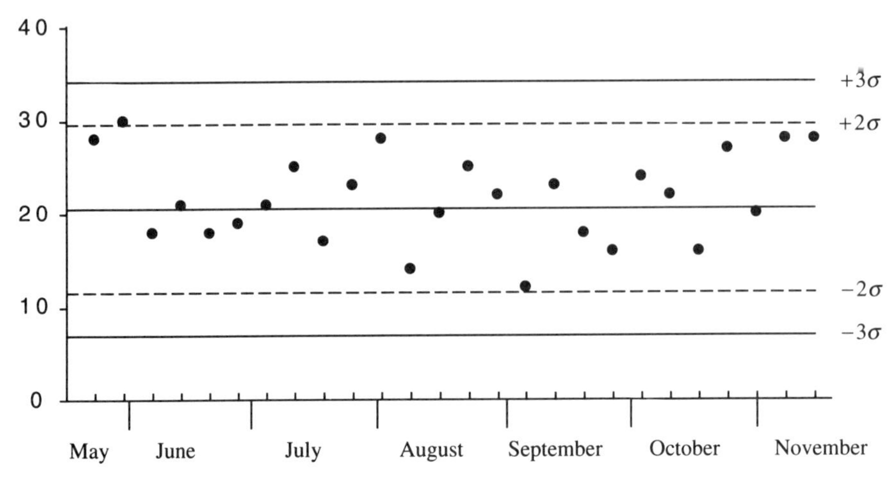

Figure 2.5 Shewhart quality control chart for weekly frequencies of injury showing "action" limits and "warning" limits.

Thus the sequence of points representing the frequency of injuries supports the hypothesis that the process is in the state of control during this period of time and no action is warranted.

2.2 COMMON CAUSES AND SPECIAL CAUSES

Quality control charts, such as that shown in Figure 2.5, are of great value in helping us distinguish what Dr. Walter Shewhart and Dr. W. Edwards Deming call *common causes* from *special* (or *assignable*) *causes*. Common causes are those associated with the overall behavior of the system when it is in a (hypothetical) *stable state of control*. Special causes are associated with deviations from this stable state. Again, as we will later discuss, this model—like all models—is imperfect but can be very useful.

Common Causes

These are the causes associated with the usual steady-state running of the process when it is in a state of control. For the health clinic example, it is easy to imagine patterns in the data, but if, as the chart in Figure 2.5 suggests, the process was indeed in a state of control, it would be a waste of time to look for assignable causes. The "high attendance at the end of May" and the "upward trend beginning in the middle of September" are probably purely random occurrences. In fact, the sequence of frequencies plotted in Figure 2.5 has the appearance expected from a random injury process with a mean rate of injuries of about 20.6 per week and if such a random process is indeed operating, this mean rate can only be changed by changing the system itself. Suppose, for example, that management decided to install new machinery with special guards with the intention of making injury less likely. Or suppose that extra lighting was installed or "nonslip" floors. Then *if* machinery, lighting, or slippery floors were important features in producing injuries, the system of common causes would be changed. Suppose these interventions by management resulted in a reduction of the mean number of injuries and the establishment of a new stable system with variation corresponding to a random sequence of frequencies with a mean of 10. Then a new *common cause system* would have been produced. If this happened, not only would there be a shift in the mean from 20.6 to about 10, but it turns out there would also be a *reduction of variation* about this lower mean. The reason for this you will see later.

Note that dealing with common causes—introducing new machinery, new lighting, nonslip floors, and so on—comes about as a result of *management decision.* Juran and Deming have said that at least 85% of problems are associated with management and only 15% or less with employees. Before the institution of quality improvement teams, this meant that employees were not able to affect the system very much. However, with the formation of these teams, employees can become, in effect, part of management and their studies can lead to fundamental change in the system itself. For example, a skilled team using simple graphical tools (e.g., see Ishikawa, 1989) may be able to discover the principal causes of injury. The teams' findings, based not on mere speculation but on data,[3] can then lead to appropriate modifications by management to, for example, the machinery or the system of lighting; such action can result in a fundamental change in the *common cause system* and so to a permanent reduction in injuries.

Special Causes

In Figures 2.6 and 2.7, we show for illustration two situations where the process would be judged to be *out of control.* In both cases it would make sense to look for special (assignable) causes. The chart in Figure 2.6 shows that up to about the middle of September, the data are

[3] Remember that factory showing a big poster with the notice: "In God we believe; everybody else, please bring data."

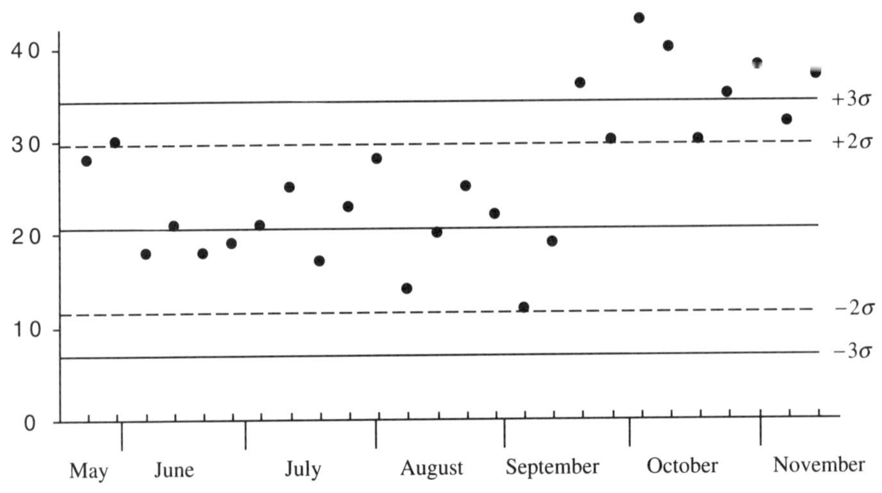

Figure 2.6 Process out of control, suggesting a real increase in mean frequency of injuries after September.

consistent with a process in a state of control with mean about 20. After that, however, of the nine further observations, six exceed the upper action limit and three more exceed the upper warning limit. Thus there seems to have been an increase in mean to about 35 injuries per week beginning in September and the reason for this should be looked into.

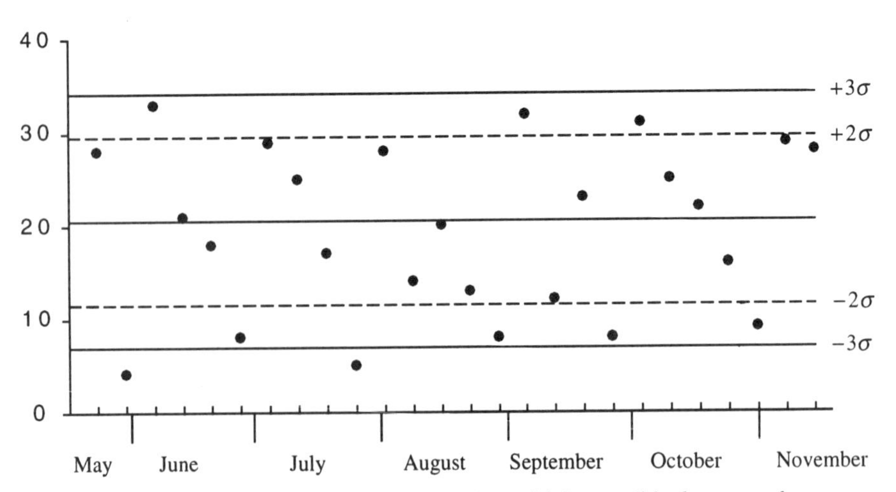

Figure 2.7 Process out of control. Variation is too high, possibly due to under-reporting of accidents in the last week of the month.

For the chart in Figure 2.7 the mean seems to be on target, but variation is unusually high with two points outside the action limits and seven more outside the warning limits. A very suspicious circumstance is that an unusually low value seems to occur in the last week of each month, which is compensated by a high value in the first week of the next month. This might be due to underreporting of accidents in the last week of the month, in a mistaken attempt to keep the reported monthly totals down.

Note that the control chart has dual functions:

1. To monitor the stable common cause system.
2. To flag possible evidences of special causes.

Both functions are important to process improvement. A newly instituted common cause system produced by process improvement should be carefully monitored to ensure that the input is not eroded but is permanently locked in. Also, significant deviations from what would be expected if the present common cause system were in operation, can call attention to the need to look for and deal with possible special (and potentially assignable) causes.

2.3 HOW DO WE CALCULATE THE CONTROL LIMITS?

Control limits are chosen so that when the process is in a state of control, only a small predetermined proportion of the points will fall outside the limits. We have supposed that when the system is in a state of control, the points on the chart will be like random drawings from a Poisson reference distribution, like the one shown to the right of Figure 2.4. Knowing this distribution, therefore, we can calculate where to put the control lines. However, direct calculation in this way is tedious and there is a simple approximation that we can often use instead.

As long as the mean frequency is not too small (as long as it is, say, 12 or more), the calculations for the action and warning limits are very easy. The distribution can then be approximated by a distribution of central importance called the normal distribution. We discuss this special distribution a little later; for the moment we merely note that the action limits and warning lines are then closely approximated by what are called 3σ (three-sigma) limits and 2σ (two-sigma) limits of the normal distribution. The letter σ (Greek *sigma)* denotes a measure of the *spread* of the distribution and is called the *standard deviation* of the distribution of the data. The appropriate standard deviation for a frequency of events varying randomly about a fixed mean value in a Poisson distribution turns out to be simply the *square root of this mean value.* Equally we can say that the variance and the mean are equal: $\mu = \sigma^2$.

In our example the mean frequency is 20.6 so the standard deviation is $\sigma = \sqrt{20.6} = 4.5$. Thus the three-sigma action limits are $20.6 \pm (3 \times 4.5) = 20.6 \pm 13.5$; that is, 7.1 and 34.1. And the two-sigma warning limits are $20.6 \pm (2 \times 4.5) = 20.6 \pm 9.0$; that is, 11.6 and 29.6. These are the limit lines drawn in Figure 2.5.

TABLE 2.1 Interpolated Action Limits and Warning Limits Calculated from the Poisson Distribution Function

Action Limit	Warning Limit	Mean Frequency	Warning Limit	Action Limit
—	—	0.5	1.89	3.25
—	—	1.0	2.94	4.75
—	—	1.5	3.91	5.88
—	—	2.0	4.83	6.93
—	—	2.5	5.69	7.93
—	—	3.0	6.50	8.91
—	—	3.5	7.24	9.86
—	0.06	4.0	7.95	10.77
—	0.23	4.5	8.76	11.66
—	0.48	5.0	9.50	12.51
—	0.83	5.5	10.18	13.31
—	1.12	6.0	10.88	14.06
—	1.36	6.5	11.62	14.89
0.07	1.69	7.0	12.30	15.73
0.19	2.06	7.5	12.94	16.52
0.38	2.31	8.0	13.68	17.26
0.66	2.65	8.5	14.34	17.97
1.02	3.05	9.0	14.96	18.79
1.17	3.31	9.5	15.68	19.56
1.37	3.66	10.0	16.34	20.27
1.68	4.06	10.5	16.95	20.96
2.04	4.34	11.0	17.65	21.75
2.22	4.71	11.5	18.29	22.48
2.47	5.09	12.0	18.91	23.16
2.83	5.40	12.5	19.59	23.88
3.11	5.79	13.0	20.21	24.63
3.34	6.16	13.5	20.84	25.32
3.66	6.49	14.0	21.50	25.97
4.04	6.91	14.5	22.09	26.73
4.25	7.24	15.0	22.75	27.43

Note that if, by fundamental modification of the system, the mean frequency of weekly accidents was reduced from 20.6 to 10, the standard deviation measuring the spread of points about this lower mean would likewise be reduced from $\sqrt{20.6} = 4.5$ to about $\sqrt{10} = 3.2$.

When the mean frequency of events is smaller than, say, 12, the approximation based on the normal distribution begins to be too crude. In this case you can use Table 2.1 to find appropriate limits. Each row of the table shows the lower action limit, lower warning limit, mean frequency, upper warning limit, and upper action limit. These have been calculated by direct interpolation in the Poisson distribution function to yield the same probabilities as those given by $\pm 2\sigma$ and $\pm 3\sigma$ limits of the normal distribution. In practice, of course, frequencies must be whole numbers. Thus, using the table if the mean frequency is 10, the action limits would be crossed only if the frequency was less than 2 or greater than 20 and the warning limits crossed if the frequency was less than 4 or greater than 16.

2.4 FOR WHAT KINDS OF DATA IS THE c-CHART USEFUL?

The c-chart that we have considered so far is good only for plotting *frequencies* of occurrence—in our example, the number of injuries per week. Frequencies usually occur as *counts per unit of time* or *counts per unit of space.* Examples are the number of telephone calls received in a minute, the number of faults in a square foot of knitted fabric, the number of rat hairs in a one-pound sample of flour, and the number of cells in one cubic millimeter of blood. It is easy to imagine examples where each one of these quantities is of concern. In each case we might need information on whether the system was a stable one, or whether in addition to random causes there were other influences that might be tracked down to assignable causes. We might also like to know when we had been successful in changing the common cause system by permanently changing the mean count in a favorable direction.

In manufacturing, c-charts are often used to monitor minor flaws and blemishes; that is, defects that are not such as would render the product useless but must nevertheless be kept low to produce a product of high quality. Examples are:

- The number of weaving flaws in a standard test sample of 10 square yards of cloth
- The number of surface imperfections on a wooden desk top

- The number of typesetting errors in 10 pages of a book
- The number of imperfect solder joints in a particular subassembly

2.5 THE NORMAL DISTRIBUTION: ITS USE FOR APPROXIMATION

The rule that enables us to set up three-sigma control lines for frequencies as

$$\text{mean frequency} \pm 3\sqrt{\text{mean frequency}}$$

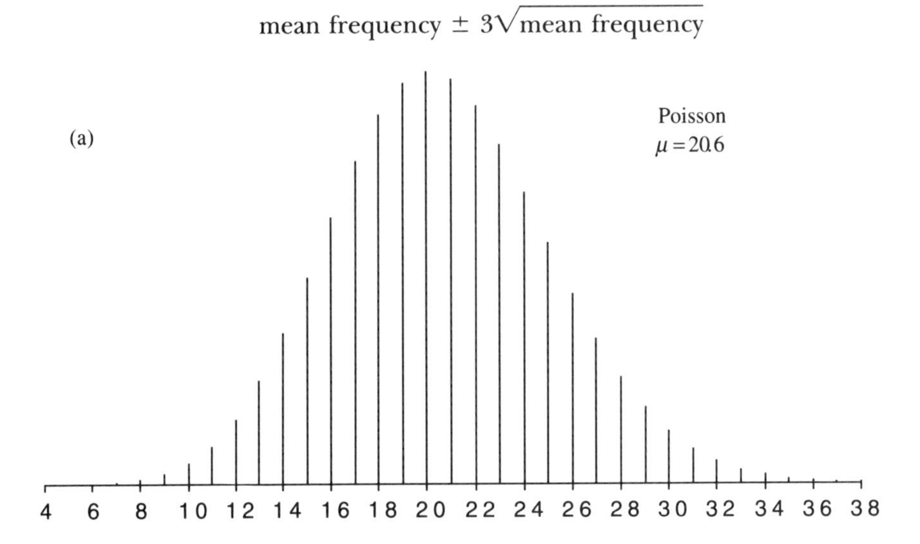

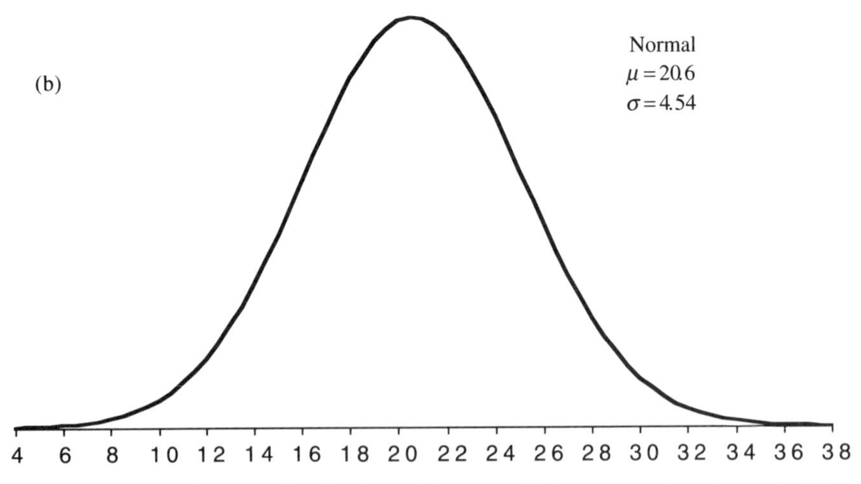

Figure 2.8 (a) A Poisson distribution with mean 20.6 representing the probabilities of different frequencies of accident. (b) An approximating normal distribution with mean 20.6 and standard deviation $\sqrt{20.6} = 4.54$.

is very convenient and simple. We have to digress a little to see where this rule comes from. Look again at the reference distribution on the right margin of Figure 2.4. This is a Poisson distribution with mean 20.6, which is shown once again in Figure 2.8a. In Figure 2.8b is an approximating smooth curve that fits it very well. This is the curve of the *normal distribution.* Its mathematical formula is given in Appendix 2C.

The normal distribution is useful for approximating other probability distributions as well as the Poisson, so we need to know more about it. The normal distribution is defined by two constants or "parameters": its mean and its standard deviation. You will see from Figure 2.9 that the normal distribution is symmetric about its mean value.

As illustrated in Figure 2.9 the standard deviation σ of the normal distribution is the distance from the point of inflection of the curve to the mean at μ. The point of inflection is the point where the gradient (slope) of the curve stops increasing and starts decreasing—the point where you start to feel a little more comfortable when you are on a roller coaster. Thus whereas the mean is a measure of *location,* which determines *where* the distribution is located, the standard deviation is a measure of *dispersion,* which determines *how widely* the distribution is spread. For illustration, Figure 2.10 shows normal distributions with $\mu = 25$ and 60 and with $\sigma = 2.5$ and 5. The normal distribution can be used as an approximating curve by suitably adjusting the constants μ and σ to match those of the distribution to be approximated. For illustration, look again at Figure 2.8a and 2.8b. The Poisson distribution in Figure 2.8a has a mean of 20.6 and a standard deviation of $\sqrt{20.6} = 4.5$. The smooth curve of Figure 2.8b is a normal distribution in which the mean and standard deviation are set equal to these same values.

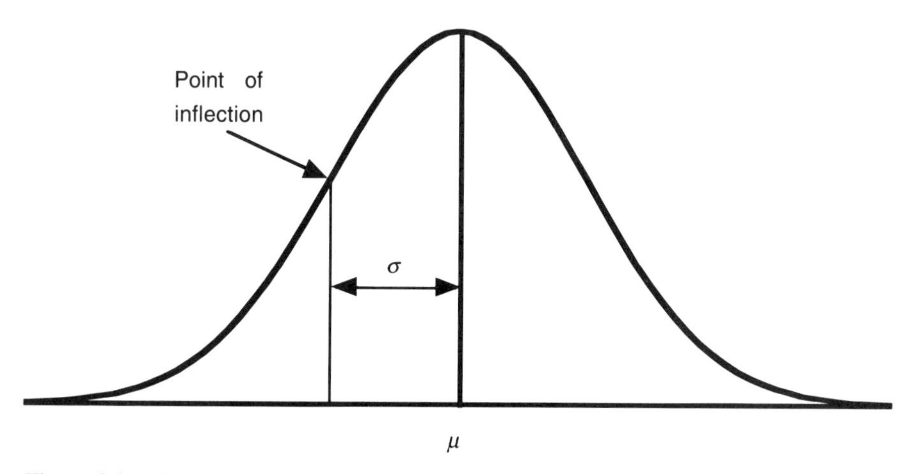

Figure 2.9 The mean, standard deviation, and point of inflection of the normal curve.

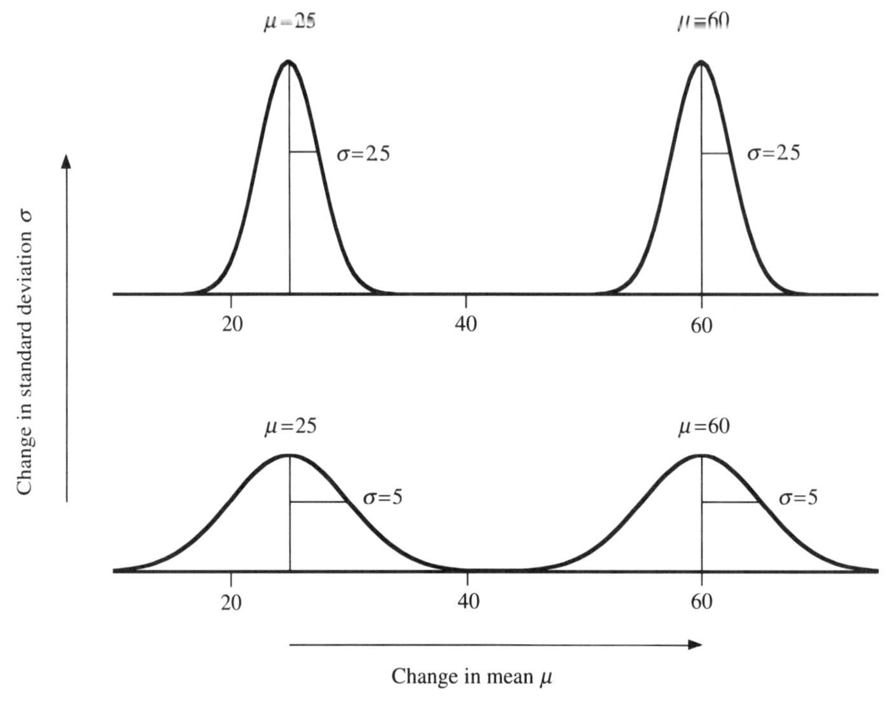

Figure 2.10 Normal distributions with means $\mu = 25$ and $\mu = 60$ and standard deviations $\sigma = 2.5$ and $\sigma = 5$.

Probabilities Associated with the Normal Distribution

Look at the normal distribution in Figure 2.11. The unshaded area, within the range of $\mu \pm \sigma$, represents a proportion 0.683 of the total area under the curve. Thus the chance of a quantity that is normally distributed lying within plus or minus one standard deviation of its mean is 68.3%, or about two-thirds. Equivalently, the probability of such a quantity falling in the two "tail" areas outside these limits is 31.7% or about one-third.

The unshaded and lightly shaded areas within the range $\mu \pm 2\sigma$ together represent a proportion 0.954 of the total area under the curve. So the chance for a quantity that is normally distributed lying within two standard deviations of the mean is 95.4% or about $\frac{19}{20}$. Equivalently, the probability of such a quantity falling in the two tail areas outside these limits is 4.6% or about $\frac{1}{20}$.

Finally, the proportion of the distribution within the range $\mu \pm 3\sigma$ represents 0.9973 of the total area. The chance of lying within three standard deviations of the mean is 99.73% or about $\frac{399}{400}$. Equivalently, the probability of falling outside these limits is 0.27% or about $\frac{1}{400}$.

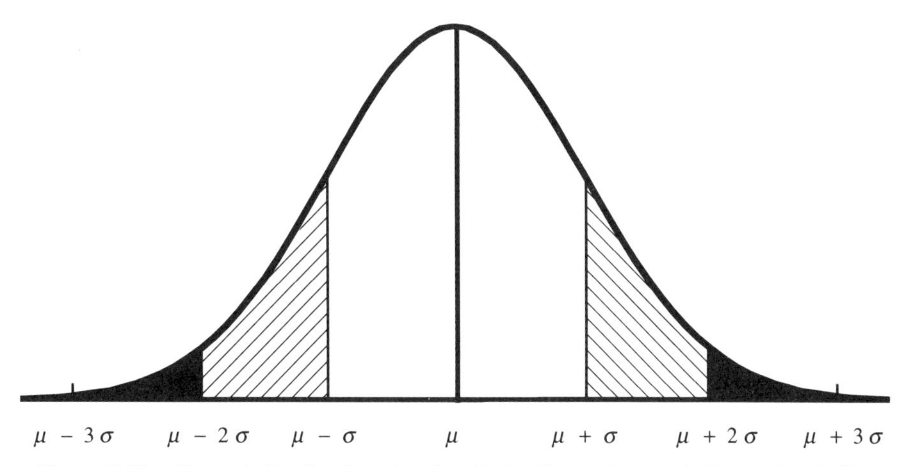

Figure 2.11 Normal distribution showing limits for $\mu \pm \sigma$, $\mu \pm 2\sigma$, and $\mu \pm 3\sigma$.

Table A at the back of this book gives an extended table of the normal distribution. It gives the probability of a normally distributed quantity deviating by as much as $\mu \pm k\sigma$ for various values of k. This is the area in *both tails* of the distribution. The probability associated with a single tail area can easily be obtained by halving the values in the table. For instance, for $k = 1$, the table gives the probability of lying outside $\mu \pm \sigma$ as 31.73%. It follows that $31.73/2 = 15.87\%$ is the probability of a normally distributed quantity *exceeding* $\mu + \sigma$.

Exercise 2.1 Suppose the measured strength of a synthetic yarn is approximately normally distributed with mean $\mu = 20.5$ and standard deviation $\sigma = 3.0$. What is the chance that the tested strength will be (a) less than 17.5, (b) between 17.5 and 20.0, and (c) greater than 26.0? □

From the above we can see why we might use the approximation $20.6 \pm (k \times 4.5)$ to calculate the $k\sigma$ control limits for the accident data. Random drawings from a Poisson distribution with mean 20.6 (which, if the process was in a state of control, would simulate the occurrence of the data) can be replaced, very nearly, by random drawings from a *normal* distribution with mean 20.6 and standard deviation $\sigma = \sqrt{20.6} = 4.5$.

2.6 ESTIMATING THE STANDARD DEVIATION FROM DATA

The standard deviation σ has so far been introduced as a measure of spread, which for the particular case of the normal distribution is the distance from the mean μ to the point of inflection of the curve. However, the standard deviation is an important measure of spread whether or not the data are normally distributed. In general, it measures the *root mean square deviation* of the data from the mean.

What is meant by this and how the standard deviation can be estimated from the data are best explained in terms of an example. Consider the sample of 26 weekly injury totals plotted in Figure 2.3 and previously listed. It was supposed earlier that we knew that the mean number of injuries determined over a long period of time was 20.6. If we treat this as the long-run mean μ, the data y and the deviations of the data from this mean $(y - \mu)$, are respectively,

y	28,	30,	18,	21,	18,	...,	28
$y - \mu$	7.4,	9.4,	−2.6,	0.4,	−2.6,	...,	7.4

The sum of squares of these deviations is

$$(7.4)^2 + (9.4)^2 + (-2.6)^2 + (0.4)^2 + (-2.6)^2 + \cdots + (7.4)^2$$
$$= 610.76.$$

If we had been able to make observations over a very long (theoretically infinite) period of time during which the process was completely stable, then the average of such squared deviations from the mean μ would be the long-run *variance* σ^2 for this process, sometimes called the "true" variance. All we have in fact are the 26 values of our sample.

However, if the process was in a state of control, then we could regard these as a random sample from the hypothetical longer series and their averaged squared deviation from the mean would provide an estimate of σ^2; namely,

$$610.76/26 = 23.49,$$

where the divisor $n = 26$ is the number of observations. Thus, assuming that we somehow knew the long-run mean $\mu = 20.6$, we could obtain an estimate for the standard deviation of

$$\sqrt{23.49} = 4.85.$$

Now almost invariably the true mean μ will not be known and we have to replace it by an estimate. This is supplied by the *sample average* $\bar{y}$, which uses the data we actually have. Thus we can write

$$\hat{\mu} = \bar{y},$$

where here and throughout this book the "hat" (applied in this case to μ) means "an estimate of." For the present case

$$\hat{\mu} = \bar{y} = (28 + 30 + 18 + 21 + \cdots + 28)/26 = 21.65$$

and the sum of squared deviations from this sample average is

$$(6.35)^2 + (8.35)^2 + (-3.65)^2 + (-0.65)^2 + \cdots + (6.35)^2 = 581.88.$$

By substituting that sample average for the true mean we have cheated a bit and it can be shown that, as a consequence, if we divide by $n = 26$, we will underestimate σ^2. This can be exactly allowed for by dividing not by $n = 26$ but by $n - 1 = 25$. Thus the *sample variance,* the estimate $\hat{\sigma}_S^2$ of σ^2 obtained entirely from the sample of data itself, is

$$\hat{\sigma}_S^2 = 581.88/25 = 23.28$$

and the corresponding *sample standard deviation* is

$$\hat{\sigma}_S = \sqrt{23.28} = 4.82.$$

The subscript S on $\hat{\sigma}_S$ is used here to denote that the estimate is obtained from the sum of squared deviations about the average.

If in general we denote a sample of n data values by

$$y_1, y_2, y_3, \ldots, y_n,$$

then the sample average is

$$\bar{y} = \sum y/n,$$

where the symbol Σ (capital sigma) indicates a sum. Also, the operation we have carried out above to obtain the sample variance can be written

$$\hat{\sigma}_S^2 = \frac{\sum (y - \bar{y})^2}{n - 1}.$$

By taking the square root of this quantity we get the sample standard deviation $\hat{\sigma}_S$. An alternative formula for the sum of squares $\Sigma(y - \bar{y})^2$ is $\Sigma\, y^2 - n\bar{y}^2$.

Exercise 2.2 Obtain the sample average and sample standard deviation for the following data:

6, 4, 3, 8, 4. □

For the accident data of Figure 2.4 the variance estimate based on the individual observations themselves is $\hat{\sigma}_S^2 = 23.28$, yielding a value for the standard deviation of $\hat{\sigma}_S = 4.82$. We saw earlier that if we are prepared to make the assumption that the data follow a Poisson distribution with a long-run mean of μ, then σ^2 will be equal to μ and $\sigma = \sqrt{\mu}$. In practice we will rarely know the value of μ, but if we replace μ by the average $\bar{y}$ of the data, then on the Poisson assumption a second estimate of σ is $\sqrt{\bar{y}}$. For this example then, $\bar{y} = 21.65$ and $\hat{\sigma}_D = \sqrt{21.65} = 4.65$ (where, by the subscript D in $\hat{\sigma}_D$, we mean an estimate that is appropriate only if the data follow the assumed Poisson *distribution*). For these data agreement between $\hat{\sigma}_S = 4.82$ and $\hat{\sigma}_D = 4.65$ is good; once more suggesting that the system depicted in Figure 2.5 is in a state of control. In making such comparisons, it should be borne in mind that estimates of the standard deviation can themselves be subject to large errors. Thus, an estimate $\hat{\sigma}_S$ has a percentage standard deviation of about $100/\sqrt{2(n - 1)}$. For example, the estimate $\hat{\sigma}_S = 4.82$ based on $n = 26$ observations itself has a percentage standard deviation of about $100/\sqrt{50} \approx 14\%$ and hence a standard deviation of about $4.82 \times 0.14 = 0.67$.

However, when the frequencies of accident are not in a state of control $\hat{\sigma}_S^2$ tends to be larger than $\hat{\sigma}_D^2$. For example, for the data in Figure 2.6, where the system is clearly out of control, $\hat{\sigma}_S^2 = 71.28$ and $\hat{\sigma}_D^2 = \bar{y} = 26.19$.[4] Equivalently, $\hat{\sigma}_S = 8.44$ and $\hat{\sigma}_D = 5.12$.

[4] In statistics texts it is shown that an approximate statistical test is provided by referring $(n - 1)\, \hat{\sigma}_S^2/\bar{y}$ to a table of the χ^2 distribution with $n - 1$ degrees of freedom. For this example the table shows that the probability that the discrepancy between the two estimates is due to chance is less than 1 in 1000.

Exercise 2.3. The data shown in Figure 2.6 for the "out of control" process are as follows:

Week	1	2	3	4	5	6	7	8	9	10	11	12	13
Patients	28	30	18	21	18	19	21	25	17	23	28	14	20
Week	14	15	16	17	18	19	20	21	22	23	24	25	26
Patients	25	22	12	19	36	30	43	40	30	35	38	32	37

Confirm that the sample average and sample variance are $\bar{y} = 26.19$ and $\hat{\sigma}_S^2 = 71.28$. □

Exercise 2.4. The data for Figure 2.7 are as follows:

Week	1	2	3	4	5	6	7	8	9	10	11	12	13
Patients	28	4	33	21	18	8	29	25	17	5	28	14	20
Week	14	15	16	17	18	19	20	21	22	23	24	25	26
Patients	13	8	32	12	23	8	31	25	22	16	9	29	28

Perform the same calculations for these data. What do you conclude? □

2.7 CHARTS FOR PROPORTIONS

Frequently we are interested in the *proportions* of manufactured articles that contain some defect. Often, however, the acceptable proportion of defects is extremely small and therefore can only be studied directly by sampling unacceptably large numbers of items from routine production. For this reason, an exaggerated test procedure is sometimes used. Thus condoms are sometimes tested by inflating a number of test specimens to a very high fixed pressure and noting the proportion that burst. Figure 2.12 shows data taken during the start-up of a machine making these articles. Studies from similar machines had shown that a high-quality product was produced if the proportion failing the test could be maintained at or below $p = 0.20$ (20%). The testing procedure consisted of taking a sample of $n = 50$ condoms every two hours from routine production and testing them. If the proportion defective was in fact equal to 0.20, the mean number failing the test would be 10 out of 50.

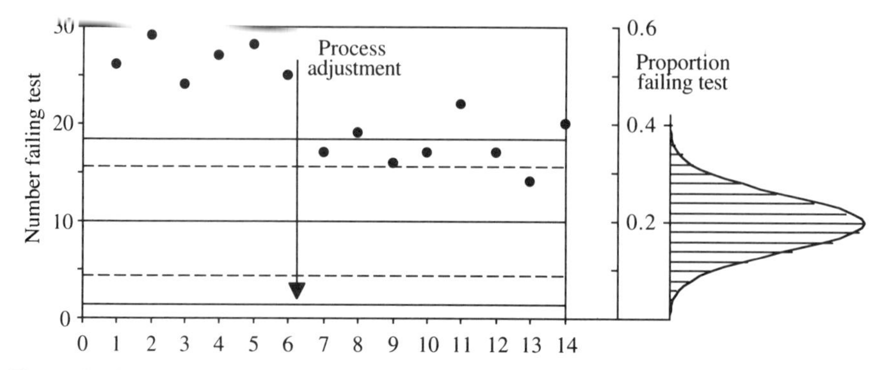

Figure 2.12 Number of condoms failing inflation tests during the start-up of a new machine. Also shown are a binomial reference distribution for $p = 0.20$ and $n = 50$, the approximating normal curve, and three-sigma action limits and two-sigma warning limits.

In Figure 2.12 the vertical scale on the right shows the *proportion* defective out of a sample of 50 tested items, while the scale on the left shows the corresponding *number* defective out of a sample of 50. The reference distribution indicated by the bars on the right shows the probabilities of getting various proportions failing if the process was in a state of control with the mean at the desired level of 0.20. Also shown are the appropriate three-sigma action limits and two-sigma warning limits. Obviously, the data show that during this start-up phase the process was badly out of control, with the number of items failing the test initially as high as 50%. A process modification made after observation number 6 brought the number of detectives down to around 40% but further measures were clearly needed to bring the process into a state of control at the desired level of $p = 0.20$.

The reference distribution indicated by the bars to the right of Figure 2.12 is called the *binomial* distribution. To understand what this distribution is, imagine a very large drum full of tickets containing a proportion of exactly $p = 0.20$ (20%) red tickets corresponding to "failures" and the rest white tickets corresponding to "successes." Suppose the tickets in the drum are well shuffled and a series of sample drawings of 50 tickets are made. Then, the binomial distribution will tell you exactly what the probability is of drawing a sample containing 0, 1, 2, . . . , 50 red tickets. The mathematical formula for the exact calculation of the binomial distribution is given in Appendix 2B.

Exact control limits corresponding to any desired degree of probability can be calculated from the binomial distribution, but once again there is a simple formula based on the approximating normal distribution (shown in Figure 2.12 by the continuous curve). To use the approximation we need to know what are the mean and the standard deviation for the binomial distribution. If the true proportion defective has some value p and a random sample of n items are taken of which number y fail to pass the test, then y will be distributed in a binomial distribution with mean $\mu = np$ and standard deviation $\sigma = \sqrt{np(1 - p)}$.

For our example the desired proportion defective is $p = 0.20$. Thus, for the process in a state of control, the mean number of failures in a sample of 50 is $\mu = 50 \times 0.20 = 10$ and the standard deviation of the number of failures is $\sigma = \sqrt{50 \times 0.20 \times 0.80} = \sqrt{8} = 2.83$. The normal distribution in Figure 2.12, which has $\mu = 10$ and standard deviation $\sigma = 2.83$, is seen to provide an excellent approximation.

For p less than 0.5 the normal approximation will be found to be reasonably good provided that the sample size n is greater than $12(1 - p)/p$. In the present example $p = 0.2$ and therefore n should be greater than $12 \times 0.8/0.2 = 48$. So, in our example, the size of $n = 50$ is adequate and we can calculate the three-sigma limits as $10 \pm (3 \times 2.83)$ and the two-sigma warning lines as $10 \pm (2 \times 2.83)$. In general then, the approximate $\pm k\sigma$ limit lines can be found from the simple formula $np \pm k\sqrt{np(1 - p)}$. With $k = 2$, we obtain "warning" limits, and if we put $k = 3$ we obtain "action" limits.

Exercise 2.5. Suppose that the proportion failing the test was maintained at $p = 0.10$ and that a sample of 120 condoms was used. (a) Could you use the normal approximation? (b) Calculate the warning and action limits for the number failing the test. □

Figure 2.13 shows the process during a subsequent period when it had been brought to a state of control with the mean number of defects in a 50-item sample varying about the desired level of 10. It might be thought that for a process of this kind, where we wish to keep the proportion of defectives *down*, only the upper limits would be of interest. Figure 2.13 illustrates a situation where this was not the case.

Note that the 10th and the 12th points on the chart *fall below the* -3σ *line.* This indicated that something unusual was occurring and actually gave rise to the finding of an assignable cause. Investigation showed that the testing procedure was responsible for these aberrant points. A fault in the air line had developed and the condoms tested at about this

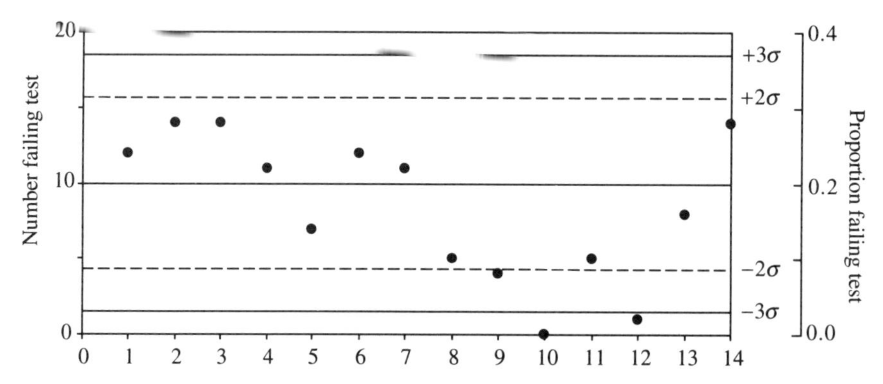

Figure 2.13 Control chart showing operation of the process after a state of control had been achieved with p close to 0.20.

time were inadvertently submitted to a much reduced air pressure, resulting in a falsely low value of the proportion defective. Action was taken therefore to ensure that this could not occur in future production.

If you use the binomial distribution with a value of n less than $12(1 - p)/p$, the approximation based on the normal distribution should *not* be used. In that case you can obtain appropriate control limits from Table 2.2.

For each combination of p and n the table shows five entries. These are the lower action limit, lower warning limit, mean np, upper warning limit, and upper action limit. The limits have been obtained by interpolation in the binomial distribution function to yield the same probabilities associated with $\pm 2\sigma$ and $\pm 3\sigma$ limits of the normal distribution. In practice, of course, the data are whole numbers; thus, for example, using the table if $n = 80$ and $p = 0.1$, the action limits would be crossed if the number defective was smaller than 1 or larger than 16 as would the warning limits, if the number defective was smaller than 3 or larger than 13.

2.8 ARE WE USING THE RIGHT REFERENCE DISTRIBUTION?

A friend of ours recently moved from Madison, Wisconsin, to San Francisco, California. He and his wife started to look for a house and were surprised by the high prices. When you evaluate a house, you unconsciously compare it with a mental reference distribution of what

TABLE 2.2 Interpolated Action Limits and Warning Limits Calculated from the Binomial Distribution Function*

n \ p	0.2					0.15					0.1					0.05					0.01				
5	—	—	1	2.69	3.84	—	—	0.75	2.16	3.41	—	—	0.5	1.81	2.89	—	—	0.25	1.00	1.99	—	—	0.05	0.55	0.99
10	—	—	2	4.38	5.91	—	—	1.50	3.68	5.03	—	—	1.0	2.83	4.19	—	—	0.50	1.85	2.97	—	—	0.10	0.80	1.70
15	—	—	3	5.89	7.84	—	—	2.25	4.87	6.75	—	—	1.5	3.77	5.46	—	—	0.75	2.44	3.85	—	—	0.15	0.90	1.90
20	—	0.19	4	7.42	9.61	—	—	3.00	5.98	8.00	—	—	2.0	4.64	6.53	—	—	1.00	2.89	4.55	—	—	0.20	0.96	1.98
25	—	0.80	5	8.82	11.16	—	0.07	3.75	7.16	9.48	—	—	2.5	5.45	7.51	—	—	1.25	3.42	4.98	—	—	0.25	1.13	2.33
30	0.01	1.36	6	10.18	12.80	—	0.37	4.50	8.28	10.74	—	—	3.0	6.17	8.43	—	—	1.50	3.84	5.71	—	—	0.30	1.41	2.64
35	0.27	2.09	7	11.58	14.33	—	0.93	5.25	9.35	11.90	—	—	3.5	6.92	9.30	—	—	1.75	4.29	6.14	—	—	0.35	1.59	2.80
40	0.92	2.72	8	12.86	15.82	—	1.29	6.00	10.40	13.05	—	0.12	4.0	7.73	10.11	—	—	2.00	4.74	6.76	—	—	0.40	1.71	2.90
45	1.30	3.39	9	14.16	17.28	0.13	1.82	6.75	11.42	14.30	—	0.32	4.5	8.46	10.95	—	—	2.25	5.06	7.20	—	—	0.45	1.81	2.97
50	2.01	4.14	10	15.49	18.74	0.40	2.27	7.50	12.43	15.47	—	0.61	5.0	9.12	11.84	—	—	2.50	5.58	7.76	—	—	0.50	1.88	3.17
55	2.44	4.89	11	16.75	20.08	0.96	2.81	8.25	13.43	16.59	—	1.02	5.5	9.85	12.70	—	—	2.75	5.91	8.16	—	—	0.55	1.94	3.45
60	3.12	5.57	12	17.97	21.56	1.21	3.29	9.00	14.41	17.69	—	1.23	6.0	10.58	13.50	—	—	3.00	6.35	8.71	—	—	0.60	2.00	3.64
65	3.70	6.32	13	19.26	22.89	1.61	3.86	9.75	15.38	18.77	0.04	1.51	6.5	11.24	14.24	—	—	3.25	6.74	9.05	—	—	0.65	2.21	3.77
70	4.30	7.12	14	20.52	24.28	2.10	4.34	10.50	16.35	19.83	0.15	1.92	7.0	11.90	14.96	—	—	3.50	7.04	9.63	—	—	0.70	2.38	3.86
75	5.04	7.89	15	21.75	25.66	2.44	4.96	11.25	17.30	20.88	0.32	2.19	7.5	12.63	15.80	—	0.02	3.75	7.51	9.96	—	—	0.75	2.52	3.94
80	5.62	8.62	16	22.94	26.94	3.02	5.43	12.00	18.25	21.92	0.58	2.49	8.0	13.28	16.58	—	0.09	4.00	7.85	10.50	—	—	0.80	2.63	3.99
85	6.28	9.39	17	24.17	28.32	3.35	6.05	12.75	19.18	22.95	1.00	2.90	8.5	13.91	17.31	—	0.17	4.25	8.21	10.88	—	—	0.85	2.72	4.23
90	7.04	10.20	18	25.41	29.65	3.91	6.54	13.50	20.11	23.98	1.14	3.20	9.0	14.62	17.98	—	0.27	4.50	8.62	11.34	—	—	0.90	2.80	4.44
95	7.65	11.03	19	26.62	30.92	4.29	7.14	14.25	21.03	25.01	1.33	3.51	9.5	15.26	18.79	—	0.39	4.75	8.92	11.76	—	—	0.95	2.87	4.60
100	8.33	11.80	20	27.81	32.25	4.82	7.69	15.00	21.96	26.04	1.63	3.94	10.0	15.89	19.54	—	0.54	5.00	9.33	12.11	—	—	1.00	2.93	4.72

*Each cell of the table shows five entries for each combination of p and n. These entries are the lower action limit, lower warning limit, mean np, upper warning limit and upper action limit.

your experience tells you it might be worth. The *appropriate* reference distributions of house prices in Madison and San Francisco are very different and, obviously, you could make serious errors of judgment if you had not taken account of this. The relevance of this to the quality problem lies in the application of the reference distribution, which decides the $\pm k\sigma$ limits of, for example, the Shewhart chart.

Overdispersion[5]

In discussing the Poisson distribution we mentioned that if the process was in a perfect state of control the mean frequency μ of accidents per week would stay constant. Now this might not be so. For example, in some weeks certain more dangerous machinery, not used at other times, might be employed. This could cause the mean μ itself to vary from one week to another. Suppose that the variance of the μ values from week to week was σ_μ^2 and that the mean of these μ values was $\bar{\mu}$. Then (e.g., see Box, Hunter, and Hunter, 1978, p. 143) the overall variance for weekly accidents would not be $\bar{\mu}$ but

$$\sigma^2 = \bar{\mu} + \sigma_\mu^2$$

and σ would be the square root of this value.[6]

Similarly, for the binomial distribution example concerning the testing of 50 condoms at two-hour intervals, the probability of bursting might vary from one test of 50 condoms to another, depending on the temperature of the room or on some other factors. Suppose this resulted in a variance σ_p^2 *for p itself* about a mean value (say, $\bar{p} = 0.2$).

[5] Overdispersion is a well-developed statistical subject and many authors have devoted attention to it. We mention a few references: Altham (1978); Breslow (1984); Cameron and Trivedi (1986); Chatfield and Goodhardt (1970); Collings and Margolin (1985); Cox (1983); Crowder (1978, 1985); Dean (1992); Dean and Lawless (1989); Fisher (1950); Haseman and Kupper (1979); Jorgensen (1987); Kupper and Haseman (1978); McCullagh and Nelder (1989); Moore (1987); Prentice (1986); Tarone (1979); and Williams (1975, 1982).

[6] In Section 2.1 we mentioned two other assumptions required to obtain the Poisson distribution. When any of these assumptions is false, overdispersion may appear. Thus *autocorrelations* between the observations in the sample may cause overdispersion (e.g., see Al-Osh and Alzaid, 1987; Barron, 1992; Cox and Solomon, 1988; Luceño, 1995a, 1996b; McKenzie, 1988; Zeger, 1988). Similarly, autocorrelation may produce extrabinomial variation (Luceño and de Ceballos, 1995). Also, the possibility of observing *simultaneous* events—for example, two or more injuries at the same time—may cause overdispersion.

Then the overall variance for the condoms failing the test would not be $n\bar{p}(1 - \bar{p})$ but[7] $\sigma^2 = n\{\bar{p}(1 - \bar{p}) + (n - 1)\sigma_p^2\}$.

Now there are two ways to think about this. One could say that the influences that changed the probability of accidents from week to week or the probability of failures of condoms from test to test are *special causes* that should be tracked down and eliminated. To do this, it would be argued, you need to use limit lines obtained from the Poisson and binomial distributions, representing processes in a perfect state of control. For example, by using Poisson limits, overdispersion was very obvious in the data of Figure 2.7 and, as we saw, this was almost certainly due to an assignable cause that could be eliminated, namely, the underreporting of accidents in the last week of the month.

However, there are some instances where overdispersion is known to occur but attempts to discover its cause have failed. It may then be decided that at least for the time being overdispersion has to be regarded as inevitable and to avoid too many false alarms while searching for other special causes,[8] the limits must be widened. To provide a check one may employ two other methods for estimating σ namely the *rational subgroup method*, and the *moving range method.*

Rational Subgroup Method

This method, due to Shewhart, amounts to looking over past records to find periods where the process is judged to have been stable. By calculating the variances during these periods and pooling the results, you can obtain an overall estimate, which we will denote by $\hat{\sigma}_G^2$.

Suppose that within, say, three such "stable" periods of n_1, n_2, and n_3 observations you calculated the variances $\hat{\sigma}_1^2$, $\hat{\sigma}_2^2$, and $\hat{\sigma}_3^2$ by the "sum of squares" method. Then the overall estimate $\hat{\sigma}_G^2$ would be obtained from

$$\hat{\sigma}_G^2 = \frac{(n_1 - 1)\hat{\sigma}_1^2 + (n_2 - 1)\hat{\sigma}_2^2 + (n_3 - 1)\hat{\sigma}_3^2}{n_1 + n_2 + n_3 - 3}.$$

Moving Range Method

Another way to obtain an estimate of σ is to use the *moving range.* This is the average absolute change from one observation to the next. By

[7] The probability might also vary *within a test of* 50 *condoms,* which would have the effect of slightly *reducing* the overall variance. This effect, however, is small and we ignore it in this discussion.

[8] Overdispersion may be caused by autocorrelations between the observations, among other things. The cause of these autocorrelations may be difficult to find, or if known, it may be impossible or undesirable to eliminate.

absolute change is meant the difference ignoring the sign. We illustrate the calculation of moving ranges by returning to the accident data plotted in Figure 2.3.

For these data the successive ranges are $|30 - 28| = 2$, $|18 - 30| = 12$, $|21 - 18| = 3$, and so on. The vertical bars refer to differences when the sign is ignored. In all, 25 values of the moving range are available. The average of these is $\overline{MR} = 5.76$. Based on the assumption that the data are roughly normally distributed and statistically independent, an estimate of the standard deviation can be obtained by dividing $\overline{MR}$ by the factor 1.128. Thus the estimate $\hat{\sigma}_M$, say, obtained from $\overline{MR}$ is $\hat{\sigma}_M = 5.76/1.128 = 5.11$. This value is slightly larger than the previous $\hat{\sigma}_D = 4.65$, but such a difference could well be due to sampling variation.

When there are a number of rational subgroups, the moving range method can be used by averaging all the ranges of individual pairs within the subgroups. It should be bourne in mind that the rational subgroup method can lead to underestimation of the appropriate σ because of a tendency to choose *unusually* uniform periods of operation as subgroups, which underestimate the typical variation of the process when in a "stable" condition. Also, the moving range method, in addition to assuming that the data are normally distributed, also assumes that successive observations are *statistically independent.* This means that the level of one observation does not affect the probability of where the next one will fall. In some circumstances, however, there is a tendency for a high value to be followed by another high value and vice versa. Successive observations are then said to be *positively autocorrelated.* This causes the moving range method to underestimate σ. There are less frequent circumstances where successive observations can be negatively correlated, which could result in overestimation of σ. Control charts for autocorrelated data are discussed in Chapters 10 and 11.

Exercise 2.6. The data for the number of condoms failing the test plotted in Figure 2.12 are as follows:

26, 29, 24, 27, 28, 25, 17, 19, 16, 17, 22, 17, 14, 20.

Estimate the standard deviation you might use for a control chart by (a) treating the data before and after the process adjustment as two rational subgroups and, (b) calculating the moving range for each subgroup and averaging. (c) Comment on the appropriateness of these two methods. (d) Would you expect the standard deviation to be the same after adjustment? □

Know Your Process

So, you may ask, which method do you recommend that we use? In this book you will find very few golden rules, because, as always, what is best to do can never depend solely on statistics and mathematics; it depends very much on circumstances and, in particular, the nature of your particular process. We suggest that, when several methods are available to estimate σ, you should look at them all, bearing in mind what you know about the system for which the estimate is to be used.

For illustration we employ the count data of Figures 2.3, 2.6, and 2.7. For the data of Figure 2.3, we saw that if we use the sample average 21.65 to estimate μ the Poisson assumption yields an estimate of $\hat{\sigma}_D = \sqrt{21.65} = 4.65$. Direct calculation from the sample variance of the 26 observations gives $\hat{\sigma}_S = 4.82$ and the moving range estimate is $\hat{\sigma}_M = 5.11$. These are all in excellent agreement. Table 2.3 shows estimates of σ given by these three methods for the data of Figures 2.3, 2.6, and 2.7. None of these sets of data is really suitable to illustrate the rational subgroup method. For completeness, however, we illustrate this method with the data of Figure 2.6, treating the first 17 observations as one subgroup and the last 9 as another. The estimated variance obtained by the sum of squares method is 24.28 for the first group and 19.75 for the second, so that the pooled estimate is

$$\hat{\sigma}_G^2 = \frac{(16 \times 24.28) + (8 \times 19.75)}{16 + 8} = 22.77.$$

So $\hat{\sigma}_G$ is $\sqrt{22.77} = 4.77$ as shown in the table. Alternatively, we can

TABLE 2.3 Estimates $\hat{\sigma}_D$, $\hat{\sigma}_S$, and $\hat{\sigma}_M$ for the Data of Figures 2.3, 2.6 and 2.7. For Figure 2.6 $\hat{\sigma}_G$ is also given.

Figure	Apparent State of Control	Average	Poisson $\sqrt{\text{Average}}$ $\hat{\sigma}_D$	Sum of Squares $\hat{\sigma}_S$	Moving Range $\hat{\sigma}_M$	Rational Subgroups $\hat{\sigma}_G$
2.3	In control	21.65	4.65	4.82	5.11	
2.6	Changing mean	26.19	5.12	8.44	5.64	4.77(5.25)
2.7	Underestimation in 4th week	19.46	4.41	9.07	11.13	

combine the moving ranges within the rational subgroups. (This can be done by omitting from the average of the 23 paired ranges the value $36 - 19 = 17$, which occurs between the two "subgroups.") This gives the value 5.25 shown in brackets in the table.

As always, to appreciate what is happening you need to *look at the data* (and think about the process) at the same time that you look at the estimates. For the data of Figure 2.3, where the process appears to be in a state of control, the estimates agree very well. So that from the data we would probably conclude that the Poisson estimate $\hat{\sigma}_D = 4.65$ explains the "in control" operation of the process very well. For the data of Figure 2.6, the high value for $\hat{\sigma}_S = 8.44$ is evidently caused by the change in mean. This change does not affect the estimate $\hat{\sigma}_G$ from "rational" subgroups. For the data from Figure 2.7 $\hat{\sigma}_S$ and $\hat{\sigma}_M$ are large because of the negative correlation produced by fourth week underestimation of casualties.

Some Things to Know About Means and Variances

A machine produces a soft drink by automatically dispensing into cans three ingredients—orange juice, lemon juice, and carbonated water. The amounts dispensed (in milliliters) and their means and standard deviations are as follows:

	Orange Juice	Lemon Juice	Carbonated Water
Amount dispensed	y_1	y_2	y_3
Mean	$\mu_1 = 200$	$\mu_2 = 100$	$\mu_3 = 300$
Standard deviation	$\sigma_1 = 2$	$\sigma_2 = 1$	$\sigma_3 = 5$
Variance	$\sigma_1^2 = 4$	$\sigma_2^2 = 1$	$\sigma_3^2 = 25$

Denote by $Y = y_1 + y_2 + y_3$ the total liquid dispensed into a particular can. A question of interest is: What is the mean μ_Y of the total contents of a can? As might be expected,

$$\mu_Y = \mu_1 + \mu_2 + \mu_3 = 200 + 100 + 300 = 600.$$

Thus the mean content of a can is 600 ml.

Another question is: What is the standard deviation of Y? This will tell us how much the total contents of the can may vary. If, as in this case might be reasonable, the amount of each ingredient dispensed is statistically independent of the others, then the variance of the total Y is just the sum of the individual variances:

$$\sigma_Y^2 = \sigma_1^2 + \sigma_2^2 + \sigma_3^2 = 4 + 1 + 25 = 30.$$

Consequently,

$$\sigma_Y = \sqrt{30} \approx 5.5.$$

Two interesting conclusions from the above analysis are as follows:

1. Most of the variation in the total contents of the can comes from variation in the amount of carbonated water. Any attempt to reduce overall variation should concentrate on this factor.
2. Assuming that Y is roughly normally distributed, you could, for example, conclude that 95% of the cans would have total content between $600 - 11 = 589$ ml and $600 + 11 = 611$ ml.

Now suppose the cost of 1 ml of each of the ingredients in hundredths of a cent is as follows:

	Orange Juice	Lemon Juice	Carbonated Water
Cost	2	4	1

Then again we might be interested in the mean μ_C and standard deviation σ_C of the overall cost C. Now

$$\begin{aligned} C &= 2y_1 + 4y_2 + 1y_3, \\ \mu_C &= 2\mu_1 + 4\mu_2 + 1\mu_3 \\ &= (2 \times 200) + (4 \times 100) + (1 \times 300) = 1100. \end{aligned}$$

Thus the mean cost of the ingredients is 11 cents.

To obtain the variance, the coefficients 2, 4, and 1 must be squared.

$$\begin{aligned} \sigma_C^2 &= 4\sigma_1^2 + 16\sigma_2^2 + 1\sigma_3^2 \\ &= (4 \times 4) + (16 \times 1) + (1 \times 25) = 57. \end{aligned}$$

Thus

$$\sigma_C = \sqrt{57} = 7.5 \quad (0.075 \text{ cents}).$$

The expression for cost, $2y_1 + 4y_2 + 1y_3$, is referred to as a *linear combination* (or a *linear aggregate*) of the *random variables* y_1, y_2, and y_3.

The Mean and Variance of a Linear Combination of k Random Variables

In general, if we have k random variables with coefficients $c_1, c_2, \ldots, c_k$ (corresponding to the individual costs in the above example, but which in other applications could be positive or negative), then the linear aggregate

$$Y = c_1 y_1 + c_2 y_2 + \cdots + c_k y_k$$

has mean

$$\mu_Y = c_1 \mu_1 + c_2 \mu_2 + \cdots + c_k \mu_k.$$

And if the y's vary independently with variances $\sigma_1^2, \sigma_2^2, \ldots, \sigma_k^2$, then

$$\sigma_Y^2 = c_1^2 \sigma_1^2 + c_2^2 \sigma_2^2 + \cdots + c_k^2 \sigma_k^2. \tag{2.4}$$

The Arithmetic Average

An example concerns the mean and variance of the sample average of n observations $y_1, y_2, \cdots, y_n$, which are supposed to vary independently and are from the same distribution that has mean μ and variance σ^2.

The sample average is

$$\bar{y} = \frac{1}{n} \sum y = \frac{1}{n} y_1 + \frac{1}{n} y_2 + \cdots + \frac{1}{n} y_n.$$

So

$$\mu_{\bar{y}} = \frac{1}{n}\mu + \frac{1}{n}\mu + \cdots + \frac{1}{n}\mu = \mu,$$

$$\sigma_{\bar{y}}^2 = \frac{1}{n^2}\sigma^2 + \frac{1}{n^2}\sigma^2 + \cdots + \frac{1}{n^2}\sigma^2 = \frac{\sigma^2}{n},$$

and

$$\sigma_{\bar{y}} = \frac{\sigma}{\sqrt{n}}.$$

Thus, for example, the average of nine independently distributed observations has standard deviation $\sigma/3$. This sample average is called the *arithmetic* average. In the next chapter we will discuss a different kind of average—an exponentially weighted average. We will use Equation (2.4) to find the standard deviation of this weighted average.

APPENDIX 2A THE POISSON DISTRIBUTION OF RANDOM FREQUENCIES

Suppose there is a very small probability of an event, such as an accident, occurring in a given small interval of time such as a second. Then, in an extended period of time, such as a week, which contains many thousands of seconds, we can expect that some events (accidents) will occur. Some weeks there will be no accident, some weeks there will be just one, some weeks there will be two, and so on.

It turns out that under certain conditions (mentioned in Sections 2.1 and 2.8 and footnotes 1 and 6) the probability of 0, 1, 2 events occurring in a week depends only on the *mean* number μ of events occurring in a week. The probability is exactly given by a distribution ascribed to the French mathematician Poisson. Below we show the general form of the distribution and illustrate the calculation for the value $\mu = 20.6$, which is the value used in the clinic example in this chapter.

Number of events y	0	1	2	3	4
Probability is proportional to:	1	μ	$\frac{\mu^2}{2}$	$\frac{\mu^3}{2 \times 3}$	$\frac{\mu^4}{2 \times 3 \times 4}$
For example, if $\mu = 20.6$,	1	20.6	$\frac{(20.6)^2}{2}$	$\frac{(20.6)^3}{2 \times 3}$	$\frac{(20.6)^4}{2 \times 3 \times 4}$

These probabilities must add up to unity so that to obtain the individual probabilities you need to divide by $(1 + \mu + \mu^2/2 + \cdots) = e^{\mu}$ ($e^{20.6}$ for our example). Some things to remember about the Poisson distribution are:

- The distribution has the property that the variance σ^2 is equal to the mean μ so that the standard deviation is $\sigma = \sqrt{\mu}$.
- The same distribution applies to rare events that occur not in time but in space, such as blemishes on a table top.
- When μ is 12 or more, the Poisson distribution is well approximated by the normal distribution, which is discussed further in Appendix 2C.
- The Poisson distribution is a limiting case of the binomial distribution when p is chosen to be very small and n very large. This is discussed further in Appendix 2B.

APPENDIX 2B THE BINOMIAL DISTRIBUTION FOR RANDOM PROPORTIONS

Suppose you have a very large box of transistors, a proportion p (say, 0.1) of which are defective and a proportion $q = 1 - p$ (say, 0.9) are not defective. Suppose they can be thoroughly mixed up so that when you take a transistor from the box, this represents a *random* drawing; that is, *each transistor has an equal chance of being drawn.* Now suppose you take a random sample of n (say, 20) transistors from the box. On average, you would expect a number $n \times p$ ($0.1 \times 20 = 2$) would be defective, but obviously you won't get exactly that number each time you draw 20 items from the box. The binomial distribution tells you the probability (in what proportion of the time) that in a sample of n (20) you should get no defectives, one defective, two defectives, and so on.

This is given mathematically by the following:

Number of defectives y	0	1	2	3	4
Probability	$C_0 q^n$	$C_1 p\, q^{n-1}$	$C_2 p^2 q^{n-2}$	$C_3 p^3 q^{n-3}$	$C_4 p^4 q^{n-4}$

where

$$C_y = \frac{n!}{y!\,(n-y)!} = \frac{n \times (n-1) \times \cdots \times (n-y+1)}{1 \times 2 \times \cdots \times y}.$$

For the transistor sample , $n = 20$, $p = 0.1$, and $q = 1 - p = 0.9$.

It is fairly easy to see where the factor C_y affecting $p^y q^{n-y}$ originates. This $p^y q^{n-y}$ is just the probability of finding, for example, that the *first* y transistors that you draw are all bad and the *last* $n - y$ are all good. But you might get y bad ones and $n - y$ good ones *in some other order,* so this is only one of the ways of getting y bad ones and $n - y$ good ones. Hence we have to multiply by the number of different ways of getting y bad ones and $n - y$ good ones. This is given by C_y. For illustration, with $n = 20$, $C_0 = 1$, $C_1 = 20$, $C_2 = (20 \times 19)/2 = 190$, $C_3 = (20 \times 19 \times 18)/6 = 1140$, so that

Number of defectives y	0	1	2	3 . . .
Probability	$(0.9)^{20}$	$20(0.1)(0.9)^{19}$	$190\,(0.1)^2(0.9)^{18}$	$1140\,(0.1)^3(0.9)^{17}$ . . .
	0.1216	0.2702	0.2852	0.1901 . . .

The mean of the distribution of the number of defectives is $\mu = np$ and the variance is $\sigma^2 = npq$, so that the standard deviation is $\sigma = \sqrt{npq}$. Provided n is not too small and p is not too extreme (not too close to 0 or to 1), the binomial distribution can be approximated by the normal distribution as we have illustrated in this chapter. For $p < 0.5$, the normal approximation works quite well if

$$n \geq 12\,(1 - p)/p.$$

If, in the expression for the binomial probability, you make n very large and *at the same time* make p very small then the distribution gets closer and closer to the Poisson distribution. For example, if you were to work out the probabilities for a binomial distribution with, say, $n = 10{,}000$ and $p = 1/1000$ (so that $\mu = np = 10{,}000 \times 0.001 = 10$), you would find they were almost exactly equal to those for a Poisson distribution with mean $\mu = 10$. Furthermore, this binomial distribution would have mean $\mu = 10$ and variance $\sigma^2 = 9.99$ ($\sigma^2 = 10{,}000 \times 0.001 \times 0.999 = 9.99$), which are almost identical, as would be expected of a distribution very close to the Poisson.

Note also that by setting $\mu = np$ in the rule for adequate approximation of the binomial by the normal distribution, we get $\mu = np \geq 12\,(1 - p)$, which as $1 - p$ gets close to 1 approaches the rule for the Poisson $\mu \geq 12$.

APPENDIX 2C THE NORMAL DISTRIBUTION

Why should random errors tend to have a distribution that is approximated by the normal rather than by some other distribution? There are many different explanations—one of which relies on the central limit effect. This central limit effect does not apply only to averages. If the overall error e is an aggregate of a number of component errors $e = e_1 + e_2 + \cdots + e_m$, such as sampling errors, measurement errors, or manufacturing variation, in which no one component dominates, then almost irrespective of the distribution of the individual components, it can be shown that the distribution of the aggregated error e will tend to the normal as the number of components gets larger.

A different and somewhat remarkable argument is due to Clark Maxwell. He was concerned with proving that the velocities of individual molecular particles would be normally distributed. We will use this argument but, for simplicity, will change the context a little and pose the problem in terms of a sharpshooter firing at a target. Suppose

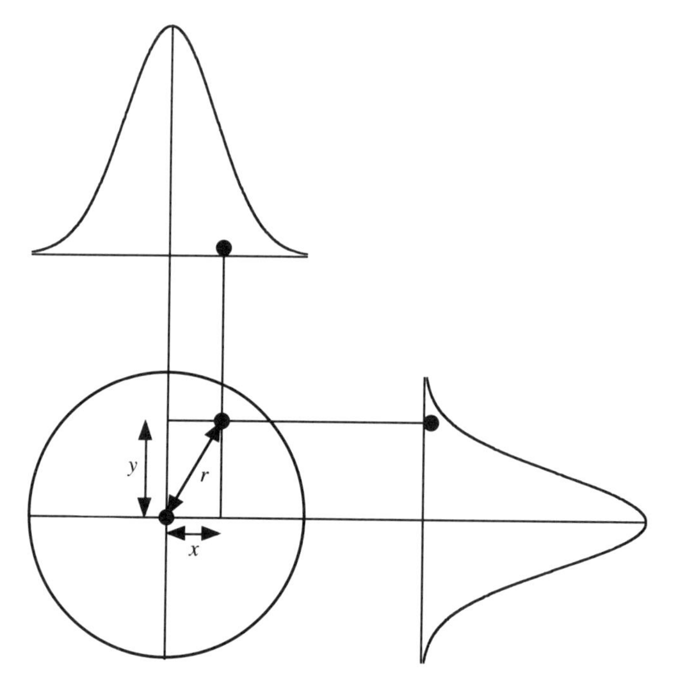

Figure 2C.1 An illustration of Clark Maxwell's derivation of the normal distribution.

neither the sharpshooter nor the rifle have any particular biases of any kind. Then, looking at Figure 2C.1, the following two postulates seem reasonable:

1. The probability of a shot hitting the target at some distance r from the bullseye is the same for every point at the same distance r from the center. That is, the probability is constant on a circumference of radius r centered at the bullseye.
2. The probability of the horizontal coordinate x having any particular value is completely independent of the probability of the vertical coordinate y having any particular value.

The remarkable fact is that, as soon as these two postulates are admitted, x and y *must* be individually normally distributed.

For those interested in mathematical matters, a very rough sketch of the proof is as follows: From postulate 1, the probability distribution of x and y must be of the form

$$p(x, y) = p(r^2) = p(x^2 + y^2).$$

But from postulate 2,

$$p(x, y) = p(x)\,p(y).$$

The only function for which $p(x)p(y) = p(x^2 + y^2)$ is the exponential function, so that $p(y)$ must be of the form ae^{by^2}, where a is a positive constant and b is a constant. Likewise for $p(x)$, so that

$$p(x, y) = p(x)p(y) = a^2 e^{b(x^2+y^2)}.$$

But the probability becomes less as x and y get larger, so that b must be negative. If we call it $-c^2$, then

$$p(y) = ae^{-c^2 y^2}.$$

If the reader will graph this function for any values of a and c, a normal curve will be obtained of the shape we have already seen in Figure 2.9.

The constants may be expressed in a more commonly used form as follows. The constant c is conveniently reexpressed in terms of the standard deviation σ, that is, the root mean square error of y. It can be shown that this requires that $c^2 = 1/2\sigma^2$. Also, since y *must* be somewhere, the constant a must be chosen to make the total area (probability) under the curve equal to 1. This requires that $a = 1/\sqrt{2\pi\sigma^2}$. Finally, the data can be centered about some mean μ other than zero by substituting $y - \mu$ for y. Putting all this together we obtain the general form of the normal distribution:

$$p(y) = \frac{1}{\sqrt{2\pi\sigma^2}} e^{-(y-\mu)^2/2\sigma^2}.$$

The normal distribution is important not only because it is a probability curve that frequently approximates the distribution of random error, but also because it provides an approximation for other distributions such as the Poisson and the binomial.

CHAPTER 3

Control Charts for Measurement Data

"When it is not necessary to change, it is necessary not to change."
A Discourse of Infallibility, VISCOUNT LUCIUS CARY FALKLAND

"Things have their due measure; there are ultimately fixed limits, beyond which, or short of which something must be wrong."
HORACE

Earlier we saw that the purpose of any quality control chart was to:

1. Monitor the common cause system.
2. Highlight *special* causes of deviation, which might then be tracked down and permanently eliminated.

A properly operating chart thus provides a continuous incentive for quality improvement and can help to ensure that gains in quality attained by quality improvement initiatives are locked in. We saw that a stable system, that is, a system in a state of statistical control, was characterized by a reference distribution. In practice, we did not need to use the whole reference distribution. Limit lines derived from the reference distribution and drawn at $\pm 3\sigma$ and $\pm 2\sigma$ were usually adequate to indicate whether or not the system was in a state of control. These limit lines could characterize the *common cause* behavior of the process, and furthermore they could indicate aberrant values pointing to *special causes* worth looking into.

For frequency data, when random variation was characterized by the Poisson distribution and the target frequency was T, the process would be in a state of control if the mean μ was equal to the target T and the observed frequency varied about the mean with standard deviation

$\sqrt{\mu} = \sqrt{T}$. When the normal approximation was adequate, the $k\sigma$ control limits were then given by $T \pm k\sqrt{T}$. For proportion defective data, when random variation was characterized by the binomial distribution and the target proportion defective was T, the process would be in a state of control if the number of occurrences in a sample of n had mean np equal to nT and standard deviation $\sqrt{np(1-p)} = \sqrt{nT(1-T)}$. When the normal approximation was adequate, the $k\sigma$ control limits were given by $nT \pm k\sqrt{nT(1-T)}$.

It was pointed out, however, in Section 2.8 that in certain practical situations control limits based on the theoretical standard deviations of the Poisson and binomial distributions could be too narrow because of "overdispersion." Overdispersion occurs, for example, when the theoretical mean, assumed fixed in the derivation of these distributions, varies somewhat from sample to sample. Although such additional variation might be regarded as due to special causes and in some cases could be eliminated using charts based on the theoretical standard deviations, there are situations where such elimination is not possible and overdispersion has to be accepted as a natural condition of the process.

We now consider the design of control charts when data are not frequencies or proportions but *measurements*. We begin with some data. Table 3.1 shows measurements of the diameter of rods taken from production. The target value T is 0.870 inch. A sample of four rods was randomly chosen and measured every hour and the table shows data taken over a particular period of 20 hours of production. The data are recorded as deviations from the target value T measured in thousandths of an inch. These data are plotted in Figure 3.1. Control charts for the sample average and the sample range, which are now described, are often referred to as $\bar{X}$ (*X bar*) and R charts. In this book, however, we will use y to represent the value of the quality characteristic because later we will need the letter X for a different purpose.

In Table 3.1 the averages $\bar{y}$ and the ranges R of the measurements of each set of four rods are also shown. The averages are plotted in Figure 3.2 as a run chart. You will see that toward the end of this period it looks as though there was an upward trend in the rod diameter, but, because of variation in the individual data values, the averages also vary, so it is hard to tell for sure. To judge whether or not the process is stable, we need to have a reference distribution that would tell us how much we would expect these averages to vary if the process was *in a state of control* about the desired target value.

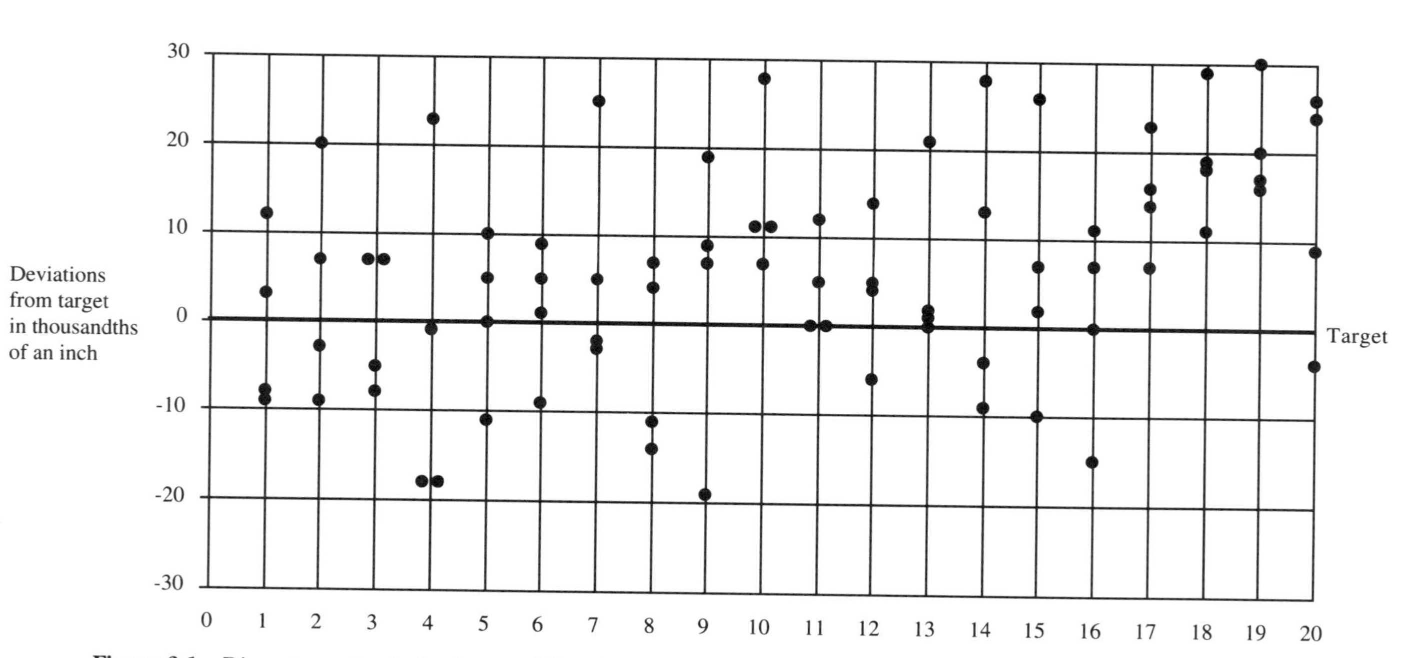

Figure 3.1 Diameters of rods in thousandths of an inch plotted as deviations from the target value of 0.870 inch.

TABLE 3.1 Measurements of the Diameter of Rods Recorded as Deviations in Thousandths of an Inch from the Target Value of 0.870 inch with Appropriate Means $\bar{y}$, Ranges R, and Moving Ranges MR

Hour	1	2	3	4	5	6	7	8	9	10
Deviations from target diameter in thousandths of an inch	−8	−3	7	−18	0	1	5	7	9	11
	3	7	−5	−23	5	−9	−2	4	−19	28
	12	−9	7	−1	10	5	25	−11	7	11
	−9	20	−8	−18	−11	9	−3	−14	19	7
$\bar{y}$, sample average	−0.50	3.75	0.25	−3.50	1.00	1.50	6.25	−3.50	4.00	14.25
R, sample range	21	29	15	41	21	18	28	21	38	21
MR, moving range for $\bar{y}$	4.25	3.50	3.75	4.50	0.50	4.75	9.75	7.50	10.25	10.00

Hour	11	12	13	14	15	16	17	18	19	20
Deviations from target diameter in thousandths of an inch	0	4	2	28	26	−15	14	19	16	−4
	0	−6	1	13	2	11	16	11	20	24
	5	14	21	−9	7	0	7	29	30	9
	12	5	0	−4	−10	7	23	18	17	26
$\bar{y}$, sample average	4.25	4.25	6.00	7.00	6.25	0.75	15.00	19.25	20.75	13.75
R, sample range	12	20	21	37	36	26	16	18	14	30
MR, moving range for $\bar{y}$	0.00	1.75	1.00	0.75	5.50	14.25	4.25	1.50	7.00	

$\bar{R} = 24.15$ $\overline{MR} = 4.99$

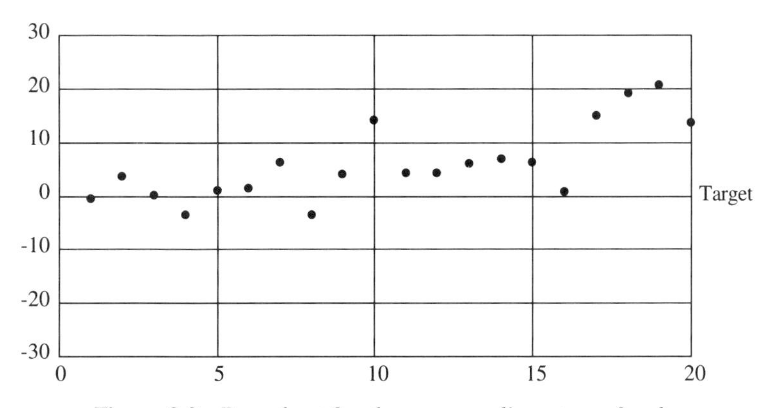

Figure 3.2 Run chart for the average diameters of rods.

Unlike data for frequencies and proportions, for measurement data the standard deviation cannot be calculated from knowledge of the mean. In fact, while in the parts industry the standard deviation is typically a very small percentage of the mean value, in the manufacture of certain antibiotics the standard deviation of the yield can be as great as 10% of the mean.

3.1 A SHEWHART CHART FOR THE SAMPLE AVERAGE

To obtain control limits for the plotted averages of Figure 3.2 we need to know how much these averages could be expected to vary if the process was stable. To do this we need to answer the questions: For a process *in a state of control,*

- What would be the shape of the appropriate reference distribution for the sample averages?
- What would be its standard deviation?

The following considerations may make it possible to answer these questions:

1. More or less regardless of the shape of the distribution of the original measurements, the shape of the distribution of *averages* looks more and more like that of a normal distribution as the sample size increases. This is called the *central limit effect* (see Appendix 2C). Because of this effect, it would often be safe to assume even for samples of size $n = 4$ that the distribution of sample averages was approximately normal.
2. If the standard deviation of individual measurements is σ_y, then, whatever the distribution of y, the standard deviation $\sigma_{\bar{y}}$ of an average $\bar{y}$ of n *independent* measurements is given by dividing the standard deviation of the original data by the square root of the sample size:

$$\sigma_{\bar{y}} = \sigma_y/\sqrt{n}. \tag{3.1}$$

   Note that this formula is true *only* for observations that are statistically independent; that is, for observations where the value of each observation does not affect the probability distribution of any of the others. On this assumption the averages of four measurements in Table 3.1 have half the standard deviation of the original data. Averages of nine would have one-third the standard deviation of the original data, and so on.

TABLE 3.2 Divisors for Average Range: To Estimate σ_y, the Average Range Is Divided by d_y; to Estimate $\sigma_{\bar{y}}$, the Average Range Is Divided by $d_{\bar{y}}$

n (sample size)	2	3	4	5	6	7	8	9	10
d_y (σ_y divisor)	1.13	1.69	2.06	2.33	2.53	2.70	2.85	2.97	3.08
$d_{\bar{y}}$ ($\sigma_{\bar{y}}$ divisor)	1.59	2.93	4.12	5.20	6.20	7.15	8.05	8.91	9.73

But what is the appropriate value to use for the standard deviation σ_y of the original data? Clearly it is the standard deviation of the measurements when the process *is in a state of control.* Unfortunately, it is very unlikely that initially, at least, the process *will* be in a state of control. In particular, as we saw, a casual look at the averages plotted in Figure 3.2 suggested a possible upward trend in the results. The following argument shows us how we might proceed.

If data taken close together in time could be represented by a white noise series like that in Figure 1.1a, then the variation in observations taken from this series one hour apart should be no more than that for observations taken one minute apart. Therefore the appropriate standard deviation σ_y would be that for measurements taken (nearly) at the same time. This could be supplied by the standard deviations of individual sets of four measurements. These could be obtained as in Section 2.6 from the sum of squares of deviations from the averages of each set of four, but an easier way to obtain them is by averaging the *ranges* of the samples of four measurements.[1]

The range within each sample of four is simply the largest value in the sample minus the smallest value. These extreme values have been underlined in Table 3.1 so as to simplify calculation of the 20 individual ranges shown in the table. The average of the ranges from these 20 samples is $\bar{R} = 24.15$. To estimate the standard deviation σ_y, we divide $\bar{R}$ by an appropriate divisor d_y given in Table 3.2.[2]

For our example, $n = 4$, and so d_y is 2.06. Thus an estimate of the standard deviation σ_y within a set of four measurements is

$$\hat{\sigma}_y = \bar{R}/d_y = 24.15/2.06 = 11.72.$$

As mentioned earlier, we use a “hat” to indicate an estimate; thus $\hat{\sigma}_y$ means an estimate of σ_y.

[1] For small values of n (say, $n < 10$) and approximately normally distributed data, the sample variance obtained by averaging ranges of sets of n observations is almost as efficient as that obtained by the sum of squares method. For a brief discussion of the “efficiency” of a statistic see p. 242.

[2] The ASTM standard uses the notation d_2 instead of d_y.

The estimated standard deviation $\hat{\sigma}_y = 11.72$ for measurements taken close together would also be an estimate of the standard deviation of individual measurements (y) taken further apart *if* the process were in a state of control. In that case, from this estimate of σ_y we could obtain an estimate of $\sigma_{\bar{y}}$, the standard deviation of the *averages* of four measurements. To get this estimate we divide by the square root of n. Thus

$$\hat{\sigma}_{\bar{y}} = \hat{\sigma}_y/\sqrt{n} = 11.72/2 = 5.86.$$

Alternatively, using Table 3.2 we can use the divisor $d_{\bar{y}} = d_y\sqrt{n}$ from

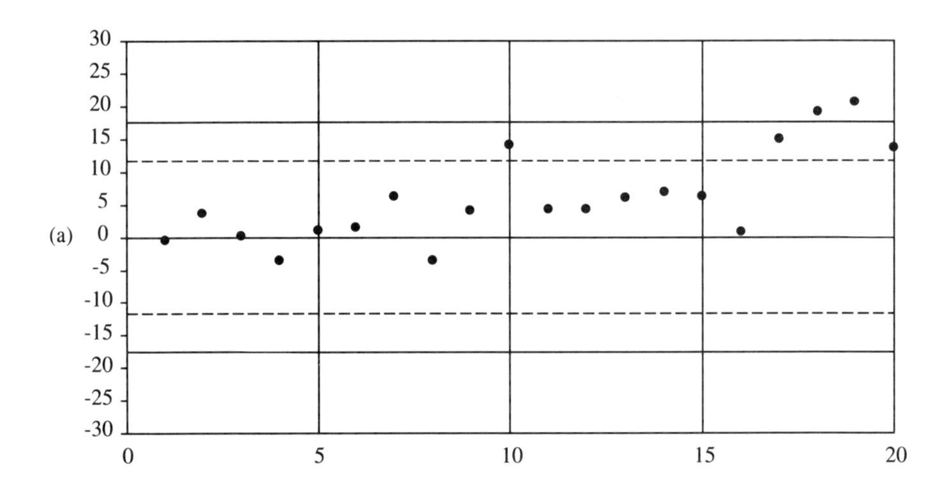

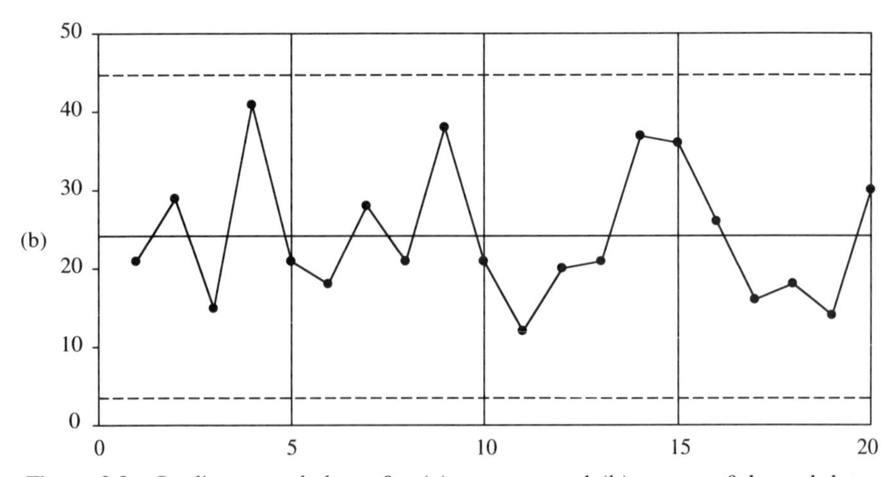

Figure 3.3 Quality control charts for (a) averages and (b) ranges of the rod data.

which $\sigma_{\bar{y}}$ can be obtained directly. Thus

$$\hat{\sigma}_{\bar{y}} = \bar{R}/d_{\bar{y}} = 24.15/4.12 = 5.86,$$

as before.[3]

Using the value $\hat{\sigma}_{\bar{y}} = 5.86$, control limits for the averages about the target value (zero) are given by $\pm(k \times 5.86)$. Thus with $k = 3$ the 3σ limits for averages of four measurements would be ±17.58 and with $k = 2$ the 2σ limits would be ±11.72. These limits are superimposed on the data for averages in Figure 3.3a to provide the final control chart.

It appears that, towards the end of the period of time under study, the process was not in a state of control and there was a drift upward in the mean diameter of the rods, possibly due to tool wear.

3.2 RANGE CHARTS

Departures of the mean from the target value are not the only kind of instability that needs to be monitored. It sometimes happens that the *spread* of the distribution of values is changing from one point in time to another. To monitor this, we can run a second chart on the ranges themselves. Because the sample ranges are calculated from limited data, they can be expected to vary a good deal even when the process is stable. However, once again, it is possible to calculate the degree of variation that would be expected to occur for a process in a state of control. Specifically, an estimate $\hat{\sigma}_R$ of the standard deviation of the ranges of n normally distributed values can be obtained by dividing the average range by a factor d_R listed below.

n Sample Size	2	3	4	5	6	7	8	9	10
d_R (Divisor)	1.32	1.91	2.34	2.69	2.99	3.25	3.47	3.68	3.86

The distribution of the range is only approximately normal, especially for small samples, but, nevertheless, $\pm2\sigma$ and $\pm3\sigma$ limits are usually used to supply action limits and warning limits. The range is always positive so that when a limit falls below zero, it is replaced by zero. For the example of the rod diameters in Table 3.1, $\bar{R} = 24.15$. The sample size is 4 and the corresponding value[4] of d_R is 2.34 so that

$$\hat{\sigma}_R = \bar{R}/d_R = 24.15/2.34 = 10.32.$$

[3] Note that an approximate rule of thumb that works quite well for n between 3 and 9 is simply $d_y \simeq \sqrt{n}$ and $d_{\bar{y}} \simeq n$.

[4] In the notation of the ASTM standard $d_R = d_2/d_3$.

Thus if the target value for the mean range is set at $\overline{R} = 24.15$ the $\pm k\sigma$ limits of the sample ranges are $24.15 \pm (k \times 10.32)$—the 2σ limits are 3.51 and 44.79 and the 3σ limits are 0 and 55.11. The 20 sample ranges for the rod data are plotted on the range chart in Figure 3.3b. No points fall outside even the 2σ control limits and we can see that although for these data the mean is out of control, the variation *about* the mean seems stable.

Thus the method that we have used here to set up mean and range charts for measurements may be summarized as follows:

1. A small group of n measurements, frequently 4 or 5, is taken at fixed intervals of time and the averages $\bar{y}$ are plotted on a control chart with limits of $T \pm k\hat{\sigma}_{\bar{y}}$, where $\hat{\sigma}_{\bar{y}} = \hat{\sigma}_y/\sqrt{n}$, and σ_y is estimated from the small groups of data taken fairly close together in time.
2. The ranges are plotted on a second chart with limits $\overline{R} \pm k\hat{\sigma}_R$.

3.3 ADDITIONAL ACTION RULES FOR SHEWHART CHARTS

Suppose that we can obtain satisfactory control limits. How should we use them? It is generally accepted that a single point lying outside the $\pm 3\sigma$ limits of a Shewhart chart should be regarded as a *signal* that some action is needed and that, in particular, an assignable cause should be sought. But it is sensible to feel almost equally suspicious of, for example, several successive points lying all on one side of the target value. Thus various additional action rules have been proposed from time to time (e.g., see Nelson, 1985; Camp and Woodall, 1987).

According to a set of widely used rules devised by Western Electric (1956), an action signal is provided by any one of the following:

Rule 1: A single point lying beyond the three-sigma limits.

Rule 2: Two out of three consecutive points lying beyond the two-sigma limits.

Rule 3: Four out of five consecutive points lying beyond the one-sigma limits.

Rule 4: Eight consecutive points lying on one side of the target value.

As an illustration, note that under Western Electric rules the chart of Figure 3.3a would signal the rod-making process to be out of control at time $t = 16$, because at this time eight successive points are above the target value. It should be remembered when we use a *set* of rules of this kind that the probability of a *false alarm* (of one or other of the rules wrongly signaling an out of control state) will be increased.

3.4 ARE WE USING THE RIGHT REFERENCE DISTRIBUTION?

This is a question we asked before for the charts of Chapter 2. In our description of the taking of sample rods that led to the data in Table 3.1, we told how each hour four separate rods were randomly chosen and each one measured. This would of course be very different from taking a single rod and measuring it four times; the standard deviation you would obtain if you did this would be that of a measurement process only and would not include the variations of diameter occurring in the rods themselves. Mistakes of this kind are commonly made when setting up control charts; thus a control chart based on a standard deviation obtained from repeated measurements of the same rod might show most of the points falling outside the 3σ limits, even though the process was in a state of control.

This, however, is not the only problem. For some systems, particularly those involved with processing liquids or gases, observations taken close together in time are likely to be highly autocorrelated so that each reading is highly dependent on previous ones. For example, for some chemical processes if you were to look at successive temperature measurements made automatically ten times every second they might look like this:

$$174.2, 174.2, 174.2, 174.2, \ldots .$$

After a few hundreds of such readings the value might change to 174.3. Obviously, the formula $\sigma_{\bar{y}} = \sigma_y/\sqrt{n}$ for independent data will not apply to a sequence of such autocorrelated data. However, it is perfectly possible that observations of temperature made less frequently, say, once every 30 minutes, would be essentially independent.

Familiarity with your process may help you to know when to expect such problems and what is a reasonable sampling interval to use for process monitoring purposes. As in Chapter 2, different ways for calculating the value of $\sigma_{\bar{y}}$ may be compared. When suitable data is available you can use the rational subgroup method by pooling estimates of standard deviations calculated during periods of operation when the process was judged to be stable (although as we said this method may tend to yield underestimates). The method of moving ranges is also available. For example, we can use the moving range to estimate $\sigma_{\bar{y}}$ directly using the rod diameter averages $\bar{y}$ given in Table 3.1. For these data the moving ranges are $|3.75 - (-0.50)| = 4.25$, $|0.25 - 3.75| = 3.50$, and so on, yielding $\overline{MR} = 4.99$. As before, on the assumption that the $\bar{y}$ values are roughly normally distributed and *statistically indepen-*

dent, an estimate of the standard deviation is then obtained by dividing by the factor 1.128 to give 4.99/1.128 = 4.42. The estimate 4.42 agrees reasonably well with that obtained earlier from the average range within samples, $\overline{R}/d_{\bar{y}} = 5.86$. If we had found that the moving range method gave a value much larger than that obtained from the within-samples method, this could suggest, for example, that the repeated measurements were not statistically independent and in particular that they might have been made on the same piece rather than on separate pieces. If the moving range method gave a much smaller value, this could be due to positive autocorrelation between the sample averages. (See also Section 5.2.)

If the assumptions concerning independence and appropriate sampling are true, the formula $\hat{\sigma}_{\bar{y}}^2 = \hat{\sigma}_y^2/n$ shows how the variance is reduced by averaging n observations. However, we are on safer ground if this can be checked experimentally—by comparing the estimate of the variance of $\bar{y}$ calculated from $\hat{\sigma}_y^2/n$ with more direct methods using rational subgroups and the moving ranges. Note that the rational subgroup method and the moving range method are available even when $n = 1$, that is, when only one single observation is taken at each sampling time.

In later sections of this chapter we discuss EWMA and Cusum charts as alternatives to the Shewhart chart. It is important to bear in mind, however, that all the stipulations we have made about Shewhart charts apply equally to these charts. In particular, for these charts to operate correctly, observations must be statistically independent and an appropriate value for the standard deviation must be used. [See, for example, Johnson and Bagshaw (1974); Alwan (1992); Palm, Rodriguez, Spiring, and Wheeler (1997).]

Finally, there is the question: Is *any* kind of monitoring chart discussed in this chapter an appropriate tool for this application?

The answer depends on the usefulness to that application of a model where it is supposed that the process can be brought to a state of control occasionally affected by special causes. As we have seen all models are necessarily approximate so that violations of assumptions does not necessarily imply the uselessness of a derived procedure. However, it is disturbing when substantial violations are found to be common. Thus, according to Alwan and Roberts (1988), "an empirical study of 235 quality control applications suggests that violations of the assumptions are the rule (87% observed) rather than the exception in practice leading to: (a) a false assurance that the process is stable, (b) a false search for special causes, (c) failure to search for special causes which can be seen with better analysis, (d) failure to see and act on systematic variation, and (e) control charts being ignored."

What can safely be said is that it is always valuable to plot a run chart of the data as it comes from the process and to exhibit it in such a way that it is forced to the attention of those running the process. Standard warning and action limits are, however, inappropriate for a basically unstable system. Instead some *process adjustment* strategy, as those discussed in later chapters of this book, may be needed to reduce variability. For the moment, however, we shall continue to suppose that the standard monitoring model provides an adequate approximation to reality.

3.5 SPECIFICATION LIMITS, TARGET ACCURACY, AND PROCESS CAPABILITY

The earlier discussion was about checking that a process was in a state of control and finding and eliminating special causes. For this purpose we were usually concerned with averages of n observations and their appropriate reference distribution and hence their control limits. These control limits have nothing to do with *specification limits,* which tell us what we would like the process to do, namely, that only very rarely would *individual* items be produced that were out of specification.

We can be in trouble either because the process has too great variability or because the process mean is off target. The *potential* ability to fit most production within the specification limits is measured by the *spec/sigma ratio*:

$$S_\sigma = \frac{\text{span of specification limits}}{\sigma_y}.$$

This ratio measures directly the span of the specification limits relative to σ_y.

We should certainly like this ratio to be comfortably large. In particular, *if* we could rely on the assumption of normality for the original data, and *if* the mean were on target, then a value of $S_\sigma = 6$ would ensure that each of the specification limits was three standard deviations away from the mean and that bad product would be produced only about once in 400 times. On this basis, it has become customary to divide S_σ by 6 to get the *process capability index* C_p:

$$C_p = \frac{\text{span of specification limits}}{6\sigma_y} = \frac{S_\sigma}{6}. \qquad (3.2)$$

However, the merits of the factor 6 depend on specific circumstances and assumptions and we feel happier in simply quoting S_σ itself.

Whether or not the potential of the process—as measured by S_σ (or C_p)—is realized is heavily dependent on the extent to which the process mean is off target. This can be measured by the *off-target ratio* S_T:

$$S_T = \frac{\mu - T}{\sigma_y}.$$

The off-target ratio S_T measures how far the mean is from the target relative to the standard deviation σ_y. The ratio can be positive or negative depending on whether the mean is above or below the target. A valuable associated criterion that measures how well the process output is centered within the specification limits is the C_{pk} index defined by

$$C_{pk} = \frac{|\text{specification limit} - \mu|}{3\sigma_y},$$

where the specification limit substituted in the formula is that closest to the mean.

Assuming the data to be normally distributed, three possible situations are illustrated in Figure 3.4. Figure 3.4a illustrates the happy situation where the process is situated nicely on target in the middle of the specification range. In this example, the span of the specification limits is 10σ so that $S_\sigma = 10$ ($C_p = 1.66$) with $S_T = 0.00$ and $C_{pk} = 1.66$.

A less than comfortable situation is illustrated in Figure 3.4b. Here the specification limits are at $T \pm 2\sigma$, $S_\sigma = 4$ ($C_p = 0.66$), $S_T = 0.00$, and $C_{pk} = 0.66$. The mean is on target and although we are doing the best we can with this process in its present state, even so, about 5% of our production will be out of specification. Obviously, we must try to reduce the standard deviation σ_y.

Finally, in Figure 3.4c the span of the specification limits is 7σ so that $S_\sigma = 7$ ($C_p = 1.16$) and the process is *potentially* capable of running reasonably satisfactorily. The problem is that the process mean is much too low. Consequently, about 16% of production is out of specification on the low side with the off-target ratio

$$S_T = \frac{\mu - T}{\sigma_y} = \frac{-2.5\sigma_y}{\sigma_y} = -2.5 \quad \text{and} \quad C_{pk} = 0.33.$$

The situation can be corrected by moving the process mean by 2.5 standard deviations up to the target.

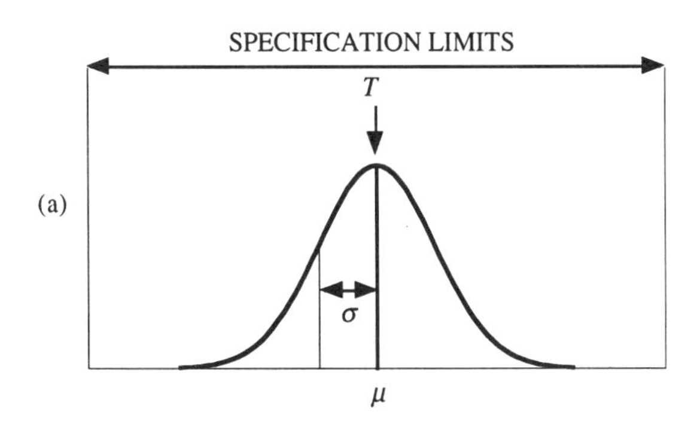

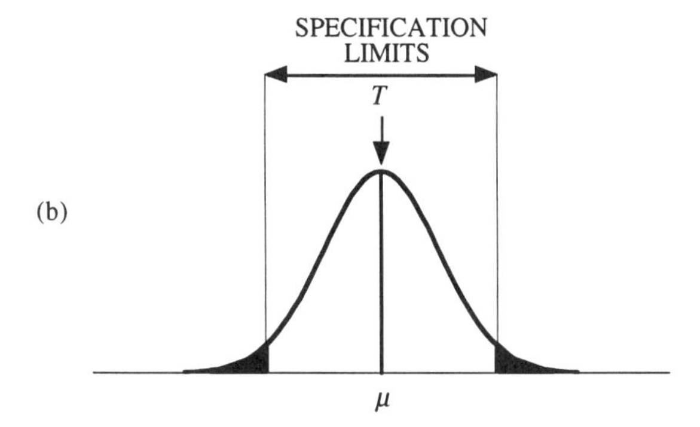

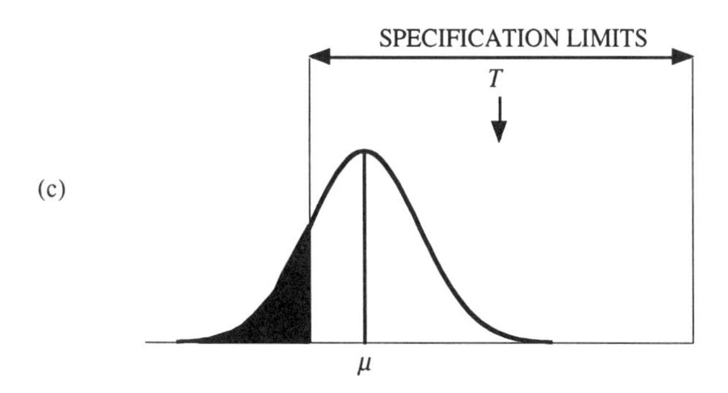

Figure 3.4 Distribution of product relative to specification limits for three different situations: (a) $S_\sigma = 10$, $C_p = 1.66$; $S_T = 0.00$, $C_{pk} = 1.66$. (b) $S_\sigma = 4$, $C_p = 0.66$; $S_T = 0.00$, $C_{pk} = 0.66$. (c) $S_\sigma = 7$, $C_p = 1.16$; $S_T = -2.50$, $C_{pk} = 0.33$.

In calculating the ratios and indices, it must be remembered that the standard deviation of *individual measurements* σ_y must be used. Because we are now not looking at averages but at individual measurements, the assumption that the distribution is normal is much more seriously strained. In particular, the fact that $S_\sigma = 6$ ($C_p = 1$) may no longer provide the expected safe margin. The implications are important (e.g., see Luceño, 1996a) and some checking of how often measurements *actually* fall outside various limits using past data is advisable. An average of recent results should be used to represent the process mean μ in calculating S_T.

Mean Square Error and Variance

The long-run average of $(y - T)^2$, the squared deviation from the target value, is called the *mean square error* or MSE. The importance of maintaining the mean on target is further illustrated by the fact that *the mean square error is equal to the variance plus the square of the deviation of the mean from the target,*

$$MSE = \sigma_y^2 + (\mu - T)^2;$$

or in terms of the off-target ratio S_T,

$$MSE = \sigma_y^2(1 + S_T^2).$$

If you want to reduce deviation from the target, then usually the first thing to do is to try to get the process mean equal to the target value, so that S_T is zero. If you can do this, then the mean square error will be equal to the process variance σ_y^2. You might then start to consider how to reduce σ_y^2 itself.

Minimum Mean Square Error (MMSE)

Later in this book we talk about MMSE control, that is, about methods of control that produce *minimum mean square error* at the process output. If such a method also forces the mean to be on target, then the method will produce minimum variance and hence minimum standard deviation.

Unbiased Estimates

The above concepts are also useful in consideration of various possible estimates of an unknown constant (parameter). If one such estimate varies about the true value of the unknown parameter with smallest mean square error, it is called a minimum mean square error (MMSE) estimate. Furthermore, if the mean value of the estimate is equal to the true value of the parameter (corresponding to the "target" in the previous discussion), then the MMSE estimate becomes a minimum variance estimate. The estimate is then said to be *unbiased.*

3.6 PROCESS MONITORING CHARTS USING AN EXPONENTIALLY WEIGHTED AVERAGE

In Section 3.3 we discussed additional action rules for Shewhart charts such as are embodied, for example, in the four "Western Electric rules." The intention is to highlight suspicious local behavior. A different approach is to use, as an addition to the Shewhart chart, a local time average as an indicator of recent performance.

It was suggested by Roberts (1959) that such a local indicator might take the form of the *exponentially discounted* average—that is, an average in which past data values are "remembered" with geometrically decreasing weight. As we mentioned in Chapter 1, such a device had previously been used for forecasting.

Suppose, for example, that data are available for some quality characteristic y up to the present time t. Then, with present and past values of y denoted by $y_t, y_{t-1}, y_{t-2}, \ldots$, an exponentially weighted average $\tilde{y}_t$ with *discount factor* equal to, say, 0.6 is

$$\tilde{y}_t = \text{constant}\ (y_t + 0.6y_{t-1} + 0.36y_{t-2} + 0.216y_{t-3} + \cdots), \quad (3.3)$$

where the *relative* weights applied to y_t, y_{t-1}, y_{t-2} are 1.0, 0.6, 0.6^2, and so on. In general, the discount factor is denoted by θ (Greek *theta*), which theoretically must have some value between -1 and $+1$, but we will assume for our present purpose it is between 0 and 1. The weights are in the relation $1, \theta, \theta^2, \theta^3, \ldots$. Now the sum of the coefficients $1 + \theta + \theta^2 + \theta^3 + \cdots$ is $1/(1 - \theta)$, so we can make the weights add up to 1 by setting the constant in Equation (3.3) equal to $1 - \theta$. Finally, the exponentially weighted average with discount factor θ is

$$\tilde{y}_t = (1 - \theta)(y_t + \theta y_{t-1} + \theta^2 y_{t-2} + \cdots). \quad (3.4)$$

For example, with $\theta = 0.6$, $1 - \theta = 0.4$ and Equation (3.3) becomes

$$\tilde{y}_t = 0.4\ (y_t + 0.6y_{t-1} + 0.36y_{t-2} + 0.216y_{t-3} + \cdots).$$

The weights are thus 0.4, 0.24, 0.144, . . . and will eventually sum to 1.

To see why we need the weights to add up to 1, consider any weighted average of present and past data, say,

$$\dot{y}_t = w_1 y_t + w_2 y_{t-1} + w_3 y_{t-2} + \cdots$$

with weights w_1, w_2, w_3, ... Thus, for example, if $\dot{y}_t$ were the ordinary arithmetic average of, say, the last ten values, then

$$\dot{y}_t = 0.1y_t + 0.1y_{t-1} + 0.1y_{t-2} + \cdots + 0.1y_{t-9}$$

and the weights are all equal to $1/10 = 0.1$.

It is the necessary property of any average that if any constant *c*, say, is added to each of its component data values, then the value of the average must also increase by *c*. For this to happen, the weights must add to 1.

Because in practice we will continually recalculate this exponential average as each new data value becomes available, it is called a *moving* average. Finally, the quantity $\tilde{y}_t$ in Equation (3.4) is called an exponentially weighted moving average, or EWMA for short. The EWMA turns out to have a number of different functions and applications, which are discussed in different parts of this book. For the moment we consider the application to process monitoring.

The constant $1 - \theta$ has special importance on its own and is given the distinguishing symbol λ (Greek *lambda)*, where $\lambda = 1 - \theta$. Note that, unless θ is close to 1 (λ is close to 0), the weights will die out (become negligible) fairly rapidly. In Figure 3.5, the weights for $\theta = 0.4$ ($\lambda = 0.6$) are plotted and also those for $\theta = 0.6$ ($\lambda = 0.4$) and

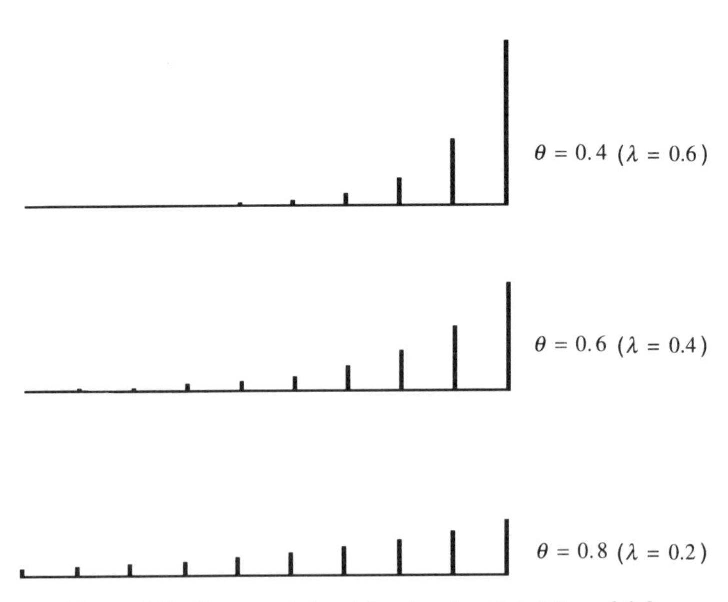

Figure 3.5 Exponential weights for $\theta = 0.4$, 0.6, and 0.8.

for $\theta = 0.8$ ($\lambda = 0.2$). From this diagram we see that using a smaller value for θ (a larger value for λ) results in weights that die out more quickly and place more emphasis on recent observations. Such a choice produces a chart that reacts quickly but does not possess the same stabilizing and averaging effect obtained for larger θ (smaller λ). A good compromise for EWMA charting seems to be a value of θ close to 0.6 ($\lambda = 0.4$).

EWMA Monitoring Chart

It would be laborious to use Equation (3.4) to repeatedly recalculate the EWMA. Fortunately, a very convenient formula for updating the EWMA, as each observation becomes available, is:

$$\tilde{y}_t = \lambda y_t + \theta \tilde{y}_{t-1}. \tag{3.5}$$

Thus, given the EWMA $\tilde{y}_{t-1}$ we can calculate $\tilde{y}_t$, as soon as we obtain the new observation y_t.

For example, let's use this formula for calculating the EWMAs for a process that had a target value of $T = 10$ for which ten successive observations were as follows:

Observation	1	2	3	4	5	6	7	8	9	10
y	6	9	12	11	5	6	4	10	12	15

Suppose we wish to calculate the EWMAs with $\lambda = 0.4$ and $\theta = 0.6$ so that the general formula is $\tilde{y}_t = 0.4y_t + 0.6\tilde{y}_{t-1}$. We do not know the value for $\tilde{y}_0$ so we will set it equal to the target value $T = 10$ and we then obtain

$$\begin{aligned}
\tilde{y}_1 &= (0.4 \times 6) + (0.6 \times 10) = 8.4,\\
\tilde{y}_2 &= (0.4 \times 9) + (0.6 \times 8.4) = 8.64,\\
\tilde{y}_3 &= (0.4 \times 12) + (0.6 \times 8.64) = 9.98,\\
&\vdots
\end{aligned}$$

Exercise 3.1. Calculate the remaining EWMAs for this series. Plot the original data and the EWMAs. □

You will see that the new value of the EWMA is an interpolation between the old EWMA and the new data point. This means that if you draw a straight line between these two numbers, the new EWMA would be at a fraction λ of the distance between $\tilde{y}_{t-1}$ and y_t. With an EWMA control chart, the $\tilde{y}_t$ values are plotted between their appropriate limit lines. If the process were in a perfect state of control and the deviations $y_t - T$ from target formed a random sequence with standard deviation σ_y, the standard deviation of the EWMA would be

$$\sigma_{\tilde{y}} = \left(\frac{\lambda}{2 - \lambda}\right)^{1/2} \sigma_y. \tag{3.6}$$

In particular, using this formula we find the following:

θ	0.80	0.70	0.60
λ	0.20	0.30	0.40
$\sigma_{\tilde{y}}/\sigma_y$	0.33	0.42	0.50

Thus given an appropriate value for λ, we can calculate three-sigma or other appropriate limits for $\tilde{y}_t - T$. It is understood, of course, that if the data were averages from samples of n items, then σ_y in Equation (3.6) would be replaced by the standard deviation of these averages.

To see where the formula (3.6) comes from, note that $\tilde{y}_t$ in Equation (3.4) is a linear aggregate of supposedly independent observations y_t, $y_{t-1}, y_{t-2}, \ldots$ each having variance σ_y^2. So formula (2.4) for the variance of a linear aggregate with coefficients $c_1 = 1 - \theta$, $c_2 = (1 - \theta)\theta$, $c_3 = (1 - \theta)\theta^2, \ldots$ leads to

$$\sigma_{\tilde{y}}^2 = \sigma_y^2(1 - \theta)^2(1 + \theta^2 + \theta^4 + \cdots).$$

But if $|\theta| < 1$ and the series on the right is continued indefinitely, then

$$(1 + \theta^2 + \theta^4 + \cdots)(1 - \theta^2) = 1,$$

and so

$$\frac{\sigma_{\tilde{y}}^2}{\sigma_y^2} = \frac{(1 - \theta)^2}{(1 - \theta^2)} = \frac{1 - \theta}{1 + \theta} = \frac{\lambda}{2 - \lambda},$$

which yields Equation (3.6).

Hunter (1989) has argued that a chart of this kind with λ close to 0.4 behaves somewhat like a Shewhart chart to which the four Western

Electric rules discussed in Section 3.3 are simultaneously applied. He suggested that a "coplot" be run consisting of a Shewhart chart for the original $\bar{y}$ values and also for the $\tilde{y}_t$ values with *their* limit lines superimposed. For illustration, calculations are shown below for the average rod diameters of Table 3.1. The EWMAs are updated using $\tilde{y}_t = 0.4\bar{y}_t + 0.6\tilde{y}_{t-1}$.

Averages and EWMAs for Rod Diameters

Average $\bar{y}_t$		3.75	0.25	−3.50	1.00	1.50	6.25	−3.50	4.00	14.25
EWMA $\tilde{y}_t$	0.000	1.500	1.000	−0.800	−0.080	0.552	2.831	0.299	1.779	6.768
Average $\bar{y}_t$	4.25	4.25	6.00	7.00	6.25	0.75	15.00	19.25	20.75	13.75
EWMA $\tilde{y}_t$	5.761	5.156	5.494	6.096	6.158	3.995	8.397	12.738	15.943	15.066

The coplot is shown in Figure 3.6. The averages of samples of four measured rods previously plotted in Figure 3.2 are indicated in Figure 3.6 by open circles. At each time t, the EWMA $\tilde{y}_t$ has also been calculated and is plotted as a solid dot. Action would be called for as soon as $\tilde{y}_t$ crossed one or other of its control limits. For $\lambda = 0.4$, these limits are just halfway between the target and corresponding Shewhart limits. This is because $\sigma_{\tilde{y}}/\sigma_{\bar{y}} = 0.5$ for $\lambda = 0.4$. In practice, no calculations need actually be made because to an adequate approximation a new value of $\tilde{y}_t$ can be obtained interpolating by eye four-tenths of the way between the previous EWMA and the last observation. It will be seen that for this particular set of data both sets of three-sigma limits

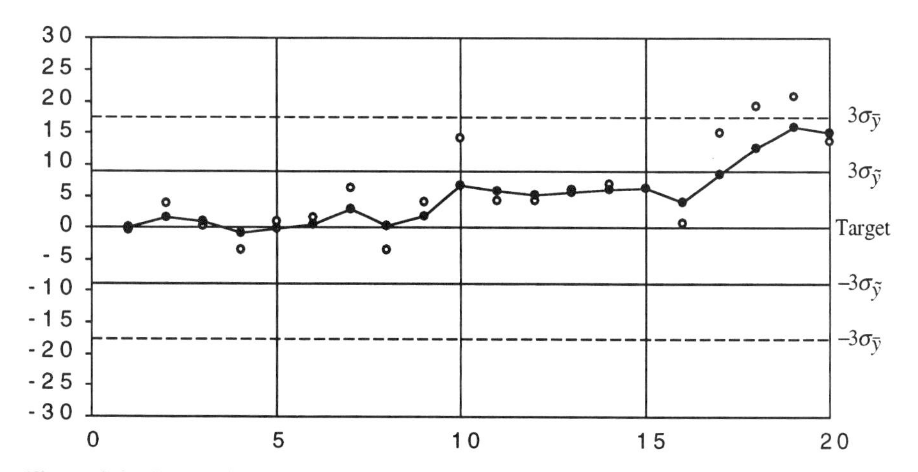

Figure 3.6 Hunter's coplot applied to averages of the rod diameter data: the averages $\bar{y}_t$ previously plotted on a Shewhart chart in Figure 3.3 are here indicated by open circles. The EWMAs $\tilde{y}_t$ of these data are indicated by solid dots. The $\pm 3\sigma$ limits for the $\bar{y}_t$ values are shown by dotted lines and for the $\tilde{y}_t$ values by solid lines.

are crossed at about the same time (although Rule 4 of the Western Electric rules would have signaled a change slightly earlier, it must be remembered that such multiple rules have a higher false alarm rate). Nothing much about the relative merits of particular procedures can be concluded from individual sets of data, but there can be little doubt that the EWMA chart is a valuable additional tool for process monitoring. (A fuller discussion is given in Chapter 10.)

As we mentioned earlier, when, as in this example, the data used to calculate the EWMAs are sample averages, the appropriate value for $\sigma_{\bar{y}}$ must be substituted in Equation (3.6), the standard deviation appropriate for the *averages.* Thus for the rod diameter the estimated standard deviation for the averages of sets of four measurements obtained from $\hat{\sigma}_y/\sqrt{n}$ was 5.86. So we obtain

$$\hat{\sigma}_{\tilde{y}} = \left(\frac{0.4}{1.6}\right)^{1/2} \times 5.86 = 2.93,$$

for this example then, the $3\sigma_{\tilde{y}}$ limits are ± 8.79.

Note that the exponentially weighted moving average chart is concerned with process *monitoring.* The rationale used is closely associated with statistical significance testing. In particular, the limit lines relate to probabilities appropriate for a process in a perfect state of control, corresponding to the *null* hypothesis of statistical hypothesis testing. The objective is to trigger the search for causes of trouble only in those instances where there is strong evidence that trouble exists.

In chapters that follow, we shall be discussing process *regulation.* In that discussion, exponentially weighted averages again play an important role but in a totally different context, associated not with statistical hypothesis testing but with statistical *estimation.* In process regulation, the appropriate model for the disturbance is that for a process *not* in a state of control, which, without adjustment, would be wandering from target. By continually estimating this disturbance it is possible to compensate for its effect. Thus the objective in these later chapters will be quite different from that of process monitoring discussed here. As we previously emphasized, both process monitoring and process regulation are of great value in appropriate contexts; however, they should not be confused, as they often are. In particular, waiting for the deviation from target to be statistically significant before making a change would usually be a very poor strategy if the objective was to keep the adjusted process as close as possible to target.

3.7 PROCESS MONITORING USING CUMULATIVE SUMS

Different processes are afflicted by different diseases. One important symptom is the occurrence from time to time of small shifts in the process mean. To detect such shifts a valuable diagnostic device is the *Cusum* due to Page (1954, 1957) and Barnard (1959).

In its simplest form, the Cusum Q is an accumulated sum of the deviations $y_1 - T$, $y_2 - T$, . . . from the process target value T. Thus

$$Q = (y_1 - T) + (y_2 - T) + \cdots + (y_t - T),$$

or more concisely,

$$Q = \sum (y_t - T).$$

A Cusum chart is obtained by plotting Q against t. To understand how the Cusum works, suppose that initially the process is in a state of control with the mean equal to the target value T. Then the deviations $y - T$ that are accumulated in the Cusum will be the sum of a set of random errors with zero mean—some positive and others negative. Suppose now that at some point in time an increase of d units occurs in the process mean, then from that point onwards, with every new observation, a quantity $+d$ will be added to Q. Thus in successive intervals d, $2d$, $3d$, . . . will be aggregated into the Cusum and the plotted value of Q will tend to slope upward. Similarly, if a decrease in the mean occurs, then the plot will tend to slope downward. In general, a change in *slope* of the Cusum can indicate a change in the *mean* of the process. Also, the magnitude of the slope change can indicate the size of the possible change in mean.

Thus the Cusum can indicate approximately *when* and *by how much* the mean of the process has shifted, providing valuable clues for the identification and elimination of sources of trouble, as the following example illustrates.

The Delinquent Calibrator Unmasked!

In a large chemical plant it was very important for the flow of a certain gas to be maintained at a constant level. Daily flow readings had a standard deviation equal to about 0.1% of the mean. Data showing deviations from the target value over an eight-month period are shown

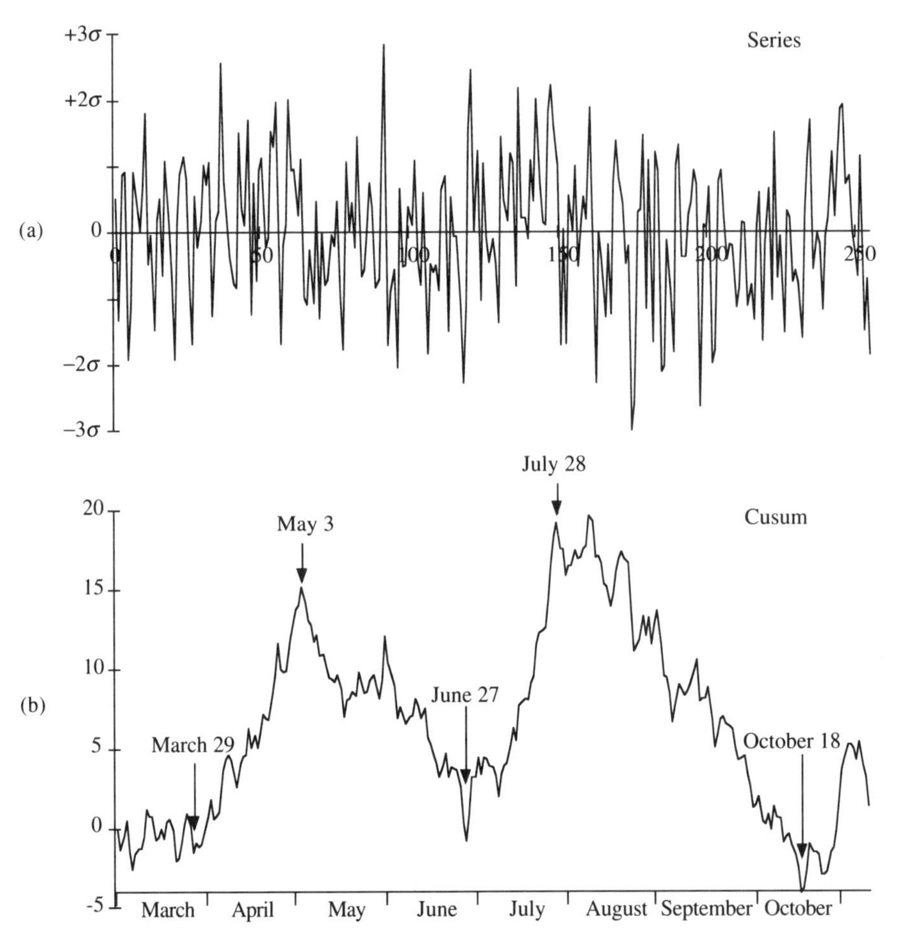

Figure 3.7. (a) Deviations from the target value of daily gas flow readings covering an eight-month period. (b) Cusum chart corresponding to this series.

in Figure 3.7a but do not immediately bring to our attention anything of interest. However, in the Cusum chart for the same data, shown in Figure 3.7b, there are a series of slope changes occurring at about March 29, May 3, June 27, July 28, and October 18. The question was therefore: "What might have happened at these times?" After some scratching of heads, it transpired that these times were close to those when the meters measuring the gas flow were recalibrated. As a result of this discovery, the calibration system was drastically modified and the problem eliminated.

Although you can get quite fancy in the study of the properties of Cusums, don't forget the utility of the basic idea shown by this example: "Cusums are a simple and valuable *diagnostic* tool for tracking down special causes."

Monitoring Surface Roughness

To clarify the ideas that follow, we have used a constructed example based on a *real* monitoring scheme in which Cusums were employed. The quality characteristic y of interest was a measure of the roughness of a surface in a manufactured product, which from the study of past records we will suppose had a standard deviation $\sigma = 2$. High values of roughness were undesirable and the lower the roughness the better. The process was operated on the basis that if the mean could be maintained at a level no higher than $T = 30$, the product was regarded as satisfactory, but if the mean roughness was as high as $T + d = 32$, then it was unacceptable. Thus the "target" value[5] is $T = 30$ and the difference we want to be able to detect is $d = 2$. Figure 3.8a shows some constructed data plotted on a Shewhart chart, where the first 15 observations are from a normal distribution having mean $T = 30$ and $\sigma = 2$ representing a satisfactory product, and the second 15 observations are from a similar distribution but with mean $T + d = 32$ and so representing unsatisfactory product.

Figure 3.8b shows the same data plotted on a Cusum chart, which strongly suggests that a change did occur at about observation 15.

How Should You Decide Whether a Real Change Has Occurred?

For the data plotted in Figure 3.8 we *know* that a change of the magnitude $d = 2$ did occur after the 15th observation. This is because we generated the numbers that way. But suppose these were real data. How could you decide whether or not the pattern of change in the Cusum implied a real change in the mean?

A change in slope of the Cusum can occur, of course, as a result of inherent variation of the data and may lead to a *false alarm*, that is, to a conclusion that the mean has changed when it hasn't. What is needed is a procedure that quickly detects real changes but only rarely produces false alarms.

[5] The long-term objective, of course, was to reduce roughness to the lowest level possible. So 30 was really not so much the target as the highest value of mean roughness that produced a product of acceptable quality. However, for the purpose of defining the Cusum monitoring scheme, we will continue to refer to $T = 30$ as the "target" value.

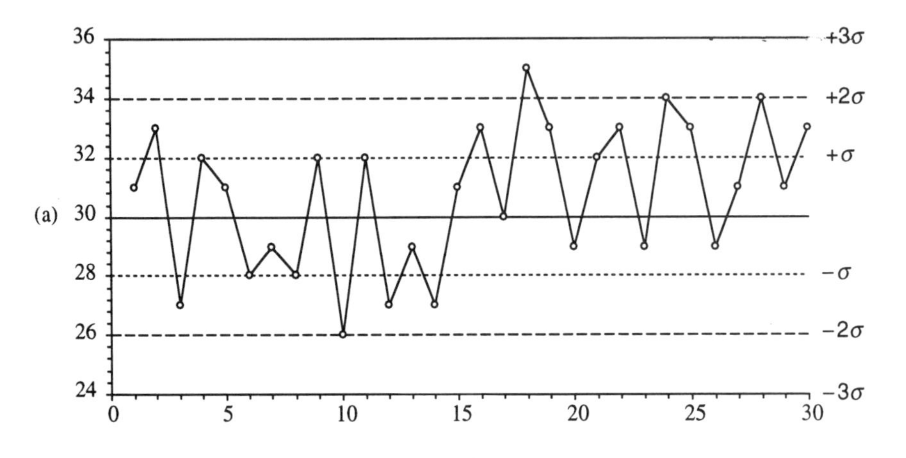

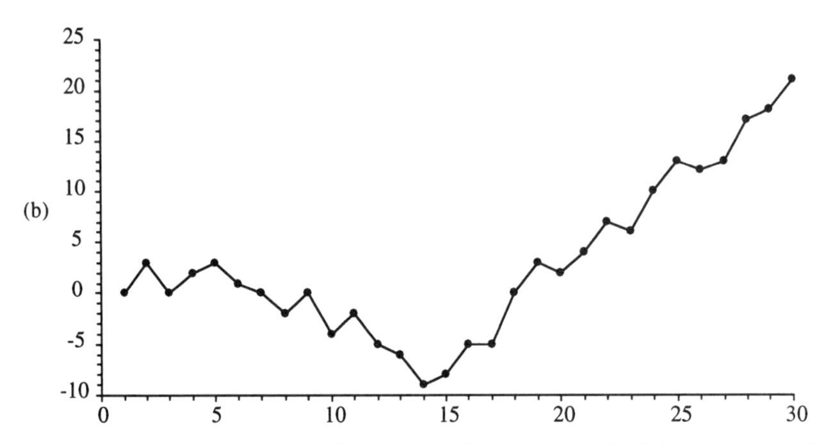

Figure 3.8. Step change example: a series with a one-standard-deviation step change occurring at observation 16 plotted on (a) a Shewhart chart and (b) a Cusum chart.

The Centered Cusum Q_*

For formal detection of a change in level we use a centered Cusum statistic Q_*, where

$$Q_* = \sum(y - T - \tfrac{1}{2}\, d).$$

The deviations here are not measured from the origin T but from an origin midway between the "acceptable" and the "rejectable" mean levels. For the example, these levels are $T = 30$ and $T + d = 32$, so

$$Q_* = \sum (y - 31).$$

The centered Cusum is essentially a handicapped score. When the mean is really on the target $T = 30$, the quantity $\frac{1}{2}d = 1$ will be subtracted from each deviation, producing a downward trend in the centered Cusum. But, if the mean level is greater than $T + \frac{1}{2}d = 31$, then the centered Cusum will tend to increase. The Cusum *test* is then as follows: accept that a genuine increase in the mean above the target has occurred as soon as the centered Cusum has increased *from its previous minimum* by more than some quality h. The quantity h is called the *decision interval.*

To see what all this means look at Figure 3.9a, which shows the centered Cusum Q_* for the data of Figure 3.8 with the decision interval set equal to $h = 10$. A local minimum of -22 occurs at the 15th observation and is not subsequently fallen short of before Q_* exceeds the value $-22 + 10 = -12$ at the 25th observation. The decision is then taken that a real increase in the mean has occurred.

How Do You Choose *h*?

Process monitoring schemes are conveniently characterized in terms of the average run lengths they produce, the *average run length* (ARL) being the average number of observations L that occur before an alarm is triggered.

Of particular importance are:

1. The average run length L_a when the process is in an acceptable state with the process mean at the target T.
2. The average run length L_r when the process is in an unacceptable state with the process mean at $T + d$.

Now a Cusum scheme is determined by the difference d you want to be able to detect, the decision interval h, and the standard deviation σ. It is most convenient to standardize d and h in terms of σ; so we actually tabulate $D = d/\sigma$ and $H = h/\sigma$ and D and H measure the detectable difference and the decision interval in units of the standard deviation. A convenient chart due to Bissell (1969) is shown in Figure 3.10. The values of D and H are shown on the horizontal and vertical axes, respectively, and the values of L_a and L_r can be read off the current lines, as indicated in the figure.

For the roughness example, $d = 2$, $h = 10$, and $\sigma = 2$; so that $D = 2/2 = 1$ and $H = 10/2 = 5$. If you follow the vertical line at $D = 1$ upward, you will see all the combinations of L_a and L_r that would

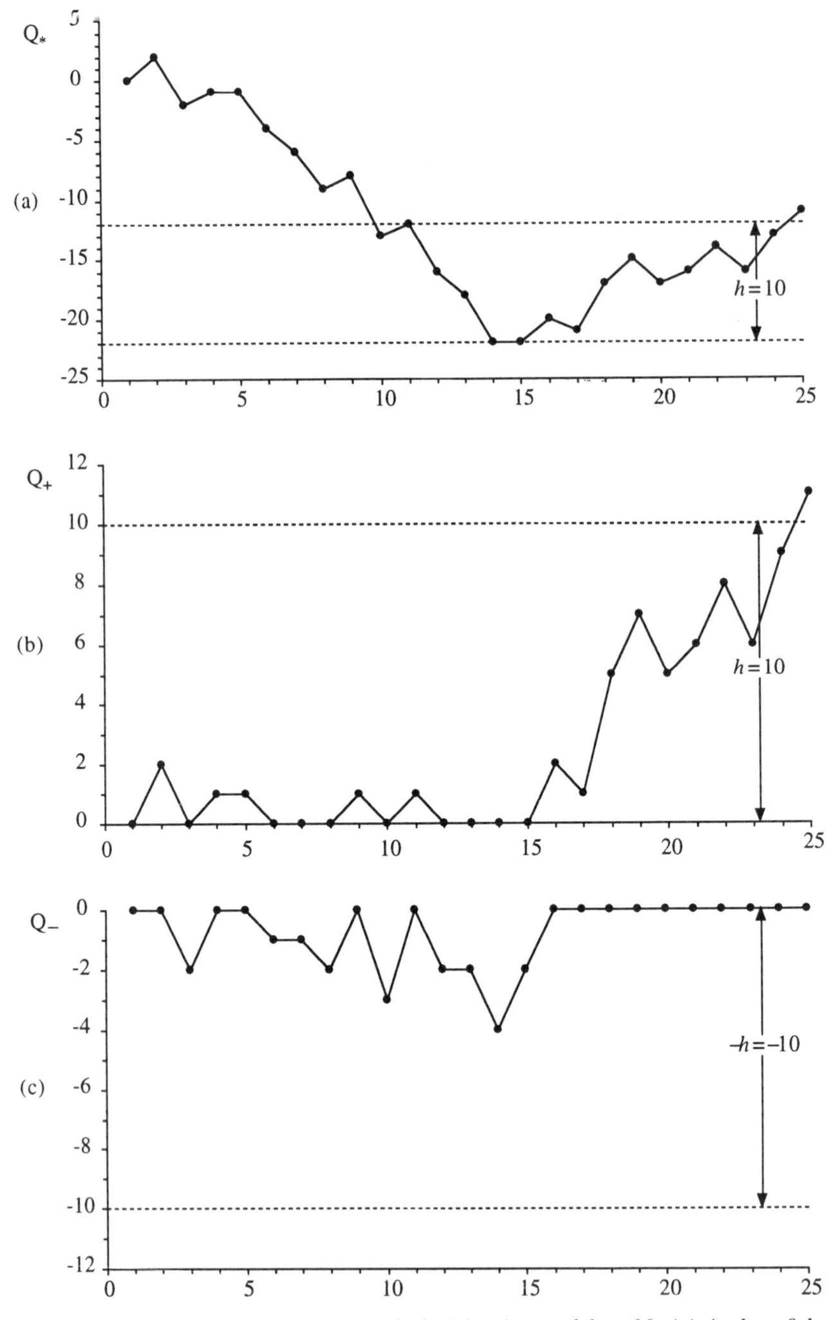

Figure 3.9 Step change example with decision interval $h = 10$. (a) A plot of the centered Cusum statistic $Q_* = \Sigma(y - 31)$; (b) a plot of the truncated centered Cusum Q_+; (c) a plot of Q_-.

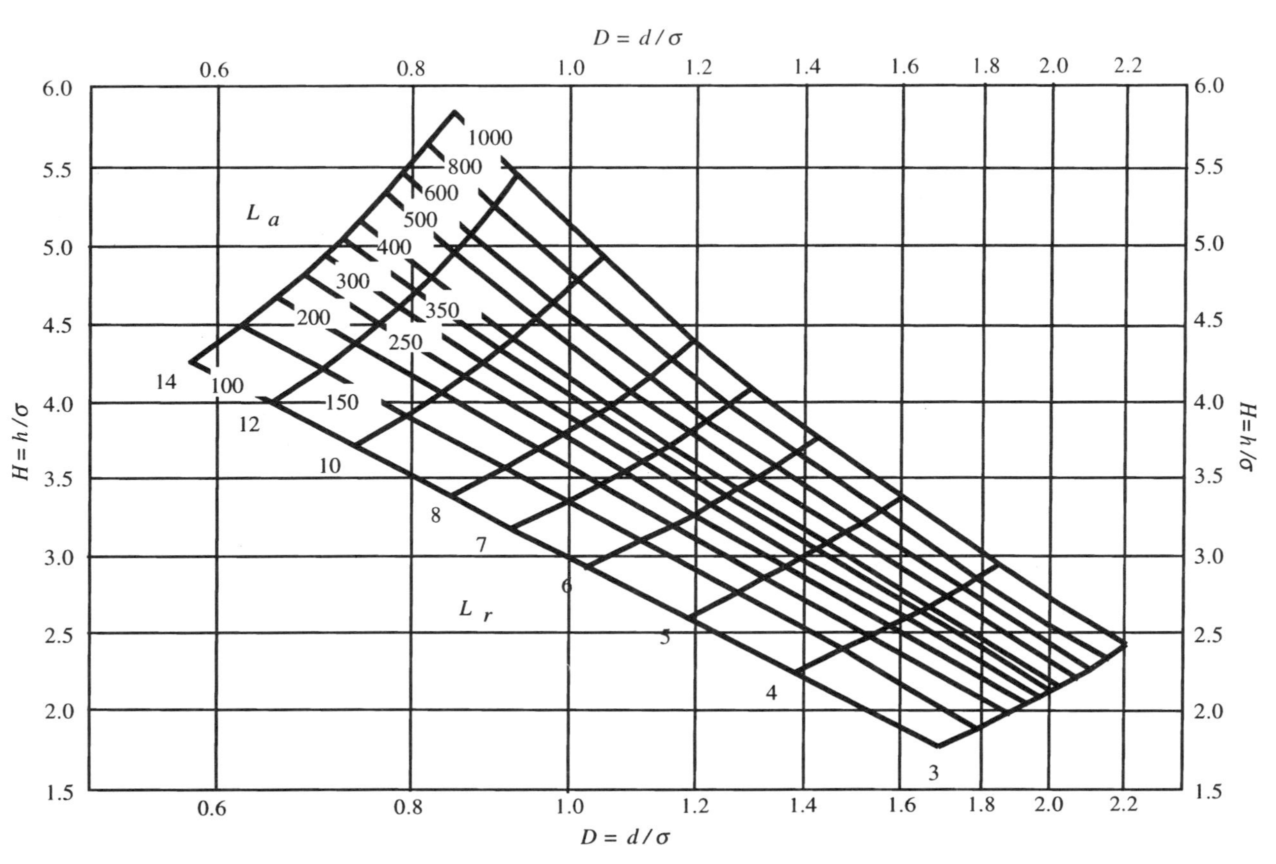

Figure 3.10 A chart for a Cusum scheme showing values of L_a and L_r for various values of $H = h/\sigma$ and $D = d/\sigma$. Adapted from Bissell (1969) by permission.

be produced for various values of H. For $h = 10$, the value chosen for illustration, the standardized decision interval is $H = h/\sigma = 5$, for which you will see that L_a is about 900 and L_r about 11. Thus if the process mean was really on the target $T = 30$, the scheme would, on the average, run for 900 observations before giving a false alarm. While if the process mean rose to $T + d = 32$, this change would be detected on the average in 11 runs. The scheme plotted in Figure 3.9a ran for ten observations before signaling. The very close agreement is fortuitous because the variation in run lengths about their means is known to be high.

A nomogram originally developed by Kemp (1962) for calculating L_a and L_r is also given in Duncan (1974, p. 472).

Scoring Method for Cusum

Although Figure 3.9a is convenient for understanding the centered Cusum, it is not of much practical use. Suppose, for example, that the process mean *did* stay on the target value $T = 30$ for 900 intervals; then the plotted points would fall on the average about 15 feet! It would be very inconvenient to require graph paper with dimensions of that kind. An equivalent, but much more practically convenient scoring method, is described below. The procedure is illustrated by the plot in Figure 3.9b, which shows the truncated Cusum Q_+. The idea is to plot the data so that any value falling below zero is recorded as a zero and a subsequent Cusum is *started from that zero value.* Once such a Cusum is initiated, either it will rise and exceed the value h, or it will, at some point, fall to zero again without reaching the value h. In the latter case, only when an increase again occurs do we need to start again to plot the Cusum beginning at the value zero.

For illustration, in the following table the first 20 values of the centered deviation $y - 31$ are shown together with the truncated Cusum denoted by Q_+. To understand what is happening, treat the centered deviations $y - 31$ in the second row of the table as the "data." The first value is 0, the next is positive so that the truncated Cusum $Q_+ = 0 + 2 = 2$ is recorded at $t = 2$. However, the next value, -4, will take the Cusum below zero. Therefore we write zero for Q_+ at $t = 3$. The next value is 1 so we begin a new Cusum with $Q_+ = 1$ at $t = 4$ and $Q_+ = 1 + 0 = 1$ at $t = 5$. But the next three values are all negative, giving three zeros of Q_+. Positive values of the deviations at $t = 9$ and $t = 11$ cause Cusums to start but to be immediately terminated. After observation 15, however, a new Cusum starts and eventually Q_+ exceeds $h = 10$ (see Figure 3.9b), indicating that the roughness measure has almost certainly increased.

t	1	2	3	4	5	6	7	8	9	10	11	12	13	14	15	16	17	18	19	20
$y - 31$	0	2	−4	1	0	−3	−2	−3	1	−5	1	−4	−2	−4	0	2	−1	4	2	−2
Q_+	0	2	0	1	1	0	0	0	1	0	1	0	0	0	0	2	1	5	7	5

Some thought and a little practice will convince you of the following:

1. The graphs in Figures 3.9a and 3.9b are identical in their effect.
2. Plotting the truncated Cusum chart of Figure 3.9b is a lot easier to do than to describe.
3. When the truncated Cusum is used, it is not even necessary to graph the data; it is sufficient to keep a numerical record of Q_+ like that shown above.
4. It is not difficult to train operators to make these calculations and/or plots.

Alternatively, a computer may easily be programmed to carry through the computations and produce a suitable warning signal as soon as Q_+ exceeds h.

Two Sided Discrepancies

For simplicity, we have introduced the topic of Cusum tests by talking about a "one-sided" situation. In the example in which the measured characteristic was roughness, it was supposed it was undesirable to have too rough a product but not undesirable for it to be too smooth. Other situations exist where we need to be made aware of deviations from the average in either direction. This can easily be done by running two truncated Cusum charts, one in each direction. Thus again using the data of Figure 3.8a, suppose that the desired level of the mean was $T = 30$ as before, but early warning was needed of deviations in *either direction.* Suppose that while a high mean of $T + d = 32$ was unacceptable, a low mean of $T + d = 28$ was also unacceptable. Then a second *centered* Cusum based on the deviations $y - 29$ could be run simultaneously. The value 29 is halfway between the acceptable quality level $T = 30$ and the unacceptable low level $T + d = 28$. Again to conserve our graph paper supply, we run a truncated Cusum Q_- for this second scheme. Below are shown the first 20 values of $y - 29$ and the corresponding values of the appropriate truncated Cusum Q_-.

t	1	2	3	4	5	6	7	8	9	10	11	12	13	14	15	16	17	18	19	20
$y - 29$	2	4	−2	3	2	−1	0	−1	3	−3	3	−2	0	−2	2	4	1	6	4	0
Q_-	0	0	−2	0	0	−1	−1	−2	0	−3	0	−2	−2	−4	−2	0	0	0	0	0

In this example, the lower limit $-h = -10$ is never approached, so there is no indication of a real reduction in the mean level of roughness at any point in these data. (See also Figure 3.9c.)

Cusum tests are sometimes conducted using a "V mask" due to Barnard (1959), which supplies the original basis for all the procedures discussed here. It may be shown that the above two-sided procedure is identical in its effect to the use of a V mask. A discussion of other aspects of Cusum techniques, including the fast initial response methods of Lucas and Crosier, 1982 will be found, for example, in Ryan, 1989.

Cusums for Poisson and Binomial Data

EWMA and Cusum techniques can also be used when the data are counts or proportions for which c-charts and binomial charts were introduced earlier. The required standard deviations may be obtained from the usual formulas. Thus, for example, an EWMA could be used to monitor the health clinic data of Figure 2.3. In that example, it was supposed that the long run mean frequency of injuries per week was 20.6. Assuming the data to have a Poisson distribution, the standard deviation was $\sqrt{20.6} = 4.5$. The limits for an EWMA chart would then be based on $\sigma_{\tilde{y}}$ obtained by inserting this value into Equation (3.6). Thus, for example, with $\lambda = 0.4$

$$\sigma_{\tilde{y}} = \left(\frac{0.4}{1.6}\right)^{1/2} 4.5 = 2.25.$$

The methods work well certainly within the limits in which the normal approximation applies.

SPC Considered More Generally

Interesting studies which underline the necessity to allow for possible overdispersion are described in Randall, Ramírez, and Taam (1996) and Ramírez and Cantell (1997). Fuller discussion of process monitoring and its relationships to many of the aspects of process improvement and management will not be attempted here. Valuable accounts of these matters will however be found in the books by, for example, Bajaria and Copp (1991), Bergman and Klefsjö (1994), Deming (1986), Hromi (1996), Imai (1986), Joiner (1994), Juran (1988), Ishikawa (1985, 1989, 1990), Kanji and Asher (1993), Montgomery (1997), Ott (1975), Pitt (1994), Ryan (1989), and Wadsworth, Stephens, and Godfrey (1986).

CHAPTER 4

Modeling Process Dynamics and Forecasting Using Exponential Smoothing

"Change alone is unchanging."
HERAKLIETOS OF EPHESOS

"And coming events cast their shadows before."
Lochiel's Warning, THOMAS CAMPBELL

4.1 INTRODUCTION

In Chapters 2 and 3 we discussed the use of control charts for monitoring the stability of processes, for indicating when we should be seeking possible assignable causes, and for locking in process improvements.

Sometimes it may *not* be possible to achieve satisfactory control without process adjustment. When this is so, it is important that adjusments be made on a logical, efficient, and systematic basis. Without such a basis, whether, when, and by how much to adjust the process rest on individual opinion and product quality is likely to suffer because different operators and supervisors may differ widely in skill and experience.

To arrive at rational schemes for adjustment we introduce two new concepts: (1) the use of simple dynamic models to approximately represent the inertia of a process to change and (2) the use of exponentially weighted moving averages specifically to forecast future values of a time series.

4.2 LINEAR INTERPOLATION AND EXTRAPOLATION

We begin by reminding the reader of methods for linear interpolation and extrapolation. Suppose you have two values y_0 and y_1 obtained at times $t = 0$ and $t = 1$; you join them with a straight line and you want to know what the value given by that line would be at some intermediate value, say, $t = 0.6$. The required value is given by the interpolation formula

$$y_{0.6} = 0.4y_0 + 0.6y_1.$$

Thus (see Figure 4.1a) if $y_0 = 3$ and $y_1 = 13$, then

$$y_{0.6} = (0.4 \times 3) + (0.6 \times 13) = 9.$$

The general expression for making an interpolation y_δ at some intermediate value δ (where δ is the Greek letter *delta*) is

$$y_\delta = (1 - \delta)y_0 + \delta y_1. \tag{4.1}$$

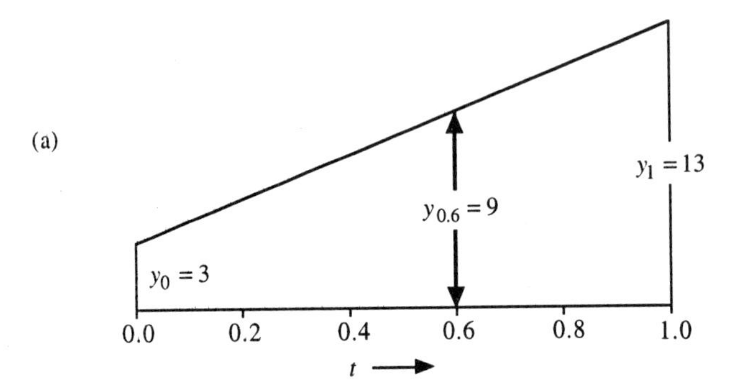

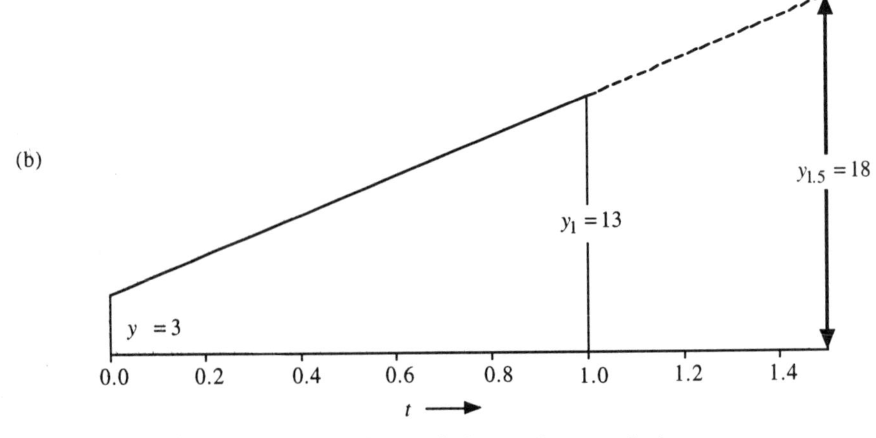

Figure 4.1 Linear interpolation and extrapolation.

The formula also works for extrapolation, that is, for values lying on an extended straight line. Thus in our example (see Figure 4.1b) the value $y_{1.5}$ obtained by setting $\delta = 1.5$ is

$$y_{1.5} = (-0.5 \times 3) + (1.5 \times 13) = 18.$$

4.3 PROCESS INERTIA (DYNAMICS)

Processes sometimes exhibit inertia. Thus it may be some time before the full effect of adjustment is experienced at the process output. How can such a phenomenon be modeled? Let's consider an example. We take a can of beans out of the refrigerator and set it on the kitchen counter at room temperature. How long will it take to warm up? Suppose that observations are made at equally spaced intervals of time $1, 2, \ldots t-1, t, t+1, \ldots$; we denote by X_t the temperature of the can's *environment* and by Y_t the temperature of the can at time t. Initially, the environmental temperature of the can is the temperature inside the refrigerator, but as soon as the can is taken out of the refrigerator its environmental temperature becomes the temperature of the room and the can will start to warm up.

Sir Isaac Newton showed that the rate at which the temperature of the can will increase is, at any stage, proportional to the distance it has still to go to reach the room temperature; that is, to the difference between the temperature of the room and the temperature of the can. The law is usually applied to continuously measured data. For discrete data available at equispaced intervals of time, we can represent Newton's law as follows. After the can is removed from the refrigerator, the increase $Y_{t+1} - Y_t$ in the temperature of the can between times t and $t + 1$ will be proportional to the difference $X_t - Y_t$ between the room temperature and the can temperature. If we denote the fixed constant of proportionality by $1 - \delta$ (where again δ is the Greek letter *delta*), we get

$$Y_{t+1} - Y_t = (1 - \delta)(X_t - Y_t). \tag{4.2}$$

For the present application $1 - \delta$ and hence δ must lie somewhere in the range 0 to 1.

This equation describes the dynamic relation between the room temperature X_t and the can temperature Y_t. Equation (4.2) is called a first-order difference equation and a system governed by such an equation represents a *first-order dynamic system.*

Now, after rearrangement, Equation (4.2) can be rewritten as

$$Y_{t+1} = (1 - \delta)X_t + \delta Y_t. \tag{4.3}$$

Comparing this with Equation (4.1), the model says that the can temperature (output) at time $t + 1$ is a linear interpolation between the room temperature (input) at time t and the can temperature (output) at time t.

For illustration, let's suppose that the constant δ in Equation (4.3) has a value of 0.5. (In practice, this constant depends on the conductivity of the can of beans and on other factors.) Then, in the examples that follow, Equation (4.3) becomes simply

$$Y_{t+1} = 0.5X_t + 0.5Y_t.$$

If the temperature in the refrigerator is 0°C and the temperature of the room stays fixed at 20°C, then (see Figure 4.2), at time $t = 1$, just before we take the can out of the refrigerator, the temperature X_1 of the can's environment and temperature Y_1 of the can are both zero. If we take the can out of the refrigerator at time $t = 2$, so that its environmental temperature suddenly changes to 20°C, there is a temperature imbalance between the can and its environment. We can find what subsequently happens to the can temperature when the environment temperature X_t undergoes this *step change* by using Equation (4.3) to compute Y_t recursively (step by step) as follows:

$$Y_2 = \{0.5 \times 0\} + \{0.5 \times 0\} = 0,$$

$$Y_3 = \{0.5 \times 20\} + \{0.5 \times 0\} = 10,$$

$$Y_4 = \{0.5 \times 20\} + \{0.5 \times 10\} = 15,$$

and so on.

In this way, the numbers Y_t in the can temperature column of Figure 4.2 can be successively calculated and we see that, when the environment temperature of the can undergoes a "step" change of 20°C, the can temperature makes an exponential (geometric) approach to this equilibrium value. Intuitively we would expect this.

The Meaning of the Constant δ

If you experiment with Equation (4.3), you will find that a small value of δ corresponds to a rapidly responding system, and a large value to

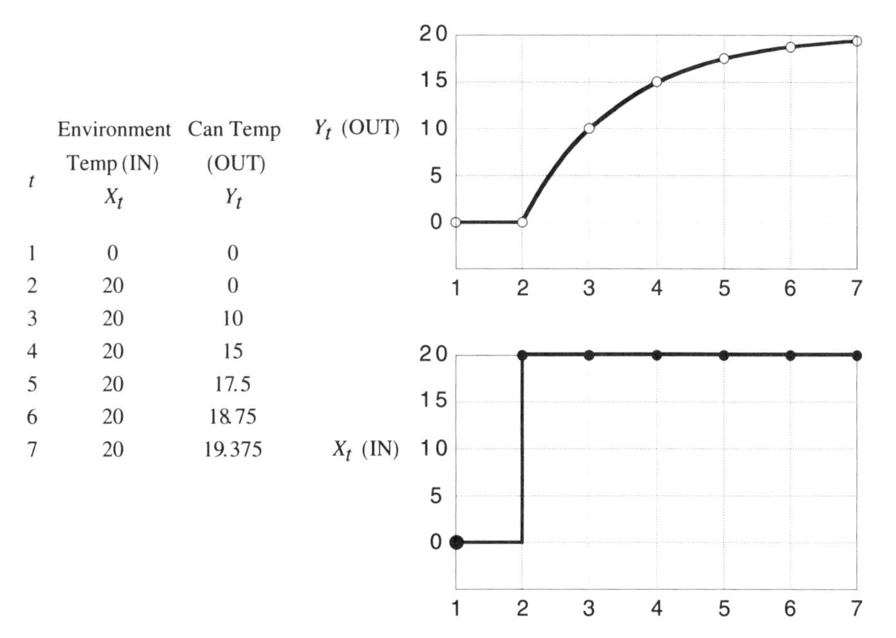

t	Environment Temp (IN) X_t	Can Temp (OUT) Y_t
1	0	0
2	20	0
3	20	10
4	20	15
5	20	17.5
6	20	18.75
7	20	19.375

Figure 4.2. Response of can temperature Y_t to a step change in the environmental temperature X_t from 0°C to 20°C.

a slowly responding system. In fact, for a step change at the input, the proportion of the total change that occurs in the first time interval is equal to $1 - \delta$. Thus, in the case of the beans example with $\delta = 0.5$, you will see that half (10°C) the eventual change in can temperature (20°C) occurs in the first time interval.

Exercise 4.1 Suppose that there are two identical water tanks, A and B, which are cubical in shape with A located above B. Tank A has a valve at its base and, when it is opened, water runs from A into B at a rate proportional to the head (height) of water in tank A. Suppose that observations of the system are made at equispaced times $t = 0, 1, 2, \ldots$, and that initially the depth X of water in tank A is 100 centimeters and that the depth Y in tank B is zero. Suppose also that the valve is opened at $t = 1$ and that the depths of water in tank B in centimeters at times $t = 0, 1, 2$ are $Y_0 = 0$, $Y_1 = 0$, and $Y_2 = 20$.

(a) Write down a difference equation that enables you to calculate subsequent values of Y.

(b) Calculate recursively $Y_3, Y_4, \ldots, Y_{10}$. □

Now Equation (4.3) also can be used to find out what would happen to the can temperature (the output) for various *patterns of change* in the environmental temperature (the input). Suppose that *initially* both the temperature of the can and the temperature of the room are equal at 20°C and imagine that we are able to change the room temperature X_t and maintain it at a series of different fixed levels during a series of time intervals. Suppose that a series of changes is made in the room temperature like those shown by solid dots in Figure 4.3 and let's use the notation X_{t+} to indicate that room temperature is maintained at a specified level during the tth time interval. Thus during the first interval it is 20°C and in the second interval it is 28°C and so on. Supposing, as before, that the constant $\delta = 0.5$. We can again use Equation (4.3) to find out what happens to the can temperature Y_t as follows:

$$Y_2 = \{0.5 \times 20\} + \{0.5 \times 20\} = 20,$$

$$Y_3 = \{0.5 \times 28\} + \{0.5 \times 20\} = 24,$$

and so on. These values are plotted as open circles in Figure 4.3. In practice, of course, the can would be heating up or cooling down in between in a series of gentle curves, but for our purpose we consider only observations made at the discrete times $t - 1$, t, $t + 1$, ...

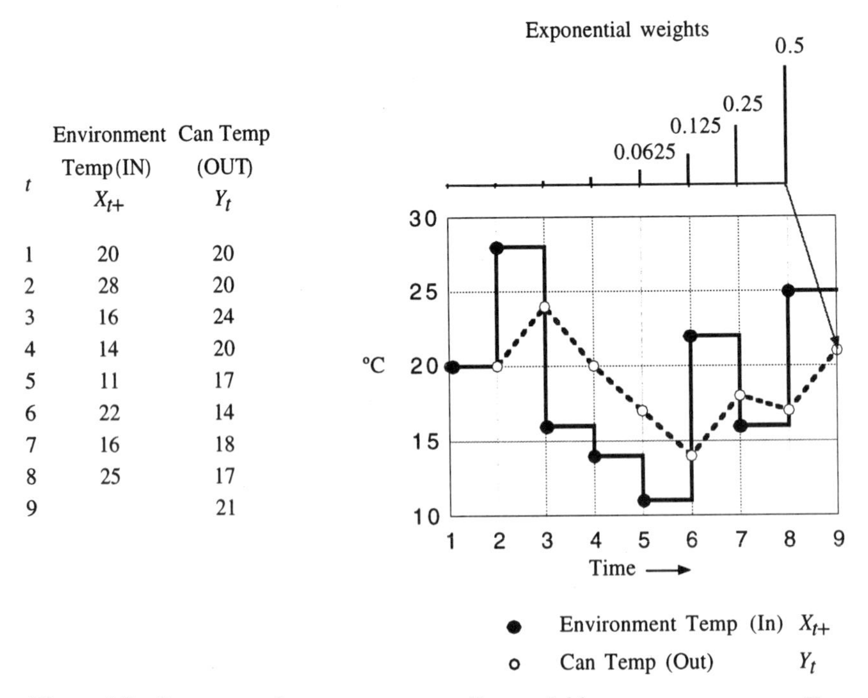

t	Environment Temp (IN) X_{t+}	Can Temp (OUT) Y_t
1	20	20
2	28	20
3	16	24
4	14	20
5	11	17
6	22	14
7	16	18
8	25	17
9		21

Figure 4.3 Response of can temperature Y_t to variable room temperature X_{t+}.

4.4 THE RETURN OF THE EXPONENTIALLY WEIGHTED MOVING AVERAGE

Now look again at Figure 4.3 and consider, for example, the can temperature Y_9 at $t = 9$. This temperature has been influenced by the room temperatures over the last several periods. We can determine how this particular can temperature depends on previous room temperatures as follows.

We calculated Y_9 by using Equation (4.3) with $t = 8$ to get

$$Y_9 = (1 - \delta)X_8 + \delta Y_8,$$

but we got Y_8 from

$$Y_8 = (1 - \delta)X_7 + \delta Y_7.$$

So substituting this equation in the previous one we have

$$\begin{aligned} Y_9 &= (1 - \delta)X_8 + \delta\{(1 - \delta)X_7 + \delta Y_7\} \\ &= (1 - \delta)\{X_8 + \delta X_7\} + \delta^2 Y_7, \end{aligned}$$

but we got Y_7 from

$$Y_7 = (1 - \delta)X_6 + \delta Y_6.$$

So again after substitution

$$Y_9 = (1 - \delta)\{X_8 + \delta X_7 + \delta^2 X_6\} + \delta^3 Y_6.$$

Continuing in this way the final term becomes negligible and we get in general

$$Y_{t+1} = (1 - \delta)\{X_t + \delta X_{t-1} + \delta^2 X_{t-2} + \delta^3 X_{t-3} + \cdots\}. \qquad (4.4)$$

Thus in our example where we supposed that $\delta = 0.5$, the effect of the operations we carried out to calculate Y_9 was the same as if we had calculated that value from

$$Y_9 = 0.5X_8 + 0.25X_7 + 0.125X_6 + 0.0625X_5 + \cdots. \qquad (4.5)$$

The quantities 0.5, 0.25, 0.125, 0.0625, . . . may be called the *weights* and these are shown as a series of vertical bars in Figures 4.3 and 4.4.

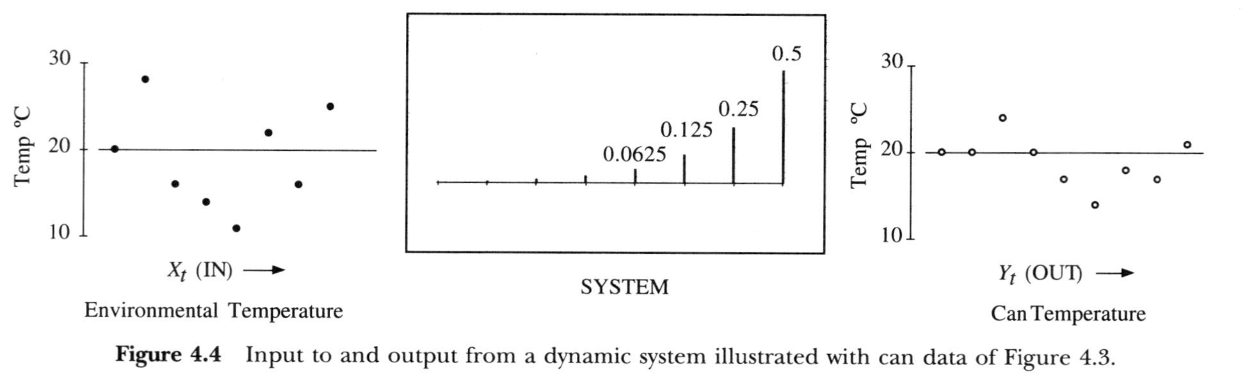

Figure 4.4 Input to and output from a dynamic system illustrated with can data of Figure 4.3.

They determine how much influence each of the previous room temperatures had in determining the temperature of the can of beans at time $t = 9$. Thus in Equation (4.4), the weights are

$$(1 - \delta), (1 - \delta)\delta, (1 - \delta)\delta^2, \ldots$$

and the equation may be written

$$Y_{t+1} = \tilde{X}_t$$

with

$$\tilde{X}_t = (1 - \delta)(X_t + \delta X_{t-1} + \delta^2 X_{t-2} + \cdots). \tag{4.6}$$

Thus the can temperature at time $t + 1$ is an exponentially weighted moving average of the previous environmental temperatures. This is, of course, the same EWMA used in Section 3.6 for process monitoring, where now δ is the *discount factor*, also called the *smoothing* constant [compare Equations (3.4) and (4.6)]. To avoid later confusion, when used to represent process dynamics we will employ the symbol δ for the "smoothing constant" rather than θ.

Now with $Y_{t+1} = \tilde{X}_t$ substituted in Equations (4.2) and (4.3), we get two general updating formulas for EWMAs.

From Equation (4.2)

$$\tilde{X}_t - \tilde{X}_{t-1} = (1 - \delta)(X_t - \tilde{X}_{t-1}),$$

and from Equation (4.3)

$$\tilde{X}_t = (1 - \delta)X_t + \delta\tilde{X}_{t-1},$$

which corresponds to the updating formula previously given in Equation (3.5).

4.5 WHAT'S THIS GOT TO DO WITH THE DYNAMICS OF AN INDUSTRIAL PROCESS?

The dynamic behavior of an industrial system is often approximated quite well by Equation (4.2), or equivalently Equation (4.3). Such a system (which could be a manufacturing process or even an office system) can then be thought of as a dynamic operation on an input X_t to produce an output Y_t as illustrated in Figure 4.4. For our example, the temperature of the can's environment is the input X_t and the output Y_t is the temperature of the can. The dynamic system in this case corresponds to the inertia in the can's ability to gain or lose heat.

System Gain

One special characteristic of the can of beans illustration was that the input X and the output Y were both temperatures measured in the same units. But for many dynamic systems the input and the output will be different entities measured in different units. For example, the output could be the speed of your car and the input the amount by which you depress the gas pedal. For a system such that a unit change in the level of the input eventually produces g units of change in the output, we will say the *system gain* is g. Then if the dynamics could be represented by a first-order system, the relation between input and output would be

$$Y_{t+1} = g\tilde{X}_t.$$

Thus the same type of relation illustrated in Figure 4.4 would hold but the output would be magnified by a factor g.

4.6 USING AN EWMA AS A FORECAST

We saw in Chapter 3 how the EWMA could be used for process monitoring and in this chapter for representing process dynamics. We now illustrate its use for forecasting. As was mentioned in Chapter 1, the use of exponential smoothing for forecasting was first arrived at empirically on the grounds that it was a weighted average with the sensible property of giving most weight to the last observation and less to the next-but-last and so on.

Thus the general idea is that, given data up to and including *time t*, which is then called the *forecast origin*, we can use the EWMA $\tilde{z}_t$ to provide an estimate $\hat{z}_{t+1}$ of the next value z_{t+1}.

In this book we will always use the "hat" notation to denote an estimate. Thus $\hat{z}_{t+1}$ is an estimate of z_{t+1}. This must not be confused with the "tilde" notation. Thus $\tilde{z}_t$ means an EWMA based on data *up to and including time t.* Depending on the application, this might or might not be used as a forecast.

We illustrate this application now with some data.

Forecasting Sales of Dingles

The second column of Table 4.1 shows the weekly sales z_t, for week $t = 21$ through week $t = 40$, of a product called "Dingles" (the other columns in that table will be explained later). These data are plotted in Figure 4.5, which shows how at week 40 an exponentially moving average of current and previous sales with a discount factor of 0.6 produces a value $\tilde{z}_{40} = 331$ that can be used as an estimate $\hat{z}_{41}$ of the sales to be expected in week 41 (331,000 boxes).

As we mentioned earlier, we use the symbol δ for the discount factor only for the application of process dynamics. In discussing the forecasting application, we will use θ as we did in Chapter 3. Thus, in general, the EWMA forecast one step ahead is

$$\hat{z}_{t+1} = \tilde{z}_t = (1 - \theta)(z_t + \theta z_{t-1} + \theta^2 z_{t-2} + \theta^3 z_{t-3} + \cdots). \tag{4.7}$$

In particular, with $\theta = 0.6$ and the *forecast origin* at $t = 40$, the exponentially smoothed value that is used as a forecast of sales (in thousands) for the week $t + 1 = 41$ is

$$\hat{z}_{41} = \tilde{z}_{40} = 0.4z_{40} + 0.24z_{39} + 0.14z_{38} + 0.09z_{37} + \cdots$$

and for the data of Table 4.1,

$$\begin{aligned}\hat{z}_{41} = \tilde{z}_{40} &= (0.4 \times 340) + (0.24 \times 350) + (0.14 \times 320) \\ &\quad + (0.09 \times 300) + \cdots \\ &= 331.\end{aligned}$$

For updating the EWMA we use, as before in Equation (3.5),

$$\tilde{z}_t = \lambda z_t + \theta \tilde{z}_{t-1}, \tag{4.8}$$

where λ (Greek *lambda*) is $1 - \theta$.

Equivalently, in terms of the forecasts, this equation becomes

$$\hat{z}_{t+1} = \lambda z_t + \theta \hat{z}_t. \tag{4.9}$$

This is the formula that is most often used for calculating EWMA forecasts. It says that as soon as the actual sales figure z_t becomes available at time t, you can interpolate between z_t and $\hat{z}_t$ to get $\hat{z}_{t+1}$. Table 4.1 shows the calculations for the sales data using the discount factor $\theta = 0.6$. To get the process started we have supposed a forecast

TABLE 4.1 Sales in Thousands of Boxes of Dingles Over a 20-Week Period, Calculation of Forecasts $\hat{z}_{t+1}$, Forecast Errors e_t for a Trial Value $\theta = 0.6$, and Sum of Squared Forecast Errors $S = 12{,}309$

t	z_t	$\hat{z}_t = 0.4z_{t-1} + 0.6\hat{z}_{t-1}$	$e_t = z_t - \hat{z}_t$
21	260	250	10
22	240	254	−14
23	220	248	−28
24	240	237	3
25	260	238	22
26	260	247	13
27	280	252	28
28	270	263	7
29	240	266	−26
30	250	256	−6
31	310	253	57
32	300	276	24
33	300	286	14
34	260	291	−31
35	290	279	11
36	320	283	37
37	300	298	2
38	320	299	21
39	350	307	43
40	340	324	16
41		331	

$$S = 10^2 + (-14)^2 + (-28)^2 + \cdots + 16^2 = 12{,}309$$

of sales for week 21 of 250.[1] Then using Equation (4.9) we get

$$\hat{z}_{22} = (0.4 \times 260) + (0.6 \times 250) = 254,$$

$$\hat{z}_{23} = (0.4 \times 240) + (0.6 \times 254) = 248,$$

and so on.

This recursive procedure gives the "one step ahead" forecasts $\hat{z}_{t+1}$, which to the nearest digit are listed in the third column of Table 4.1, and in particular the forecast $\hat{z}_{41}$ of 331,000 boxes for week 41. In Figure 4.5, each new forecast $\hat{z}_{t+1}$, indicated by a horizontal line, is an interpolation 0.6 of the way between the new observation z_t and the old forecast $\hat{z}_t$. There is no point in attempting to be too exact in these calculations, because rounding errors do not accumulate, and we have

[1] After a few time periods the forecasts are very insensitive to the choice of this initial value, so a guess is often good enough.

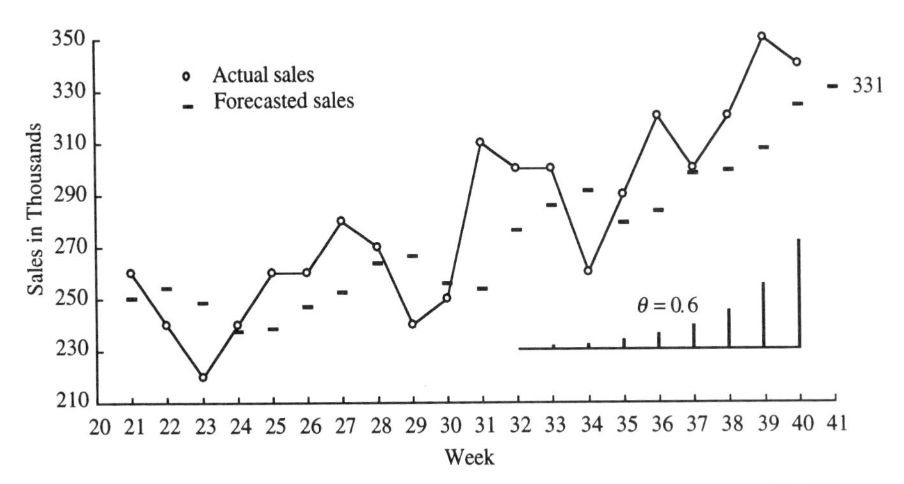

Figure 4.5 Sales z_t, in thousands of boxes, from week $t = 21$ to $t = 40$. Using a discount factor $\theta = 0.6$, the exponentially weighted average $\tilde{z}_{40} = 331$ calculated in week 40 is used as a forecast $\hat{z}_{41}$ of sales in week 41.

given the forecasts to the nearest whole number. Note also that, although exponential averages are easily calculated using the interpolation formula (4.9), for many applications, simply plotting the values as in Figure 4.5 and interpolating by eye would be good enough.

Alternative Forms of the Relationships for EWMAs

Exponentially weighted moving averages—EWMAs—have a number of different applications and interpretations. We now pause for a moment to assemble in Table 4.2 three equivalent relationships, which we denote by A, B, and C (see Appendix 4A). These are written in terms of both the EWMAs $\tilde{z}_t$ and the equivalent one step ahead forecasts such as $\hat{z}_{t+1}$, and also for the forecast errors e_t, where, for example, $e_t = z_t - \tilde{z}_{t-1} = z_t - \hat{z}_t$.

Note that the three relationships in Table 4.2 are identities true *for any set of numbers* $z_t, z_{t-1}, z_{t-2}, \ldots$ and any value of θ between -1 and

TABLE 4.2 Three Important Relationships for Exponentially Weighted Moving Averages

If $\tilde{z}_t = \hat{z}_{t+1} = \lambda(z_t + \theta z_{t-1} + \theta^2 z_{t-2} + \cdots)$ with $\lambda = 1 - \theta$ and $e_t = z_t - \tilde{z}_{t-1} = z_t - \hat{z}_t$, then

A: $\tilde{z}_t = \lambda z_t + \theta \tilde{z}_{t-1}$ or $\hat{z}_{t+1} = \lambda z_t + \theta \hat{z}_t$

B: $\tilde{z}_t - \tilde{z}_{t-1} = \lambda(z_t - \tilde{z}_{t-1})$ or $\hat{z}_{t+1} - \hat{z}_t = \lambda e_t$

C: $z_t - z_{t-1} = e_t - \theta e_{t-1}$

+1. They do *not* involve any assumption about underlying time series models. This last statement is of great importance and is illustrated in the following exercise based on the data in Exercise 3.1.

Exercise 4.2. To verify the relationships in Table 4.2, complete the entries in the following table using $\lambda = 0.4$ ($\theta = 0.6$):

	A			B		C	
t	z_t	$\hat{z}_t$	e_t	$\hat{z}_{t+1} - \hat{z}_t$	λe_t	$z_t - z_{t-1}$	$e_t - \theta e_{t-1}$
1	6	10.00	−4.00	−1.60	−1.60		
2	9	8.40	0.60	0.24	0.24	3	3
3	12	8.64	3.36	1.34	1.34	3	3
4	11	9.98					
5	5	10.39					
6	6	8.23					
7	4	7.34					
8	10	6.00					
9	12	7.60					
10	15	9.36					
11		11.62					

Solution: Using $\theta = 0.6$, we have the following:

	A			B		C	
t	z_t	$\hat{z}_t$	e_t	$\hat{z}_{t+1} - \hat{z}_t$	λe_t	$z_t - z_{t-1}$	$e_t - \theta e_{t-1}$
1	6	10.00	−4.00	−1.60	−1.60		
2	9	8.40	0.60	0.24	0.24	3	3
3	12	8.64	3.36	1.34	1.34	3	3
4	11	9.98	1.02	0.41	0.41	−1	−1
5	5	10.39	−5.39	−2.16	−2.16	−6	−6
6	6	8.23	−2.23	−0.89	−0.89	1	1
7	4	7.34	−3.34	−1.34	−1.34	−2	−2
8	10	6.00	4.00	1.60	1.60	6	6
9	12	7.60	4.40	1.76	1.76	2	2
10	15	9.36	5.64	2.26	2.26	3	3
11		11.62					

□

How Good Is the EWMA Forecast?

Look again at the solution for Exercise 4.2. Note that we could substitute *any* numbers we liked for the series $z_1, z_2, \ldots, z_n$, for the starting value $\hat{z}_1$, and for a value for θ between -1 and $+1$. (The arbitrary numbers are underlined in the solution to the exercise.) We could then calculate values for $\hat{z}_2, \hat{z}_3, \ldots, \hat{z}_n$ and hence for $e_1, e_2, \ldots, e_n$. However the original numbers were chosen, all the demonstrated relationships A, B, and C would be exactly true. But, it would not be necessarily true that at time $t - 1$, say, $\tilde{z}_{t-1} = \hat{z}_t$ would supply a forecast of z_t that was of much value. Whether $\tilde{z}_{t-1}$ is a *good* forecast of z_t depends on whether the series being forecast can, to a reasonable approximation, be represented by a particular model called the IMA model and whether an appropriate value is used for θ. We discuss these questions in the next chapter and merely note here that, for the Dingles data, the conditions are met reasonably well.

Pete's Rule

The formulation of the updating process given by formula B in Table 4.2 helps intuitive understanding of the logic of exponential forecasting. This equation, with $\lambda = 1 - \theta = 0.4$, is

$$\hat{z}_{t+1} - \hat{z}_t = 0.4(z_t - \hat{z}_t). \tag{4.10}$$

Now imagine a situation where the sales figure for week 34 has just become available. By looking at the series plotted in Figure 4.5 you will see that at this point in time Dingles sales are about 30,000 boxes below forecast. Pete, the production manager, is sitting in his office when Sam, the sales manager, bursts in and says "Pete, the sales are down this week. They are 30,000 boxes below forecast! Let's lower the forecast for next week by 30,000." Now the production schedule is based on the forecast and Pete wishes neither to produce an unnecessary large inventory nor to be unable to meet next month's demand. So he might say, "Sam, we both know that a lot of the big fluctuations we get in sales both up and down are really due to chance causes. I have a rule that I use consistently and that I find works very well. I always discount 60% of the current deviation from forecast whether it's up or down, so let's make an update of the forecast by lowering it by just 40% of what you suggest, namely, by 12,000 boxes."

Pete's rule is $\hat{z}_{t+1} - \hat{z}_t = 0.4(z_t - \hat{z}_t)$—precisely the same as Equation (4.10). So although he may not be aware of it, Pete is using an exponentially weighted moving average forecast with a discount factor (smoothing constant) θ of 0.6. If you look again at Figure 4.5 you will see, incidentally, that Pete's sales forecast for week 35 would have been a lot closer than Sam's.

4.7 EFFECTS OF CHANGING THE DISCOUNT FACTOR

In the above illustration a discount factor of $\theta = 0.6$ was used yielding, for example, a forecast $\hat{z}_{41}$ of sales in week 41 of 331,000 boxes. Figure 4.6 illustrates the effect of using other discount factors. With $\theta = 0.4$ we get a forecast of 338,000 and with $\theta = 0.8$ of 313,000. If we had to guess an appropriate value for θ there would be some obvious pros and cons to consider. With the gentle smoothing provided by a large value of θ like 0.8, the exponential average reaches back a long way into the series and so is more effective in smoothing out the noise. On the other hand, it will be much *slower to react* to a change in level of the series than an EWMA with a smaller value of θ. With $\theta = 0.8$, for example, the forecast $\hat{z}_{41}$ is lower than that with $\theta = 0.6$ or $\theta = 0.4$ because it gives less weight to more recent happenings when sales are higher. A different compromise is best for different kinds of series. Noisy, slowly moving, series would favor larger values of θ, while smoother, fast-moving series would favor smaller values of θ. We shall

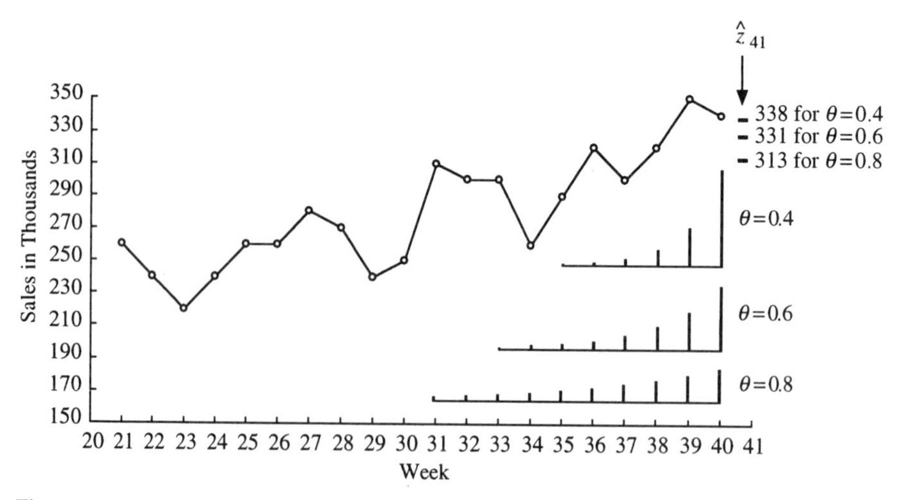

Figure 4.6 Exponential weights for $\theta = 0.4$, 0.6, and 0.8 leading to forecasts $\hat{z}_{41}$ of 338,000, 331,000, and 313,000 boxes.

see below that we don't actually need to guess, for if we have a record of past sales, we can use it to calculate the best value of θ for the particular product we wish to forecast.

4.8 ESTIMATING THE DISCOUNT FACTOR[2]

Suppose we had a series of weekly sales of Dingles stretching back for a number of weeks (100 values would be good and 200 even better). Then we could use these to find out which value of θ gave the smallest forecast errors. We illustrate the calculation with the z_t values of Table 4.1, although it must be understood that in practice more than 20 values would be needed to get a reasonably good estimate of θ.

Having obtained in succession the values of $\hat{z}_t$ for $t = 21, 22, 23, \ldots,$ 40 we can calculate the forecast errors $e_t = z_t - \hat{z}_t$ shown in the fourth column of Table 4.1. Thus

$$e_{21} = 260 - 250 = 10, \quad e_{22} = 240 - 254 = -14, \ldots$$

and so on. In this way we get S, the sum of the squared errors, for the trial value $\theta = 0.6$ from

$$S = 10^2 + (-14)^2 + (-28)^2 + 3^2 + \cdots + 16^2 = 12{,}309.$$

By carrying through the calculation for various trial values of θ between 0 and 1 (say, 0, 0.1, 0.2, . . . , 1.0) and drawing a graph of S versus θ, we can find the value ($\hat{\theta}$, say) that minimizes the sum of squares of the forecast errors.

Figure 4.7 shows such a plot of S versus θ for 100 values of the Dingles series. You will see that Pete was quite close to guessing right. The minimum value of the sum of squares occurs at about $\hat{\theta} = 0.7$ (or equivalently $\hat{\lambda} = 0.3$), where the minimum sum of the squares was $S_{\min} = 75{,}300$.

On the horizontal scale of Figure 4.7 we show both θ and $\lambda = 1 - \theta$. The quantity λ may be called the *nonstationarity constant.* The symbols λ and θ are, of course, equivalent in the sense that if you know one you

[2] A more exact discussion of estimation is given in Chapter 12 of this book and, for example, in BJR.

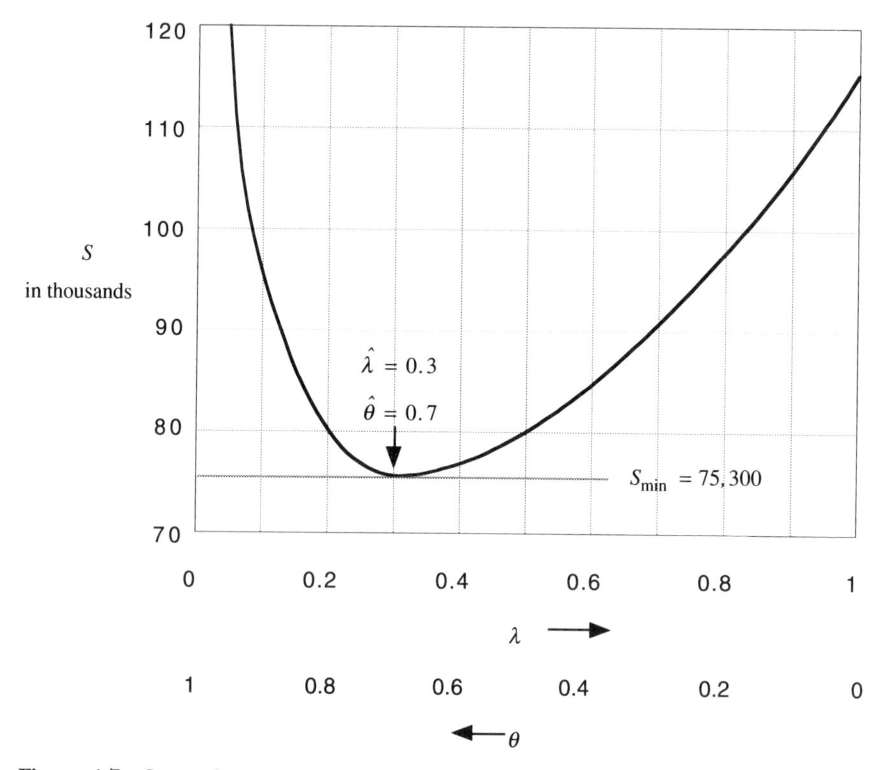

Figure 4.7 Sum of squares curve for 100 weekly sales figures showing a minimum close to $\hat{\theta} = 0.7$ or equivalently to $\hat{\lambda} = 1 - \hat{\theta} = 0.3$.

know the other, but λ is important because, as we shall see, it directly describes the degree of *nonstationarity* of a powerful class of time series, which we discuss in the next chapter. The estimated value of λ for the Dingles series is $\hat{\lambda} = 1 - \hat{\theta} = 0.3$.

The flatness of the sum of squares curve near its minimum is characteristic and is partly responsible for the high degree of robustness of forecasts of this kind. You can see, for example, that if you had used Pete's guess of $\theta = 0.6$ ($\lambda = 0.4$) this would have given an error sum of squares only about 2% higher than the minimum value.

4.9 STANDARD DEVIATION OF FORECAST ERRORS AND PROBABILITY LIMITS FOR FORECASTS

The estimated standard deviation of the forecast errors can be obtained from $\hat{\sigma}_e = \{S_{min}/(n - 1)\}^{1/2}$. For this example $\hat{\theta} = 0.7$, $\hat{\lambda} = 0.3$,

$S_{min} = 75{,}300$, and $n = 100$; so

$$\hat{\sigma}_e = \sqrt{75300/99} = 27.6.$$

If, as is often reasonable, we make the assumption that the forecast errors are approximately normally distributed, this means that you can use the normal distribution to make probability statements. For example, at time t we could say: "There is a probability of about 2/3 that next week's sales will lie within $\hat{z}_{t+1} \pm 27.6$."

4.10 WHAT TO DO IF YOU DON'T HAVE ENOUGH DATA TO ESTIMATE θ

If you are just getting started with forecasting, and you have very little data with which to estimate θ, it is helpful to know that in practice λ often seems to lie between 0.2 and 0.4 and the value $\lambda = 0.3$, which we estimated for the Dingles series, is often a good compromise. In practice, you can start off using such a guessed value and when you have more data you can check it out with the sum of squares curve.

Use of the EWMA for Process Adjustment and Regulation

In Chapter 3, control charts based on the EWMA were introduced for process monitoring. In the chapters that follow, we shall be discussing process *regulation* by feedback adjustment. In that discussion, exponentially weighted averages also play an important role but in a totally different context. Here they are associated not with statistical hypothesis testing but with statistical *estimation*. In process regulation, it is supposed that the system is *not* in a state of control and without adjustment would be wandering from target. The EWMA provides a valueable means of *estimating* the level of the disturbance and, by compensating its effect, of keeping the adjusted process close to target. Both process monitoring and process adjustment are of great value and importance in appropriate contexts. However, they should not be confused. In particular, waiting for the deviation from target to be statistically significant before making a change is usually a very poor strategy.

APPENDIX 4A SOME USEFUL RELATIONSHIPS AMONG EWMAs AND THE DATA

The following relationships are set out in Table 4.2:

$$\text{A:}\quad \tilde{z}_t = \lambda z_t + \theta \tilde{z}_{t-1}; \tag{4A.1}$$

$$\text{B:}\quad \tilde{z}_t - \tilde{z}_{t-1} = \lambda(z_t - \tilde{z}_{t-1}), \quad \text{or writing } z_t - \tilde{z}_{t-1} \text{ as } e_t,$$

$$\tilde{z}_t - \tilde{z}_{t-1} = \lambda e_t; \tag{4A.2}$$

$$\text{C:}\quad z_t - z_{t-1} = e_t - \theta e_{t-1}. \tag{4A.3}$$

Equation (4A.1) is true because

$$\begin{aligned} \lambda z_t + \theta \tilde{z}_{t-1} &= \lambda z_t + \lambda(\theta z_{t-1} + \theta^2 z_{t-2} + \cdots) \\ &= \lambda(z_t + \theta z_{t-1} + \theta^2 z_{t-2} + \cdots) \\ &= \tilde{z}_t \end{aligned}$$

Equation (4A.2) is true because

$$\tilde{z}_t = \lambda z_t + \theta \tilde{z}_{t-1}$$

and

$$\tilde{z}_{t-1} = \lambda \tilde{z}_{t-1} + \theta \tilde{z}_{t-1},$$

and subtracting the second equation from the first gives (4A.2).

Equation (4A.3) is true since

$$\tilde{z}_{t-1} - \tilde{z}_{t-2} = \lambda e_{t-1}$$

and adding $e_t - e_{t-1}$ to both sides of this equation gives (4A.3).

CHAPTER 5

Time Series Models for Process Disturbances

"Where order in variety we see,
and where, though all things differ, all agree."
Windsor Forest, ALEXANDER POPE

5.1 INTRODUCTION

This chapter is about some time series models useful in the further study of control problems.[1] The simplest type of a time series, illustrated in Figure 1.1a, is a sequence of values a_t, a_{t-1}, a_{t-2}, . . . , which behaves like a series of *statistically independent* drawings from some distribution (usually supposed approximately normal) having mean zero and with fixed standard deviation σ_a. This is called a white noise series and is denoted by $\{a_t\}$. Thus a model for a quality characteristic y with its mean μ on target and in a perfect state of control is

$$y_t - T = a_t \tag{5.1}$$

with the a_t values a white noise series. In what follows, however, we need time series models that describe the behavior of processes when they are not in a perfect state of control. In particular, we need models for process disturbances that can occur when a process is left to wander without adjustment.

We shall use $z_t = y_t - T$ to mean the *disturbance*, that is, the deviation from some target value T that would occur *if no attempt at control were*

[1] This rather limited discussion of time series models is sufficient to understand the chapters on feedback control that follow. If you would like a more general account, you can find one in Chapter 12 of this book. A deeper understanding may be gained by consulting a book on time series such as Box, Jenkins, and Reinsel (1994).

made. Then, for example, a time series model for a process in perfect control will be $z_t = a_t$, according to Equation (5.1). In general, a *time series model* is an equation that relates in some way or other a sequence $\{z_t\}$ of values of the disturbance to the values $\{a_t\}$ of a white noise series. Now the values in a white noise series are statistically independent and hence unpredictable from previous values; by contrast, wandering behavior in a process disturbance implies *lack* of statistical independence between adjacent observations and some degree of predictability for a guessed value, from previous values. Therefore we first consider ways of modeling series where the observations are not statistically independent.

5.2 A STATIONARY TIME SERIES MODEL IN WHICH SUCCESSIVE VALUES ARE CORRELATED

A simple way to represent dependence of successive values of a disturbance is by means of *autoregressive* models. Just as a *regression* model might relate z_{t+1} to some other variable, x_{t+1}, autoregressive models relate z_{t+1} to one or more previous values $z_t, z_{t-1}, z_{t-2}, \ldots$ of the same series. Thus

$$z_{t+1} = 0.9z_t + a_{t+1}$$

is a *first-order autoregressive model*, in which the deviation from target z_{t+1} is made to depend partly on the previous deviation z_t and partly on a random error a_{t+1}. The autoregressive coefficient (0.9 in this example) must lie between -1 and $+1$. It measures the degree of correlation between successive values of the series. For illustration, Figure 5.1a shows part of a series generated[2] by the first-order autoregressive model $z_{t+1} = 0.9z_t + a_{t+1}$ and Figure 5.1b shows a plot of each deviation against the previous deviation, showing the *autocorrelation* in this series.

The autoregressive coefficient is usually denoted by the Greek letter ϕ (*phi*) so that, in general, the first-order autoregressive model, also called an AR(1) model, is

$$z_{t+1} = \phi z_t + a_{t+1}.$$

[2]To generate such time series, we begin with a white noise series $\{a_t\}$, which can, for example, be obtained from a table of random normal deviates. We can then start the generating process off by writing $z_1 = a_1$; then $z_2 = 0.9z_1 + a_2$, $z_3 = 0.9z_2 + a_3$, and so on.

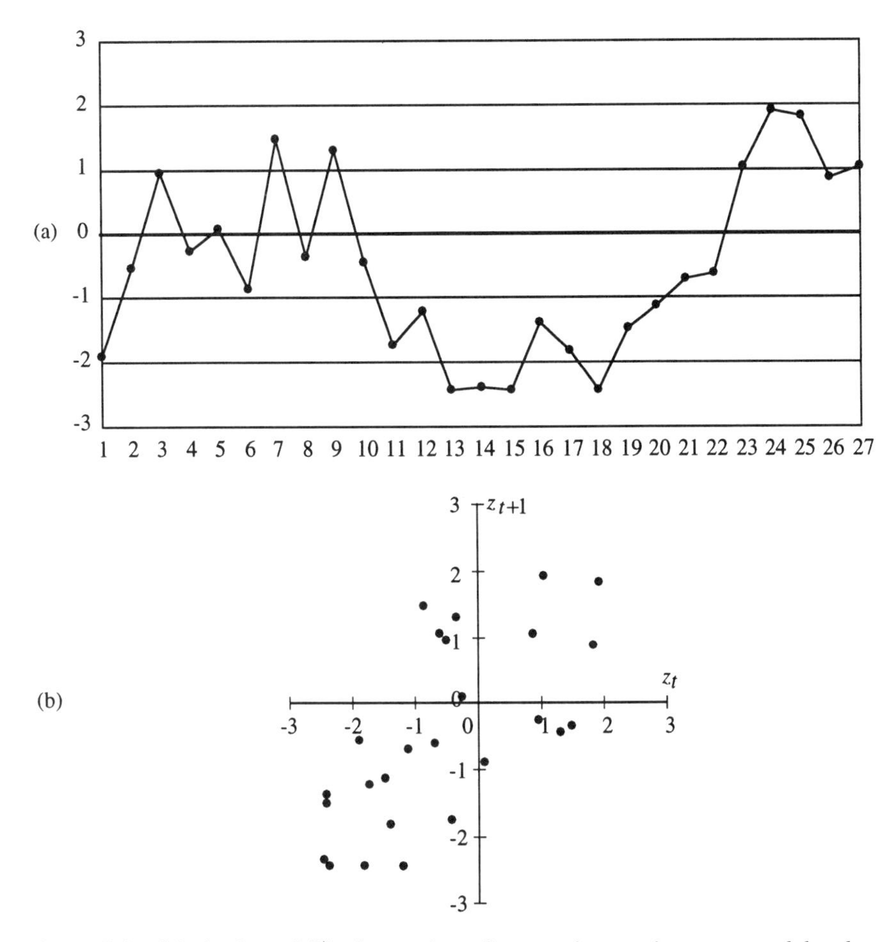

Figure 5.1 (a) A plot of 27 observations from a time series generated by the autoregressive model $z_{t+1} = 0.9z_t + a_{t+1}$. (b) A plot of each value of the time series against the previous value.

This model not only produces a correlation $\phi = 0.9$ between successive values of the series $\{z_t\}$ but also a correlation $\phi^2 = 0.81$ between values two steps apart, a correlation $\phi^3 = 0.729$ between values three steps apart, and so on.

Major Effects of Statistical Dependence: An Illustration

The monitoring techniques discussed in earlier chapters made frequent use of the assumption that, for a process in a state of statistical control, successive deviations $z_t, z_{t-1}, \ldots$ from the target would be statistically independent. The autoregressive model can be used to illustrate that violations of this assumption can have serious consequences. For illustration, we consider the estimation of the standard deviation from the average moving range $\overline{MR}$. It will be recalled that, on the assumption of statistical independence and normality, such an estimate is provided by $\sigma_z = \overline{MR}/1.128$. Now consider instead a series generated by an AR(1) model, and suppose that $\hat{\sigma}_z$ and $\hat{\sigma}_a$ are, respectively, unbiased estimates of the standard deviations of the series $\{z_t\}$ itself and of the generating white noise series $\{a_t\}$. Then if the autoregressive parameter ϕ were zero, z_t would be equal to a_t and $\hat{\sigma}_z/\hat{\sigma}_a = 1$. But if ϕ were not zero then

$$\hat{\sigma}_z = \frac{\overline{MR}}{1.128} \times \frac{1}{\sqrt{1-\phi}}, \quad \hat{\sigma}_a = \frac{\overline{MR}}{1.128} \times \sqrt{1+\phi},$$

$$\text{and} \quad \frac{\hat{\sigma}_z}{\hat{\sigma}_a} = \frac{1}{\sqrt{1-\phi^2}}.$$

For example, for the series illustrated in Figure 5.1 for which ϕ was equal to 0.9

$$\hat{\sigma}_z = \left(\frac{\overline{MR}}{1.128}\right) \times 3.16, \quad \hat{\sigma}_a = \left(\frac{\overline{MR}}{1.128}\right) \times 1.38,$$

$$\text{and} \quad \frac{\hat{\sigma}_z}{\hat{\sigma}_a} = 2.29.$$

Thus the usual calculation would, on the average, underestimate σ_z by a factor greater than three and would not supply an unbiased estimate of σ_a either. Even for small amounts of autocorrelation that are not easily detectable, the effect can be appreciable. For instance with $\phi = 0.3$, the reader may confirm that σ_z is underestimated by a factor of 1.20.

5.3 A NONSTATIONARY TIME SERIES MODEL

The property of stationarity implies that the generated time series varies in a stable manner about the fixed mean. It can make excursions from this mean but it will always return. Note, for example, the behavior of the series in Figure 5.1a. Starting at observation number 10 it makes a wide excursion away from the mean but returns by observation number 22. Such a model[3] can describe the output from a process in a state of *autocorrelated control* referred to in Chapter 1 and illustrated in Figure 1.1b. However, you cannot safely employ such a model to describe what a process would do *if it were totally uncontrolled,* because it implies that if you "sat on your hands" and did nothing you could assume that it would *of its own volition* continually return to the

[3] With the autoregressive parameter ϕ usually less extreme than the value $\phi = 0.9$ used here for illustration.

neighborhood of its mean. Unfortunately, in the real world you cannot rely on any such supposition. In practice, many processes, if they were *not appropriately controlled,* would *permanently* drift away from the target—possibly with disastrous results. We need then to consider how *nonstationary* behavior, such as is illustrated in Figure 1.1c, might be modeled. Rather unexpectedly, one way to discover how to do this is to answer the question presented below.

When Is a Forecast a Good Forecast?

Suppose at the time t you are interested in predicting the value of a series at time $t + 1$, one step ahead. The series might be for an economic indicator, the quality of a supplier's product, or sales data like that for "Dingles" in the last chapter. Suppose over a period of time some forecasting method or other has been used and you have a record of the forecasts $\hat{z}_{t+1}, \hat{z}_t, \hat{z}_{t-1}, \ldots$, that were made and the actual values $z_{t+1}, z_t, z_{t-1}, \ldots$ that subsequently occurred and hence the forecast errors $e_{t+1}, e_t, e_{t-1}, \ldots$ committed by the method. At this point the actual method of forecasting used is immaterial; it might be by guesswork; it might be by using a regression model linking the $\{z_t\}$ to other variables; it might be by using an econometric model or whatever. Now suppose by making plots like Figure 5.1b of e_{t+1} against e_t, e_t against e_{t-1}, and so forth, and by making corresponding calculations you found that these past records showed that the forecast *errors* were in some way autocorrelated. Then such autocorrelation would imply that forecast could be made of the forecast errors, which could improve the original forecast. In such a case, obviously, the forecast method could not be the best. An important qualification for a "best" method of forecasting, therefore, is that the forecast errors cannot be predicted from previous errors. In particular, this would be true if they formed a white noise series $\{a_t\}$.

Forecasting a Time Series One Step Ahead

Before proceeding, it is worth making a brief digression to consider the consequence of standing this argument on its head. Suppose we know the values $z_t, z_{t-1}, z_{t-2}, \ldots$ of a modeled time series. Then to obtain the best estimate of the next value in the series we omit the white noise element a_{t+1} from the modeled value of z_{t+1}.

For example, if we believe that a particular series can be modeled by a first-order autoregressive process, then at time $t + 1$

$$z_{t+1} = \phi z_t + a_{t+1}. \tag{5.2}$$

From this, it follows that the best estimate of z_{t+1} that can be made at time t is

$$\hat{z}_{t+1} = \phi z_t,$$

obtained by omitting a_{t+1} from the right-hand side of Equation (5.2). This makes sense because a_{t+1}, which at time t has not yet happened, varies independently of all previous elements in the series. The above rule is universally applicable for finding the one-step-ahead forecast for any linear time series model, such as is considered in this book.

We now return to the consideration of modeling nonstationary behavior and we begin by considering once more the exponentially weighted moving average (EWMA).

Qualification of the EWMA as a Forecast

A number of arguments suggest that the EWMA might provide a useful forecast of the next value in a time series:

- Exponential discounting of the past data by the EWMA agrees with common sense (e.g., see Section 1.3).
- Over a number of decades the EWMA has often been successful in forecasting many actual time series.[4]
- The rationale of Pete, the production manager, for discounting part of the discrepancy between the current sales and previous forecast (see Section 4.6), which directly leads to the EWMA, is sensible.
- The EWMA predicts the output from a first-order dynamic system (see Sections 4.3 and 4.4).

For What Model Is an EWMA a Good Forecast?

It is therefore appropriate to ask the question[5]: "What time series model would justify the use of the EWMA as a forecast of the next value in a time series?"

[4] In general, of course, the EWMA is not always the best forecast but it often does quite well in predicting nonstationary series.
[5] This question was asked and answered a long time ago by Muth (1960) using a more rigorous argument than that given here.

Now we saw earlier that, for *any* series of data, $z_t, z_{t-1}, \ldots$, and for *any* choice of the smoothing constant θ, an EWMA $\tilde{z}_{t-1}$ could be updated by using the formula

$$\tilde{z}_t = \lambda z_t + \theta \tilde{z}_{t-1}. \tag{5.3}$$

So, if $\tilde{z}_t$ is used as a forecast $\tilde{z}_{t+1}$ of z_{t+1}, the corresponding updating formula for the forecast is

$$\hat{z}_{t+1} = \lambda z_t + \theta \hat{z}_t \tag{5.4}$$

and the forecast error is

$$e_{t+1} = z_{t+1} - \hat{z}_{t+1}. \tag{5.5}$$

It follows (see formula C in Table 4.2 and Appendix 4A) that for *any* set of data $\{z_t\}$ and for *any* choice of θ

$$z_{t+1} - z_t = e_{t+1} - \theta e_t. \tag{5.6}$$

The intuitive argument used before, therefore, tells us that an EWMA with smoothing constant θ is the best forecast for a time series generated by this model when the forecast errors $\{e_t\}$ it produces are a white noise sequence $\{a_t\}$. In fact, it can be shown that the EWMA then provides a forecast with minimum mean square error (MMSE for short). MMSE means that the long-run average of the squares of the e_t values is the smallest possible. Now, in this case, we can also say that the *mean* (long-run average) of the e_t values themselves is zero, because $e_t = a_t$ and the white noise a_t values have mean zero. In such circumstances, the MMSE property also implies (see Section 3.5) that these forecast errors will have the smallest possible variance—σ_e^2 is minimized—and hence that their standard deviation σ_e is minimized. The appropriate model is therefore

$$z_{t+1} - z_t = a_{t+1} - \theta a_t. \tag{5.7}$$

This is a *nonstationary* time series model and turns out to be of special importance in the representation of nonstationary phenomenon like sales data and business and economic data. It is of particular importance in the study of process regulation and adjustment so we need to discuss its properties in some detail. We shall call it an *integrated moving average* time series model or an IMA for short.

Figure 5.2 shows five specimen series of 100 values each generated[6] by IMA models with $\lambda = 0.0, 0.1, 0.2, 0.3$, and 0.4 ($\theta = 1.0, 0.9, 0.8, 0.7$, and 0.6). In each case the limit lines shown in the figure are for $\pm\sigma_a$, where σ_a is the standard deviation of the white noise errors $\{a_t\}$ used to generate the series. When $\lambda = 0$ ($\theta = 1.0$), as in Figure 5.2a, the disturbance z_t is white noise, and what we see is a typical record for a process in a perfect state of control. Larger values of λ produce disturbances that display an increasing degree of instability, which can closely mimic the behavior of many industrial processes when no control is applied. It was shown long ago (Bachelier, 1900) that stock market prices tend to behave as if they were generated by an IMA model with λ close to 1 (θ close to zero). For such a series, the model becomes $z_{t+1} = z_t + a_{t+1}$. Thus each new value is simply the last value plus an independent error—a series called a *random walk.* Industrial time series do not show the same degree of volatility as stock prices and much smaller values of λ (usually between 0.2 and 0.4) are appropriate.

A characteristic of nonstationary disturbances, such as those generated by Equation (5.7), is that if they are left unadjusted, there is no guarantee that they will of their own volition return to the vicinity of the target value in a finite time. Indeed, such a series *does not* possess a *long-run* true mean value μ (nor a finite *long-run* variance). However, the exponentially smoothed value $\tilde{z}_t$ provides a local average that estimates the location of the disturbance at a given time t.

The Moving Range with an IMA Model

The moving range estimate of the standard deviation $\hat{\sigma}_a = \overline{MR}/1.128$ appropriate for a white noise series $\{a_t\}$ can seriously underestimate σ_a for a series generated by the IMA model of Equation (5.7). An unbiased estimate of σ_a is then

$$\hat{\sigma}_a = \frac{\overline{MR}}{1.128} \times \sqrt{\frac{2}{1 + \theta^2}}.$$

[6] To generate such a time series with, for example, $\theta = 0.6$, we can rewrite Equation (5.7) as $z_{t+1} = z_t + a_{t+1} - 0.6a_t$. Using a table of random normal deviates for $\{a_t\}$, we have calculated the successive values of the series in Figure 5.2 by supposing that $z_0 = 0$ and using the equation recursively.

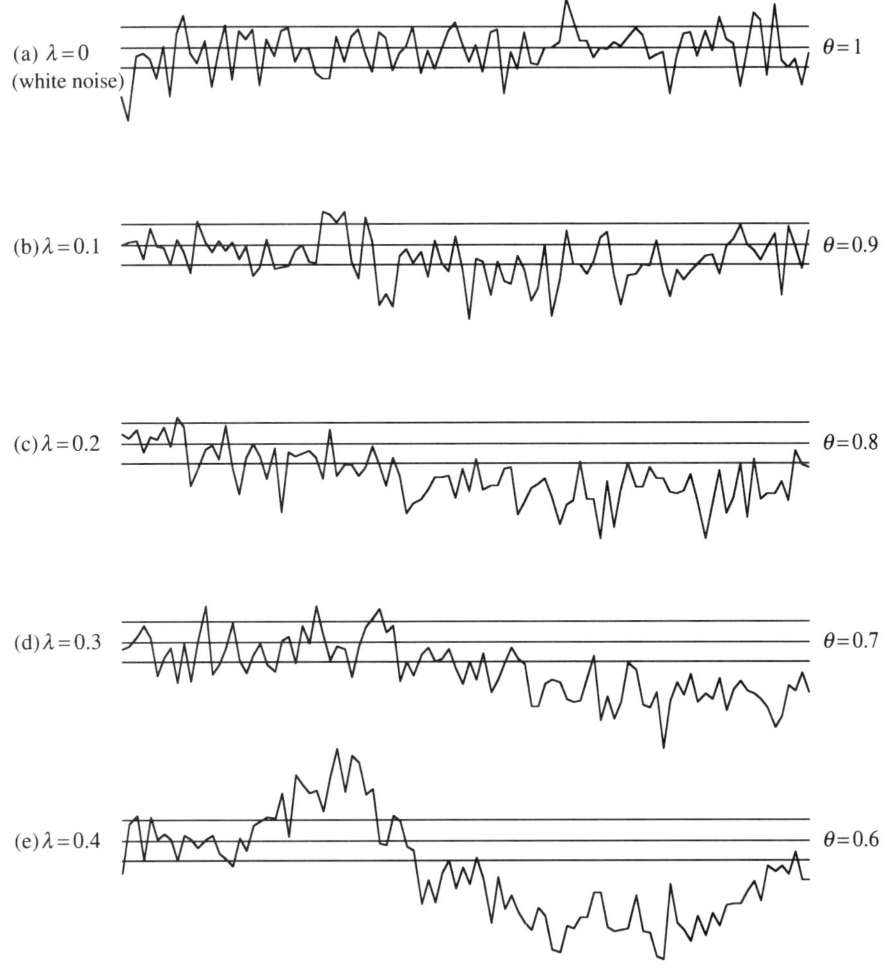

Figure 5.2 Five time series generated by the IMA model (5.7) with θ equal to 1.0, 0.9, 0.8, 0.7, and 0.6 or, equivalently, $\lambda = 1 - \theta$ equal to 0.0, 0.1, 0.2, 0.3, and 0.4.

5.4 UNDERSTANDING TIME SERIES BEHAVIOR WITH THE VARIOGRAM

A good way to understand the characteristics of various time series models including nonstationary models is by considering the variance of the difference between two values of the disturbance taken m intervals apart (see, e.g. Jowett, 1952; Cressie, 1988). We call this V_m. Thus $V_m = \text{var}(z_{t+m} - z_t)$, where $\text{var}(z_{t+m} - z_t)$ means the variance of $z_{t+m} - z_t$. Better still, we can study V_m/V_1, which tells us how much *bigger* the variance is for values m steps apart than for values one step apart. If we plot V_m/V_1 against m, we will call the resulting graph (see Figure 5.3) the *variogram.*[7]

A white noise sequence is characterized by Equation (5.7) with $\theta = 1$ ($\lambda = 0$). For such a sequence $\mathrm{var}(z_{t+m} - z_t) = \mathrm{var}(a_{t+m} - a_t) = 2\sigma_a^2$ so V_m/V_1 stays equal to 1 no matter how large m is. This says that for a process in a perfect state of control with uncorrelated errors it would make no difference whether observations were taken a hundred steps apart or one step apart, the variance of the difference would be the same. (This is a sobering thought and leads one to wonder whether any naturally occurring system could ever be in a perfect state of control.) This special case when $V_m/V_1 = 1$ for all values of m is shown in the figure by the horizontal line labeled white noise. The dotted line is for the stationary autoregressive model considered before, $z_{t+1} = 0.9z_t + a_{t+1}$. We see that for this series although the variogram increases quite rapidly at first, it then flattens out, implying in the case illustrated that the variance of values a thousand intervals apart would essentially be no larger than that for values thirty intervals apart. Note that this happens even though, for illustration, we have used a time series with a very high degree of correlation (0.9) between adjacent values. The same phenomenon will occur for any *stationary* model that varies about a fixed mean value. Such models therefore seem inappropriate for representing *uncontrolled* disturbances in the real world; for this is a world where, if left to themselves, machines *continue* to go out of adjustment, tools *continue* to wear out, and people *continue* to miscommunicate and forget. In all such circumstances, it is reasonable to expect that, *if no corrective action is taken,* observations spaced further apart will differ more and more. So a realistic model to represent such a situation would be one in which a *steady increase* in V_m/V_1 occurred as m increased. Such a model will represent a *nonstationary* disturbance. In particular, we could represent such a situation by a model in which V_m/V_1 increased *linearly* with m. As was pointed out by Box and Kramer (1992) and is illustrated in Figure 5.3, the IMA model of Equation (5.7) has precisely this property.[8]

[7] For *stationary* series, an equivalent representation uses the autocorrelation function (or correlogram), which displays the *correlation* of values of the time series d steps apart. The variogram has the advantage that it can also represent the behavior of many *nonstationary* time series. For a fuller discussion, see Chapter 12 in this volume or a book on time series such as BJR.

[8] The slope of this variogram is $\lambda^2/(1 + \theta^2)$. See Equation 5.16.

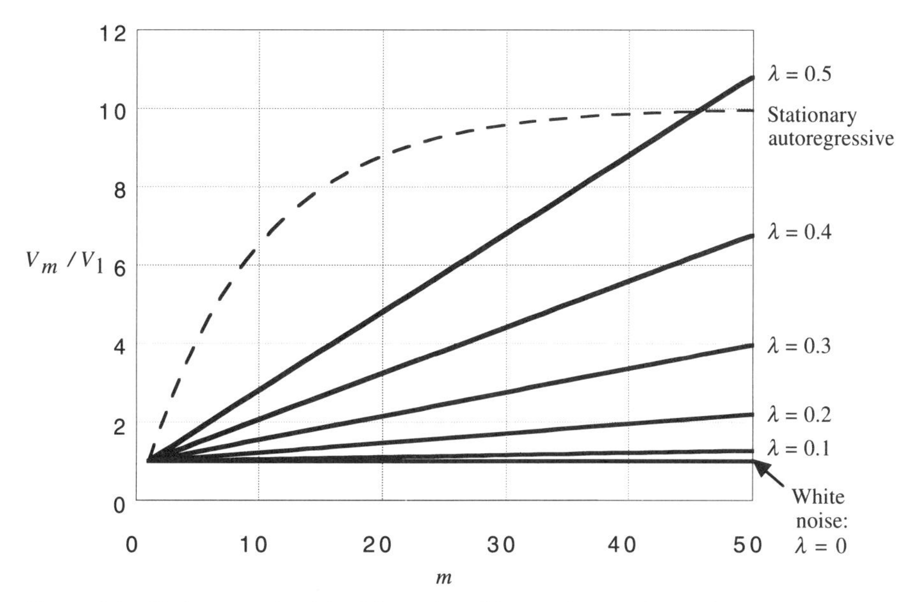

Figure 5.3 Variograms for (a) white noise, (b) a stationary first-order autoregressive model, and (c) a nonstationary IMA model for various values of λ.

As mentioned earlier, we call the quantity $\lambda = 1 - \theta$ the nonstationarity parameter.[9] It determines the degree of instability of the IMA model as evidenced by the behavior of the variograms in Figure 5.3. Some understanding of the implications of the different values of λ can be gained by studying Figure 5.3 in conjunction with Figure 5.2.

In summary, we see that the IMA model of Equation (5.7) is (1) a model for which the exponential smoothing produces the optimal forecast; (2) a model for which the variance of the difference in measurements taken m steps apart increases linearly with m; and (3) a model that can represent the situation where uncontrolled systems do not stay fixed but drift away from their initial conditions.

Unfamiliarity with this last idea happens partly because statistics is usually taught as if things stayed fixed. Also, perhaps, because any single individual person experiences only one realization of his/her

[9] Meaningful models can be obtained for θ in the range -1 to $+1$ but, for our purpose, the interesting range is usually from 0 to 1. Equivalently, $\lambda = 1 - \theta$ can be between 0 and 2, but the corresponding interesting range is between 0 and 1.

passage through life, the "roads not taken" are not experienced and seldom considered. However, opportunities do occur for glimpsing diverging reality.

One such is interlaboratory standardization studies with which many quality practitioners will be familiar. In a typical study of this sort, identical samples are sent to, say, six different labs A, B, C, D, E, and F located at different sites. These labs are all supposed to be using the same testing procedure and usually all have identical copies of a manual spelling out the details. Nevertheless, the results of repeated tests on identical samples often look something like this[10]:

Lab A	Lab B	Lab C . . .
8.9, 8.7, 9.0	6.3 6.7 6.4	10.6 10.3 10.7

When such results are shown to the participants, typically they are greatly perplexed. Thus the people from Lab B may say, "These results by Lab A and Lab C are preposterous; look at our results: 6.3, 6.7, 6.4! They are *extremely* consistent and the correct result is clearly close to 6.5. How can these people be getting values in the neighborhood of 9 and 10?" Similar remarks will be heard from Labs A and C.

One way in which such questions can be answered and standardization attempted is to arrange for representatives from different laboratories to get together and watch each other carry out the procedure. (This is particularly recommended if you are the organizer of such a gathering and are in need of a little light entertainment.)

Suppose, for example, the procedure is a chemical determination. A person from Lab B watching a person from Lab A perform the test may say something like this: "Why are you drying the sample? I've looked in the manual and it doesn't say anything about drying the sample." To which the person from Lab A responds: " Why? *Of course* you dry the sample; that's the standard procedure with a test of this kind. Where did you study your chemistry?" and so on.

It may be possible with this kind of approach to resolve the problems and get reasonably consistent results. But it is again common experience that unless such cooperative tests are continually repeated—say, every six months—the results from the various labs may again drift apart to an unacceptable extent.

[10] In case it might be supposed by some reader unfamiliar with such studies that this example is farfetched, we mention that in a recent study of cholesterol measurements (where a difference of 20 units may be regarded as clinically significant) systematic biases from lab to lab have been demonstrated of as much as 60 units! (See Hassemer, Wiebe, and Kramer, 1989.)

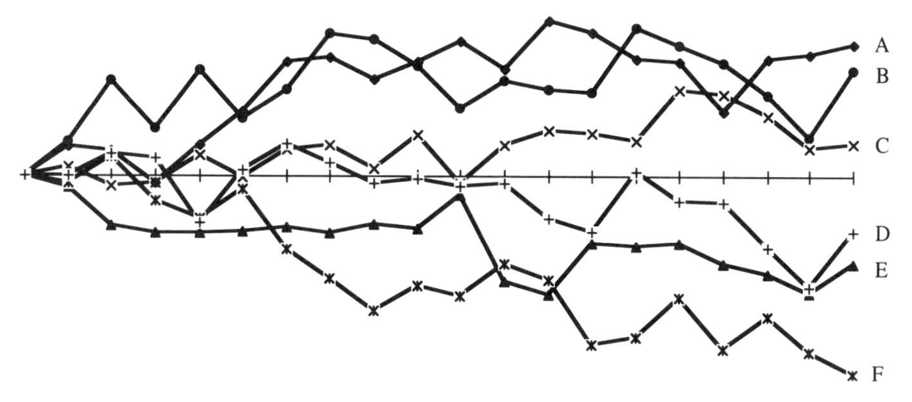

Figure 5.4 Six realizations A, B, C, D, E, and F of an IMA time series model with $\lambda = 0.4$.

This drifting apart occurs in very much the same way that the six realizations A, B, C, D, E, and F, drift apart in Figure 5.4. Each of these is, in fact, generated by the same time series model (an IMA with $\lambda = 0.4$) but with different sets of random a_t values.

If you made the necessary studies you would undoubtedly find the same thing happening with "standard" office procedures, "standard" medical procedures, "standard" legal procedures, and so forth. On a longer term basis, the same phenomenon is demonstrated by the development of the Romance languages, such as Spanish, French, Italian, Portuguese, Catalan, Romanian, and so forth all from the same basic Latin.

5.5 STICKY INNOVATION GENERATING MODEL

Now the IMA can be shown to be equivalent to a random walk model with added independent noise (e.g., see BJR). This has an illuminating interpretation as the *sticky innovation model* (Box and Kramer, 1992).

Consider a process disturbance z_t defined as the deviation $y_t - T$ from some reference target T that would occur *if no attempt at control were made.* Such a disturbance typically arises from complex causes, which may be unknowable, but a model that can approximate its behavior may be arrived at by thinking of the disturbance as containing two parts: a *transitory* part α_t and an *evolving* nontransitory part S_t. The former is associated with the *noise* and the latter with the *signal.* Thus

$$z_t = \alpha_t + S_t. \tag{5.8}$$

The noise contribution α_t is called transitory because it is associated only with the tth observation and is independent of observations taken at every other time. Measurement and observational errors are typical contributions to α_t, which will be represented by random drawings from a distribution with mean zero and variance σ_α^2.

The evolving nontransitory part S_t represents accumulated innovations that enter the system from time to time and *get stuck* there. Such "sticky" innovations can arise from a multitude of causes including wear, corrosion, and human miscommunication. Thus a car tire hits a sharp stone and *from that point onward* the tread is slightly damaged; a tiny crater, caused by corrosion, appears on the surface of a drive shaft and *remains* there; certain details in the standard procedure for taking blood pressures in a hospital are forgotten over time and the routine *from then on* is changed. Every person, thing, or system is subject to such influences, which continuously drive a natural increase in entropy (disorder). The quality practitioner's job is to defeat this process.

Let's represent the "sticky" innovations that accumulate in the system and occur at times $\ldots, t-1, t, t+1, \ldots$ by $\ldots, u_{t-1}, u_t, u_{t+1}, \ldots$ and represent these by random drawings from a second distribution[11] with mean zero and variance σ_u^2 (see Figure 5.5). Suppose that at time $t = 0$, the process is on target except for the transitory error α_0; then, at times $0, 1, 2, \ldots, t, \ldots, t+m, \ldots$, the disturbance as described by this model is

$$
\begin{aligned}
z_0 &= \alpha_0, \\
z_1 &= \alpha_1 + u_1, \\
z_2 &= \alpha_2 + u_1 + u_2, \\
&\;\;\vdots \\
z_t &= \alpha_t + u_1 + u_2 + \cdots + u_t, \\
&\;\;\vdots \\
z_{t+m} &= \alpha_{t+m} + u_1 + u_2 + \cdots + u_t + u_{t+1} + \cdots + u_{t+m}, \\
&\;\;\vdots
\end{aligned}
\tag{5.9}
$$

The disturbance z_t will then have a continuously increasing variogram like those in Figure 5.3.

[11] In Figure 5.5 these distributions are shown as normal distributions, but normality is not needed for the validity of the argument.

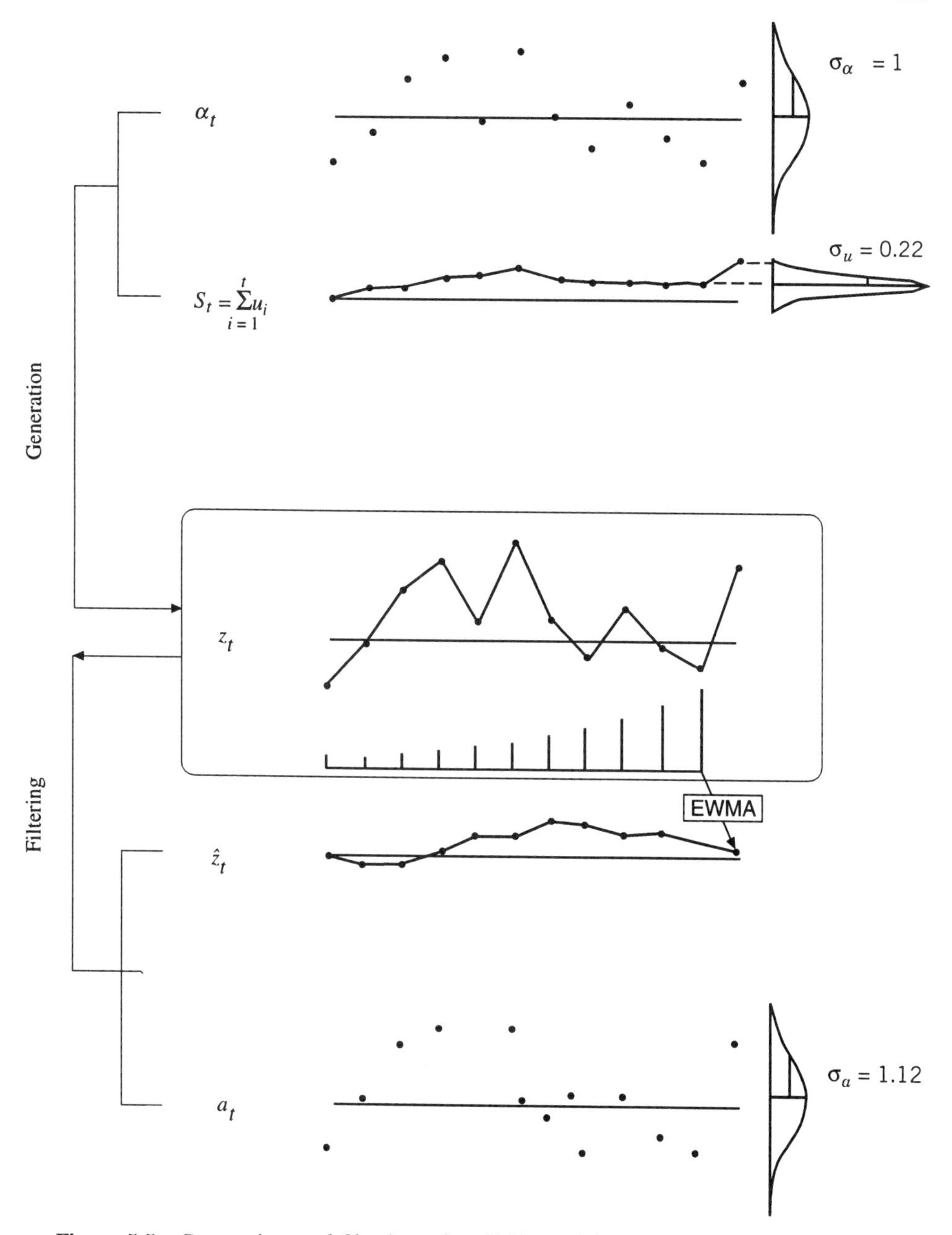

Figure 5.5 Generation and filtering of an IMA model. The diagram shows the contaminating noise α_t, the random walk signal S_t, the constructed disturbance $\alpha_t + S_t = z_t$, the extracted signal $\hat{z}_t$, and the residual noise a_t.

Specifically, if we subtract the equation for z_t from that for z_{t+m} we get

$$z_{t+m} - z_t = \alpha_{t+m} - \alpha_t + (u_{t+1} + \cdots + u_{t+m}) \tag{5.10}$$

and the variance $V_m = \text{var}(z_{t+m} - z_t)$ is the sum of the variances of the independent α_t values and u_t values, so that

$$V_m = 2\sigma_\alpha^2 + m\sigma_u^2. \tag{5.11}$$

Thus, for this model, V_m increases linearly with m and specifically

$$\frac{V_m}{V_1} = 1 + \frac{\sigma_u^2}{2\sigma_\alpha^2 + \sigma_u^2}(m - 1). \tag{5.12}$$

Now while the sticky innovation model and the IMA model are equivalent, the first contains parameters σ_u and σ_α and the second λ and σ_a. To find the relationships between these two sets of parameters we need to obtain an expression for V_m/V_1 directly from the IMA formulation of the model.

At times $t + 1, \ldots, t + m$, the IMA model is

$$\begin{aligned} z_{t+1} - z_t &= a_{t+1} - \theta a_t, \\ z_{t+2} - z_{t+1} &= a_{t+2} - \theta a_{t+1}, \\ &\vdots \\ z_{t+m} - z_{t+m-1} &= a_{t+m} - \theta a_{t+m-1}. \end{aligned} \tag{5.13}$$

If we add these equations with appropriate cancellation, we get

$$z_{t+m} - z_t = a_{t+m} - \theta a_t + \lambda(a_{t+m-1} + \cdots + a_{t+1}). \tag{5.14}$$

Thus

$$V_m = \{(1 + \theta^2) + \lambda^2(m - 1)\}\sigma_a^2 \tag{5.15}$$

and

$$\frac{V_m}{V_1} = 1 + \frac{\lambda^2}{1 + \theta^2}(m - 1), \tag{5.16}$$

which depends only on λ or equivalently on θ and describes the linearly increasing variogram in terms of the IMA formulation of the model.

Equating Equations (5.12) and (5.16) we get

$$\frac{\sigma_u}{\sigma_\alpha} = \frac{\lambda}{\sqrt{\theta}} = \frac{\lambda}{\sqrt{1-\lambda}}$$

and then, equating Equations (5.11) and (5.15) particularized for $m = 1$, we get

$$\frac{\sigma_u}{\sigma_a} = \lambda, \quad \frac{\sigma_\alpha}{\sigma_a} = \sqrt{\theta}. \tag{5.17}$$

The "sticky innovation" model (5.8) can be thought of as the *generating* model for the disturbance. On the other hand, the model $z_t = \tilde{z}_{t-1} + a_t$ with $\tilde{z}_{t-1} = \hat{z}_t$, an EWMA with appropriate smoothing constant θ, can be thought of as a model that describes the *filtering* of the series. The idea is made clear in Figure 5.5 which shows:

1. The generation of the disturbance z_t as the combination of α_t and S_t, where α_t is the transitory contaminating noise and S_t is the signal—a random walk, produced as the sum of the sticky innovations $u_1, u_2, \ldots, u_t$.
2. Optimal *filtering* is achieved by an EWMA with θ chosen so that $\lambda^2/\theta = \sigma_u^2/\sigma_\alpha^2$; the value $\hat{z}_t$ regarded as an estimate of S_t has error a_t, which has standard deviation σ_a.

Note that both the contaminating noise, represented by the α_t values, and also the errors committed in estimating the signal, represented by the a_t values, are white noise series. Figure 5.5 illustrates the situation where the series z_t is generated with $\sigma_\alpha = 1$ and $\sigma_u = 0.22$ so that $\lambda = 0.2$. An EWMA with $\lambda = 0.2$ is then used to filter the series, producing an estimate $\tilde{z}_{t-1} = \hat{z}_t$ of the signal S_t with white noise error a_t, which has standard deviation $\sigma_a = 1.12$. Note that in this example σ_a is only 12% greater than σ_α—the standard deviation of the initial contaminating noise α_t.

The ratio $\sigma_a/\sigma_\alpha = 1/\sqrt{\theta}$ will be called the *noise amplification factor* of the (optimal) filtering process. Shown in Table 5.1 are various values of σ_u/σ_α that yield different values of λ and the corresponding amplification factors σ_a/σ_α. This amplification factor is a measure of how good is the best possible filtering process for separating the signal from the noise. It is the ratio of the standard deviation of the error a_t in estimating the signal to that of the contaminating noise α_t in which the real signal is buried. For perfect, but unrealizable, signal extraction this

TABLE 5.1 Values of the Ratio σ_u/σ_α Yielding Various Values of λ with the Corresponding "Noise Amplification Factor" σ_a/σ_α

σ_u/σ_α	λ	σ_a/σ_α
0.11	0.1	1.05
0.22	0.2	1.12
0.35	0.3	1.20
0.51	0.4	1.29
0.71	0.5	1.41

factor would be 1.00. Note that, when λ is not larger than, say, 0.4, it is possible to filter out much of the "signal" corresponding to the cumulated sticky innovations as demonstrated by the moderate size of the amplification factor σ_a/σ_α. For example, for $\lambda = 0.2$, the amplification factor is 1.12 and for $\lambda = 0.4$ the amplification factor is 1.29.

Exercise 5.1. Figure 5.5 is based on the values for α_t and u_t given in the following table:

α_t	−1.2	−0.413	1.023	1.574	−0.157	1.751	0	−0.925	0.374	−0.59	−1.22	0.944
u_t	0	0.304	0.076	0.215	0.076	0.202	−0.278	−0.063	−0.013	−0.076	0.076	0.557

Confirm the subsequent calculations used for drawing the figure. □

The above calculations are made on the assumption that the optimal value of the smoothing constant θ (and hence of λ) is known. However, the procedure is remarkably robust to fairly large deviations in the choice of θ. Thus in practice a large proportion of the nonstationary part of the disturbance generated by the sticky innovation model can be filtered out by an EWMA filter with a value $\hat{\theta}$ estimated from a series of moderate size. The EWMA that forecasts the next observation is an estimate of the systematic signal and can be used in a feedback system to cancel out a large component of the disturbance. We shall see in the next chapter how this can be done.

Robustness of the EWMA

From the above it is evident that an appropriately chosen EWMA provides an efficient filter for estimating a random walk signal in a time series modeled by an IMA. However, since IMA series can be generated from a random walk signal plus noise, a reasonable question is how good (robust) is the EWMA for estimating other kinds of signals that might conceivably characterize an industrial process.

A Disturbance Model Due to Barnard

An interesting model for a process disturbance suggested by George A. Barnard (1959) is such that the local process mean μ_t is subject at random times to random sized jumps from one level to another. Specifically, it is supposed that the length of the periods between jumps is determined by random drawings from a Poisson distribution (see Chapter 2) with a mean equal to L time periods. Also, the size of the jumps is determined by random drawings from a normal distribution with mean zero and standard deviation σ_J.

Figure 5.6 shows one such realization for a signal S_t constructed in this way. In this illustration the number of periods between jumps has a mean $L = 25$, the size of the jumps has standard deviation $\sigma_J = 2$, and the contaminating noise α_t has $\sigma_\alpha = 1$. Figure 5.6 is set out in the same manner as Figure 5.5 and shows:

1. The contaminating white noise α_t.
2. The signal S_t.
3. The constructed disturbance $z_t = S_t + \alpha_t$.
4. The estimated signal $\hat{z}_t$ obtained from an EWMA whose smoothing parameter $\hat{\theta} = 1 - \hat{\lambda}$ is obtained by fitting an IMA model to the series.
5. The residual noise a_t.

In this particular realization, using estimates obtained from the data, we find $\hat{\lambda} = 0.32$ and $\hat{\sigma}_a/\hat{\sigma}_\alpha = 1.22$. The latter may be compared with the value for the amplification factor calculated from Equation (5.17), $1/\sqrt{1 - \hat{\lambda}} = 1.21$, which is appropriate for a series generated by an IMA model. Results from ten such further realizations from the same model, arranged in order according to the size of $\hat{\lambda}$, are shown in Table 5.2a.

A More Elaborate Model

To further test the robustness of the EWMA filter, a somewhat more elaborate model for the signal S_t was constructed as follows. It was supposed that the process mean μ_t could be in different equiprobable states over different time periods of randomly chosen lengths. In state 1, the mean stayed constant at some fixed level. In state 2 the mean

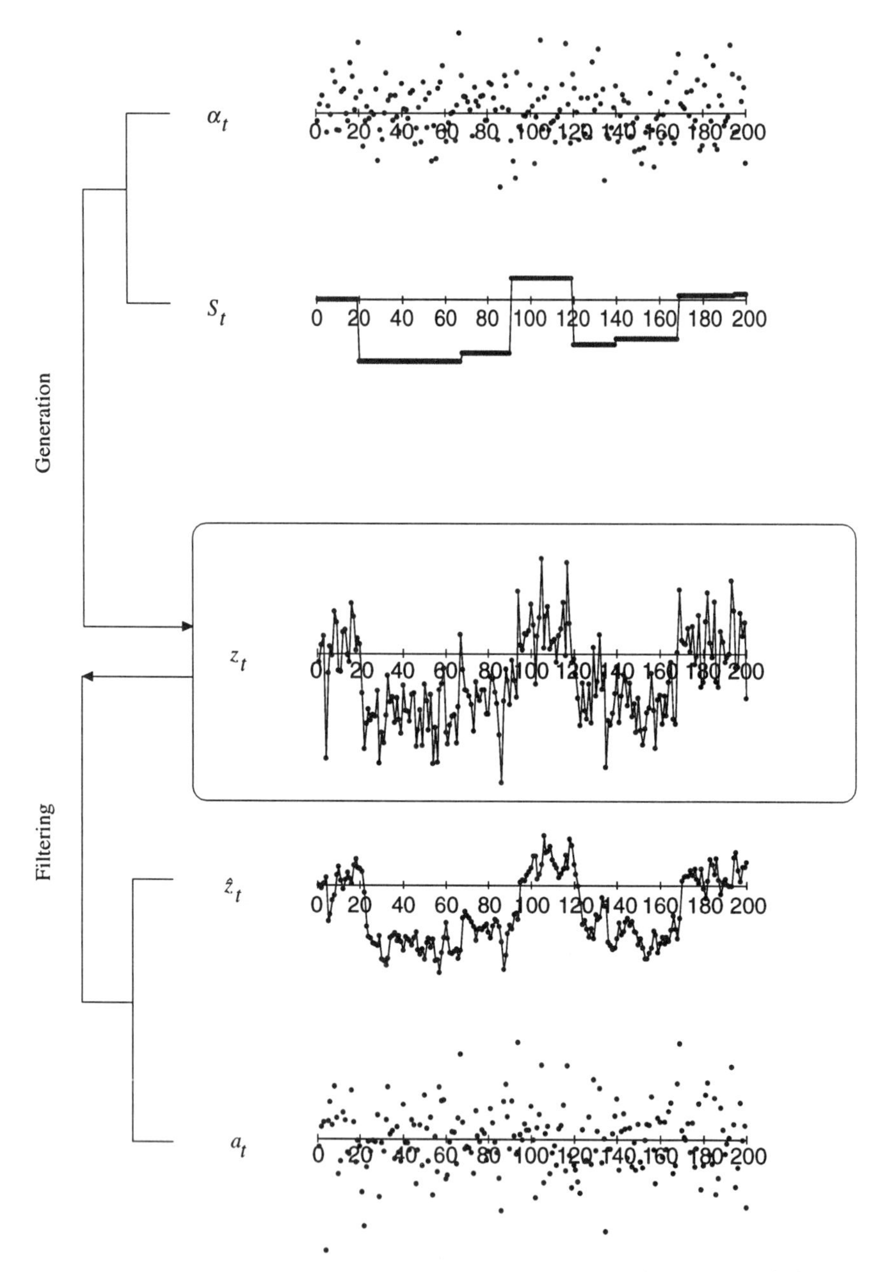

Figure 5.6 One realization of Barnard's model. The number of periods between level jumps is a Poisson random variable with mean $L = 25$; the size of the jumps are normally distributed with mean zero and standard deviation $\sigma_J = 2$; and the contaminating noise α_t has a normal distribution with mean zero and standard deviation $\sigma_\alpha = 1$. The arrangement of the diagram is similar to that of Figure 5.5.

TABLE 5.2 Estimated Values for $\hat{\lambda}$ and $\hat{\sigma}_a/\hat{\sigma}_\alpha$ Compared with Noise Amplification Factor $1/\sqrt{1-\hat{\lambda}}$ Appropriate for the IMA Model: Ten Realizations for (a) Barnard's Model and (b) a Generalization of Barnard's Model

	(a) Barnard's Model		
	$\hat{\lambda}$	$\hat{\sigma}_a/\hat{\sigma}_\alpha$	$1/\sqrt{1-\hat{\lambda}}$
	0.16	1.06	1.09
	0.27	1.18	1.17
	0.29	1.30	1.19
	0.31	1.16	1.20
	0.32	1.22	1.21
	0.33	1.14	1.22
	0.33	1.26	1.22
	0.35	1.23	1.24
	0.35	1.07	1.24
	0.43	1.25	1.32
Average	0.31	1.19	1.21
	(b) Generalizetion of Barnard's Model		
	$\hat{\lambda}$	$\hat{\sigma}_a/\hat{\sigma}_\alpha$	$1/\sqrt{1-\hat{\lambda}}$
	0.18	1.06	1.10
	0.21	1.05	1.13
	0.23	1.09	1.14
	0.24	1.19	1.15
	0.25	1.11	1.15
	0.28	1.12	1.18
	0.29	1.11	1.19
	0.32	1.30	1.21
	0.33	1.12	1.22
	0.39	1.23	1.28
Average	0.27	1.14	1.17

experienced a random jump in level. In state 3 the mean experienced a random change in slope. The duration of these states was provided by random drawings from a Poisson distribution with a mean of L time periods. Also, the eventual level changes (either sudden or gradual) are realizations from a normal distribution with mean zero and standard deviation σ_J. The standard deviation of the contaminating noise α_t is σ_α.

Figure 5.7 illustrates one realization from this model with $L = 25$, $\sigma_J = 2$, and $\sigma_\alpha = 1$. As before, it shows (1) the contaminating noise α_t;

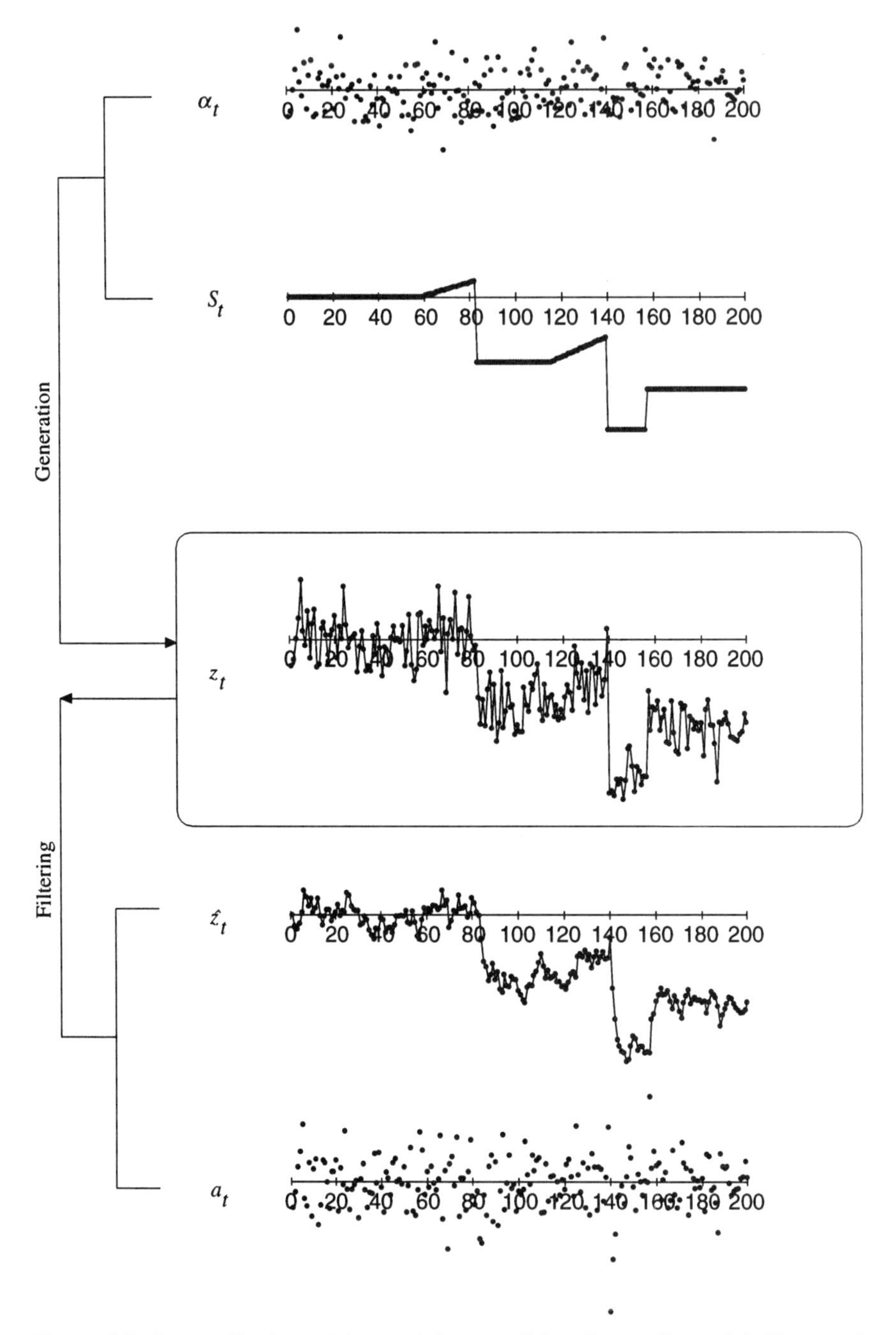

Figure 5.7 One realization of the model generalizing Barnard's model. The number of periods between switches in state is a Poisson random variable with mean $L = 25$; the size of the level or slope jumps has a normal distribution with mean zero and standard deviation $\sigma_J = 2$; and the contaminating noise α_t has a normal distribution with mean zero and standard deviation $\sigma_\alpha = 1$. The arrangement of the diagram is similar to that of Figure 5.5.

(2) the constructed signal S_t; (3) the resulting disturbance $S_t + \alpha_t$; (4) the estimated signal $\hat{z}_t$; and (5) the residual noise. Table 5.2b shows empirical results from ten realizations of this model. In the illustrated example, $\hat{\lambda} = 0.39$ and $\hat{\sigma}_a/\hat{\sigma}_\alpha = 1.23$, which compares with a value $1/\sqrt{1 - \hat{\lambda}} = 1.28$ estimated from $\hat{\lambda}$ assuming that the series was generated by an IMA time series model.

We see that, in these examples, where the series $\{z_t\}$ is not generated by an IMA model that, nevertheless, the fitted EWMA filter provides a surprisingly good estimate of the signal S_t.

CHAPTER 6

Process Adjustment Using Feedback Control: Manual Adjustment Charts

"Give me a fruitful error any time, full of seeds, bursting with its own corrections."

Comment on Kepler, VILFREDO PARETO

6.1 INTRODUCTION

Figure 6.1 shows 100 observations of a thickness measurement y_t of a very thin metallic film. These observations, listed as Series A at the end of the book, were made at equally spaced intervals of time at one stage in the manufacture of a computer chip *when no adjustment was applied.* It was desired to maintain this quality characteristic as closely as possible to the target value $T = 80$. Thus the process disturbance, as previously defined, is the deviation $z_t = y_t - 80$. As will be seen, with *no* adjustments made, major excursions from this target value occurred. In particular $\pm 3\sigma$ control limits calculated from the moving range showed the process to be badly out of control.

The first line of attack against such disturbances must be a dedicated effort to try to remove their causes and so bring the process as nearly as possible into a state of statistical control as described in Chapters 2 and 3. For some systems, however, in spite of considerable effort of this kind, a residual disturbance remains. This may arise from naturally occurring phenomena such as variation in ambient temperature, humidity, and feedstock quality, that cannot be eliminated economically; or it may occur from causes that currently are unknown. In either circumstance, some system of process adjustment may be necessary.

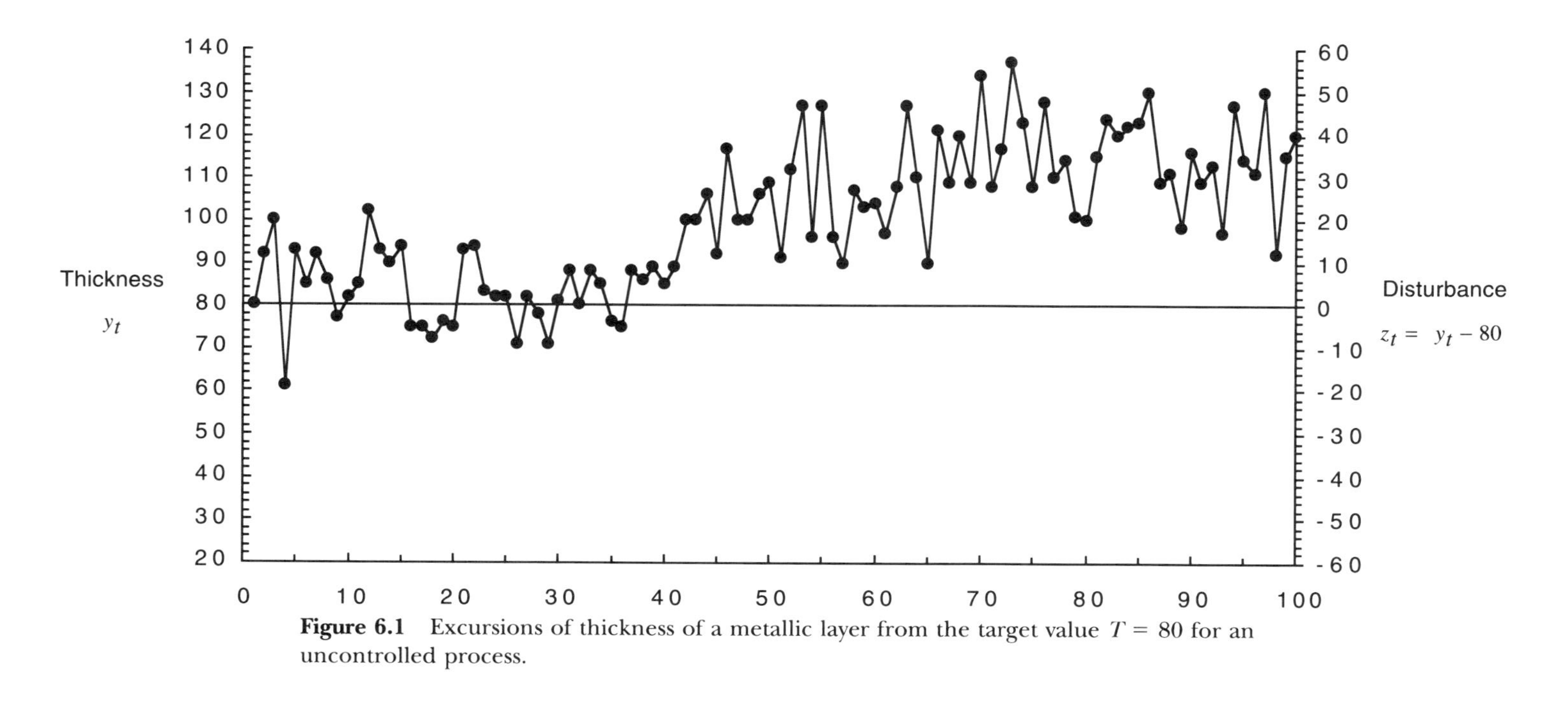

Figure 6.1 Excursions of thickness of a metallic layer from the target value $T = 80$ for an uncontrolled process.

The two approaches may be illustrated by considering the dilemma of someone who lives in Madison, Wisconsin, where the ambient temperature (the disturbance) can vary from −30°F to +95°F. The problem of keeping the body at a comfortable temperature can be solved by eliminating the special cause, for example, moving to California. But if this is not possible, convenient, or economical, it can be dealt with by installing central heating and air conditioning in the home with the temperature adjusted by a process of feedback control. Specifically, the desired target temperature is set by a thermostat on the wall and suitable heating or cooling is induced by deviations from this temperature. That feedback control is important and need not to be regarded as esoteric or unfamiliar is evidenced by the fact that for millions of people such control is the only thing that renders large areas of North America conveniently habitable.

It has most often been in the process industries that problems of irremovable disturbances have been dealt with by *feedback control*[1] and this type of control has come to be regarded as the domain of engineering process control (EPC). In this book we describe feedback techniques of most use in SPC. These are of two distinct kinds. In this chapter and in Chapter 7 we discuss discrete *proportional integral* (PI) control adapted for the use of the quality practitioner. These direct borrowings from EPC are mainly useful when the only cost to be considered is that of being off-target. However, in many SPC applications, the cost of actually making a process adjustment and sometimes the cost of surveillance (sampling and testing the product) must be considered. Chapters 8 and 9, therefore, address these rather different problems. They describe feedback techniques especially developed to take account of the additional costs and to balance them economically.

We talk here of feedback control rather than *automatic* control or *computer* control because the methods themselves have nothing to do with the way in which they are put into effect. Feedback control *can* be actuated automatically using computers, transducers, and so forth, but it need not be and we shall be particularly concerned with *manual* adjustment by suitable charts. Discussion in terms of manual adjustment charts has the advantage that it links up the subject with a number of familiar SPC ideas. Once these concepts are understood, the quality practitioner can put them into effect by any means he/she chooses. For example, rather than employ charts, the practitioner may wish to use an available process computer or a programmed calculator to determine the necessary control action.

[1] In EPC, feedforward control and other methods are also used, but we don't discuss them in this book.

Manual feedback control is, in fact, often employed in SPC, but because it is seldom acknowledged that this is what is being done, such control tends to be haphazard and inefficient. In this chapter, we introduce some basic ideas by illustrating how a feedback control chart can be developed to manually control the metallic film disturbance of Figure 6.1.

6.2 A MANUAL ADJUSTMENT CHART TO CONTROL METALLIC THICKNESS

Feedback regulation presupposes that there exists some *compensatory variable* (X, say) that can be manipulated to adjust the level of the output. For the system of Figure 6.1, this compensatory variable X is the metallic *deposition rate* that can be changed to alter the level of the output—the thickness y. Figure 6.2 shows a manual adjustment chart (Box and Jenkins, 1963) appropriate for this example. We first show how it would operate in controlling the disturbance of Figure 6.1 and consider later how it is designed.

To use this adjustment chart the operator records the latest value of thickness and then reads off on the adjustment scale the appropriate amount by which he/she should now increase or decrease the deposition rate. For example, the first recorded thickness of 80 is on target, so no action is called for. The second value of 92 corresponds on the left-hand scale to a deposition rate adjustment of -2. Thus the operator should now reduce the deposition rate by 2 units from its present level. Note that successive recorded values seen on the chart are the thickness readings that occur *after adjustment.* The underlying disturbance shown in Figure 6.1 will not be seen by the operator.

By comparing Figures 6.1 and 6.2, the chart is seen to be highly effective. While for the 100 unadjusted thickness measurements in Figure 6.1 the average squared deviation from target is 684.8, for the adjusted disturbance of Figure 6.2 it is only 124.2. Thus a more than fivefold reduction in mean square error is produced by this simple form of control, yielding a standard deviation for the adjusted thickness of about 11.1. Consider now how we arrived at this chart.

6.3 DESIGN OF THE CHART

In the adjustment chart of Figure 6.2, the output quality characteristic y to be controlled is the thickness of the film and the compensating

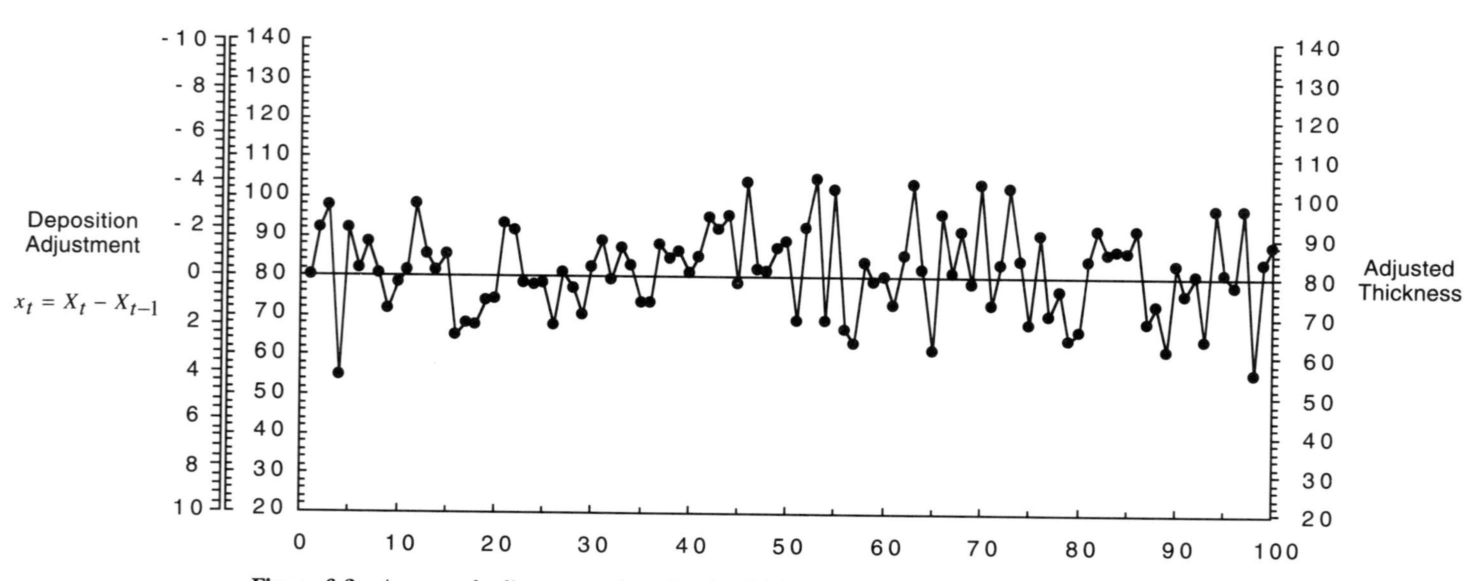

Figure 6.2 A manual adjustment chart for the thickness of a metallic layer, which allows the operator to read off the appropriate change in deposition rate needed at any stage.

variable X used for adjusting the process is the deposition rate. The chart is based on the following information:

1. A change made in the deposition rate X will produce all of its effect on the thickness y within one time interval.
2. An increase of 1 unit in the deposition rate X induces an increase of 1.2 units in the film thickness y.

In general, a system for which all the effect of a change in the compensating variable will be realized at the output in one time interval will be called a *responsive system.* Also, the factor measuring the change in the output produced by one unit of change in the compensating variable will be called the *process gain* and denoted by g. For this example $g = 1.2$, so it might at first be supposed that the appropriate adjustment would be to reduce the deposition rate by 1 unit for every 1.2 units that the thickness exceeded the target value of 80. If at time t, the *output error* that is, the deviation from the target *after adjustment,* is e_t, and if the corresponding change made in the level of the compensating variable (called the *adjustment*) is $x_t = X_t - X_{t-1}$, this would imply an *adjustment equation*

$$1.2x_t = -e_t \quad \text{or} \quad x_t = -\frac{1}{1.2}e_t. \tag{6.1}$$

The minus sign is needed in these equations because x_t needs to compensate the deviation e_t from target. We will call this the *full adjustment.* It corresponds to what Deming called "tinkering."

However, the adjustment put into effect by the chart in Figure 6.2 is only one-fifth of this full adjustment. The adjustment equation employed is thus

$$1.2x_t = -0.2\ e_t \quad \text{or} \quad x_t = -0.2\frac{1}{1.2}e_t. \tag{6.2}$$

The *damping factor* (0.2 in this example) is, in general, denoted by G. The form of the adjustment equation is consequently

$$gx_t = -Ge_t \quad \text{or} \quad x_t = -\frac{G}{g}e_t. \tag{6.3}$$

As we mentioned before, this idea of making an adjustment roughly proportional to the current error is not exactly new. Figure 6.3 shows

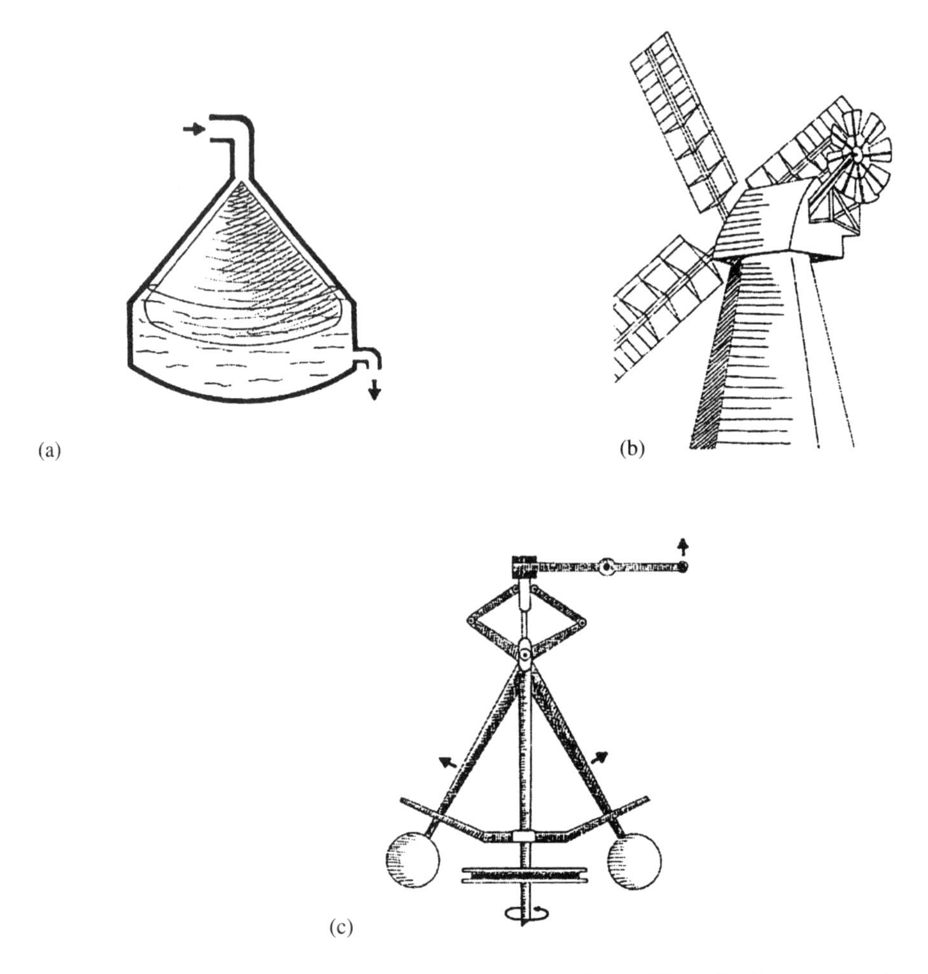

Figure 6.3 (a) Ktesibios's valve; (b) Lee's automatic "fan-tail"; and (c) James Watt's governor.

(a) a reconstruction due to Diels (1924) of the valve devised by Ktesibios (ca. 275 B.C.) to control the flow of an Egyptian water clock; (b) a rear "fan-tail" invented by Edmund Lee in 1745 and widely used in windmills to cause them to rotate so that they faced into the wind; and (c) the governor developed by James Watt (1736–1819) to control the speed of his newly invented steam engine. A comprehensive discussion of the origins of feedback control and, in particular, of the devices mentioned above will be found in Mayr (1970).

A water clock worked by allowing water to drip into a vessel. The height of the water in the vessel indicated the time. If the clock was to keep good time, an even flow had to be maintained. It is believed that a valve of the type shown was devised by Ktesibios to do this. An increase in the water level in the chamber caused the float to rise and hence to appropriately reduce the flow.

To work efficiently a windmill must face into the wind. To bring this about many windmills were designed so that the sails emanated from a cap on the top of the mill that previously were rotated by hand using a long "tail pole." The vanes on Lee's automatic "fan-tail" were at right angles to the sails and connected by gearing to wheels that could rotate the cap of the mill. A change in wind direction caused the fan to rotate and move the sails to face squarely into the wind.

Watt's governor was designed to keep his steam engine running at a constant speed in spite of changes in steam pressure and load on the system. Rotation of the governor caused two pendulums to swing outward, which, by a system of levers, regulated the steam valve.

There is even more to this simple form of control than first meets the eye and we consider now its relationship (1) to the engineers' PID control and (2) to exponential smoothing.

6.4 RELATION TO PID CONTROL

A very successful form of feedback control used by engineers for *continuous* processes employs what is called "proportional integral derivative" (PID) control. Suppose e_t is the error (deviation from target) at time t, and X_t is the *level* of the compensating variable at time t. Then X_t might be made *proportional* to e_t, to its *integral* with respect to time, or to its *derivative* with respect to time. The "three-term controller" uses a linear combination of all these modes of action. The PID *control equation* is consequently of the form

$$gX_t = k_0 + k_P e_t + k_I \int e_t dt + k_D \frac{de_t}{dt}, \tag{6.4}$$

where k_P, k_I, and k_D are constants.

Frequently, simpler forms of control are effective. One common form is *integral control,* which uses only the integral term in Equation (6.4). Another, called *proportional integral* (PI) control, uses both the proportional and integral terms.

Now in this book we discuss the control not of continuous systems but of *discrete* systems, where measurements and adjustments are made at equispaced intervals of time $t, t-1, t-2, \ldots$.

The discrete *adjustment equation* we have used for the metallic layer data is

$$gx_t = -Ge_t,$$

where $x_t = X_t - X_{t-1}$ is the *change* made at time t from the previous *level* X_{t-1} of the compensating variable X. Since this action is repeatedly applied at every step, we can write the series of adjustment equations at and before time t as

$$\begin{aligned} g(X_t - X_{t-1}) &= -Ge_t, \\ g(X_{t-1} - X_{t-2}) &= -Ge_{t-1}, \\ &\vdots \\ g(X_1 - X_0) &= -Ge_1. \end{aligned}$$

The corresponding *control equation*, which determines the actual *level* of the compensating variable X at time t, is obtained by adding these equations together to get

$$\begin{aligned} gX_t &= gX_0 - G(e_t + e_{t-1} + \cdots + e_1) \\ &= gX_0 - G\sum_{i=1}^{t} e_i. \end{aligned}$$

Now X_0 and G are both constants and, writing $gX_0 = k_0$ and $-G = k_I$, the control equation becomes

$$gX_t = k_0 + k_I \sum_{i=1}^{t} e_i. \tag{6.5}$$

This is the discrete form of integral control in which the integral in Equation (6.4) is replaced by the cumulative sum in Equation (6.5). In the next chapter we will find that to allow for process inertia we may need to use a control equation of the form

$$gX_t = k_0 + k_P e_t + k_I \sum_{i=1}^{t} e_i.$$

This is the discrete form of proportional integral (PI) control.

6.5 RELATION TO EXPONENTIAL SMOOTHING; PREDICTIVE CONTROL

Suppose as before that the disturbance at times $\ldots, t-1, t, t+1, \ldots$ is defined as the output deviations from target $\ldots, z_{t-1}, z_t, z_{t+1}, \ldots$ that would occur if *no control action of any kind* were taken. Also assume that, as before, (1) we have a responsive adjustment system in which the effect of an adjustment is fully experienced at the output by the end of one time interval, and (2) one unit of compensation produces g units of change at the output. Then, *at time t*, if we somehow knew what the level z_{t+1} of the disturbance one step ahead was going to be, we could achieve perfect control by continually setting X_t so as to just cancel it out. So the control equation would be

$$gX_t = -z_{t+1}. \tag{6.6}$$

We cannot see into the future so that Equation (6.6) is not a realizable form of control. But suppose $\hat{z}_{t+1}$ is *any* estimate (forecast, prediction) of z_{t+1} that *can be made at time t*. Then a realizable form of control is obtained by continually setting the level of the input compensating variable so that at time t it cancels the predicted value $\hat{z}_{t+1}$, and

$$gX_t = -\hat{z}_{t+1}. \tag{6.7}$$

Equivalently, the adjustment $x_t = X_t - X_{t-1}$ to the compensating variable would be such that

$$gx_t = -(\hat{z}_{t+1} - \hat{z}_t). \tag{6.8}$$

Any "predictive" form of control of this kind ensures that the deviation from target at time $t + 1$ is not the value of the disturbance itself, but rather the *error* in forecasting that disturbance one step previously.
This forecast error may be written as

$$e_{t+1} = z_{t+1} - \hat{z}_{t+1},$$

so that the adjustment Equation (6.3) leads to

$$gx_t = -Ge_t = -G(z_t - \hat{z}_t). \tag{6.9}$$

Combining Equations (6.8) and (6.9) we see that the forecast $\hat{z}_{t+1}$ we are using in Equations (6.7) and (6.8) must satisfy the equation

$$\hat{z}_{t+1} - \hat{z}_t = G(z_t - \hat{z}_t),$$

or, equivalently,

$$\hat{z}_{t+1} = Gz_t + (1 - G)\hat{z}_t. \tag{6.10}$$

This is a first order difference equation and is identical to formula A of Table 4.2 provided that we set $\theta = 1 - G$. Consequently, Equation (6.10) has the solution

$$\hat{z}_{t+1} = \tilde{z}_t = (1 - \theta)\{z_t + \theta z_{t-1} + \theta^2 z_{t-2} + \ldots\},$$

and Equation (6.7) becomes

$$gX_t = -\tilde{z}_t. \tag{6.11}$$

Therefore, the adjustment Equation (6.3), which produces discrete integral control, also implies that such control is equivalent to setting X_t so as to cancel an *EWMA* forecast $\tilde{z}_t$ of z_{t+1}. Equation (6.11) also implies that the smoothing constant for this EWMA forecast is $\theta = 1 - G$ and is entirely determined by the value of the damping factor G in the adjustment Equation (6.3). Note that this argument holds whether this particular kind of forecast with this particular choice of θ is adequate or not, and hence whether or not Equation (6.3) provides good control. As we shall see later, these issues depend on the true nature of the disturbance and the degree of robustness of the adjustment system.

6.6 WHAT IS THE ADJUSTMENT EQUATION REALLY DOING?

As illustrated in Figure 6.4, the control that is being put into effect by Equation (6.3) and by the manual adjustment chart in Figure 6.2 is really rather sophisticated. It ensures that the total compensation gX_t, which is the sum of all the incremental adjustments gx_t applied up to time t, is equal to $-\tilde{z}_t$, where $\tilde{z}_t = \hat{z}_{t+1}$ is an EWMA of the current and previous levels of the disturbance with smoothing constant $\theta = 1 - G$. Because control action is continually taken to cancel out the forecast $\hat{z}_{t+1}$, we do not see this disturbance on the chart, but only the errors in forecasting it. In Figure 6.4 the accumulated adjustments $-\hat{z}_{t+1} = -\tilde{z}_t = gX_t$ are shown by horizontal lines. The chart continuously adjusts the process so that at each stage the forecasted value one step ahead will be on the target. The deviations of the adjusted process from target, shown on the adjustment chart of Figure 6.2, are thus the forecast errors $e_t = z_t - \tilde{z}_{t-1}$.

> This can lead to some rather surprising conclusions. For example, suppose that $z_t, z_{t-1}, z_{t-2}, \ldots$ are wind directions averaged over successive short intervals of time and that small rotations of the cap of the mill are approximated by straight lines, then, Lee's fan-tail regulator will cause the windmill to point in the direction that is (approximately) an EWMA of recent wind directions.

Note that the argument employed so far uses no particular time series model to describe the disturbance and applies for any value of the damping factor G between 0 and 1. Look again at Figure 6.4. The illustration is for the specific choice $G = 0.2$ for the damping factor, which produced the control shown in Figure 6.2. The corresponding exponential weights with smoothing constant $\theta = 1 - G = 0.8$ are also shown in Figure 6.4. Now, probably Ktesibios, the English millers,

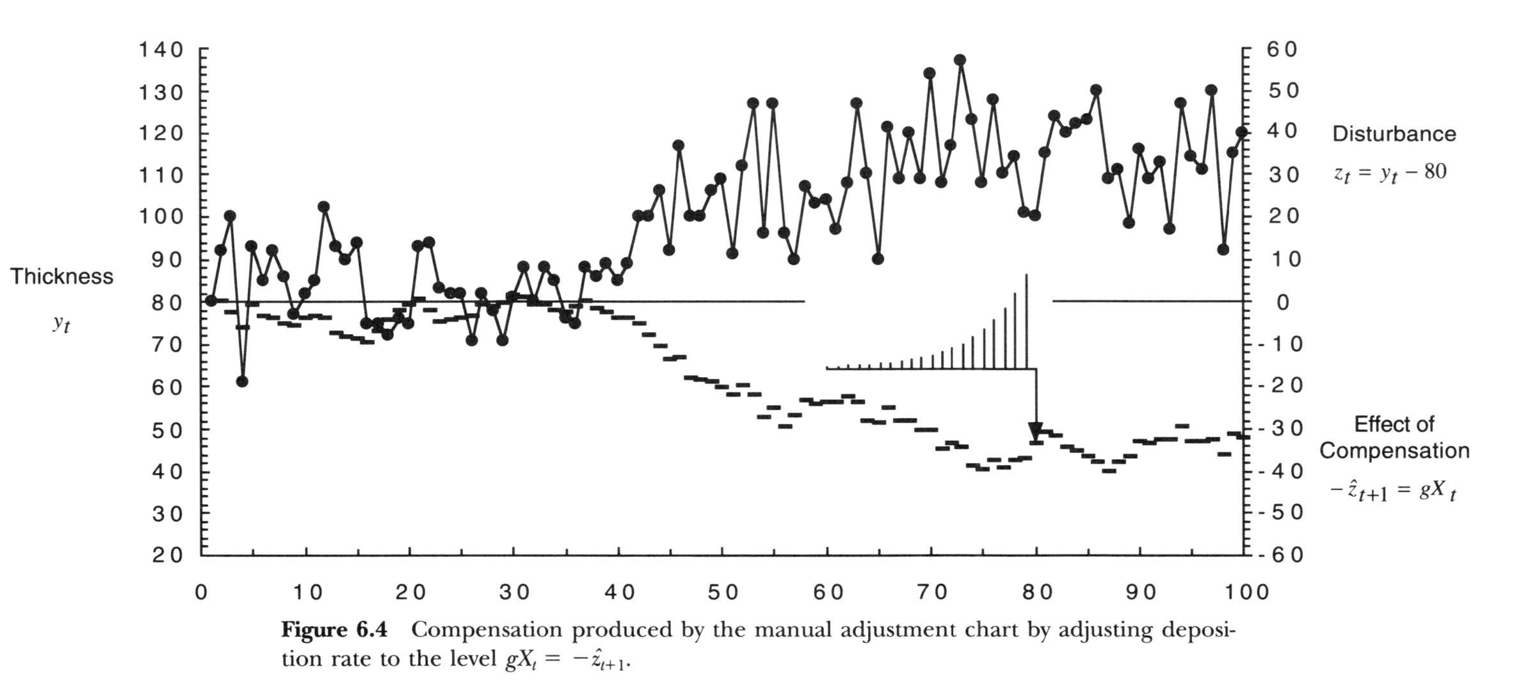

Figure 6.4 Compensation produced by the manual adjustment chart by adjusting deposition rate to the level $gX_t = -\hat{z}_{t+1}$.

and James Watt did not ponder too long about the value of G that they would use. Study of Figure 6.4 shows why they did not need to. This form of control is obviously extremely robust, for changing the damping factor G to some other value between 0 and 1 would simply mean that a different smoothing constant was used for the EWMA forecast. Thus, moderate changes in G would not have much effect—the error e_t might be slightly larger but the improvement brought about by replacing the deviation from target z_t by its forecast error e_t could still be considerable even for only a moderately good forecast.

So using the concept that the adjustment should be proportional to the last error, an idea long ago suggested by common sense, turns out to be equivalent to setting gX_t at each stage so as to cancel a forecast of the disturbance. This forecast uses an EWMA with smoothing constant $1 - G$, where G is the proportion of the full adjustment that is put into effect.

Equivalently, the scheme turns out to be a discrete form of the engineers' "integral" control

$$gX_t = k_0 + k_I \sum_{i=1}^{t} e_i$$

with $k_I = -G$.

6.7 CHOICE OF G TO MINIMIZE THE OUTPUT VARIANCE σ_e^2

For the metallic film example we used the adjustment equation $gx_t = -Ge_t$ with damping factor $G = 0.2$ and we saw that this form of control was equivalent to continually compensating the disturbance z_t by an EWMA $\tilde{z}_{t-1}$ with smoothing constant $\theta = 1 - G$. The choice of G is now considered more carefully. For the metallic film example we have a sample of 100 values of the uncontrolled series $\{z_t\}$. These were plotted in Figure 6.1. One way to choose G would be to estimate it from these data, as we did for the "Dingles" example in Chapter 4. A plot of the sums of squares of the forecast errors obtained from a series of trial values of G is shown in Figure 6.5. The minimum value is about $\hat{G} = 0.24$, which is near enough to the value 0.2 we used in the feedback control chart of Figure 6.2.

We therefore know that, at least for these data, this choice of G produces close to the smallest average squared error for the control scheme. Note that this argument *does not assume a particular form of model for the disturbance*; it simply rests on the empirical fact that, for *this*

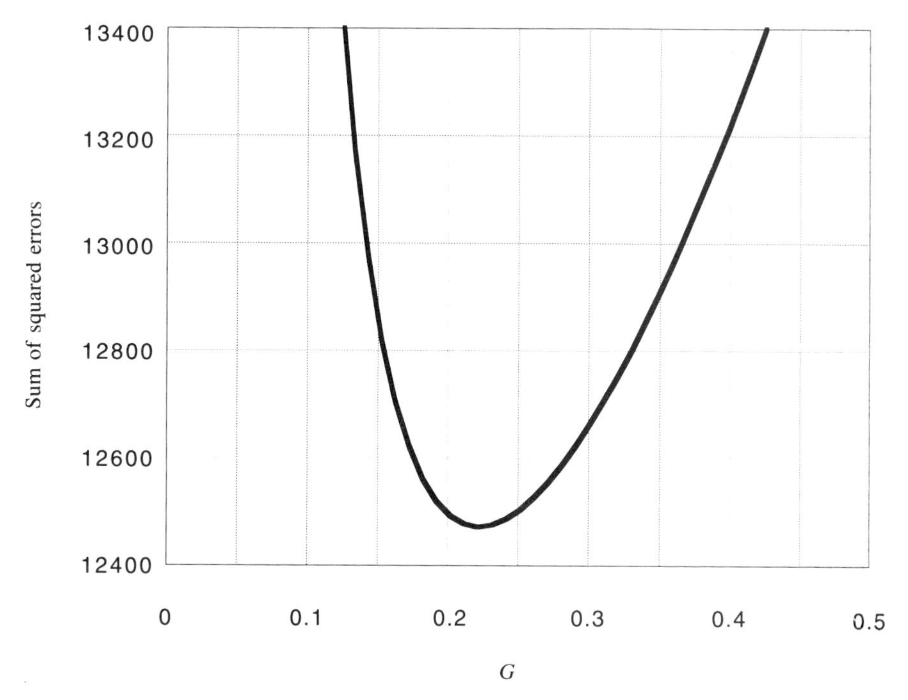

Figure 6.5 Plot of the values of the sum of squares of the forecast errors for the thickness data obtained from a series of trial values of G.

sample of 100 values of the disturbance, a value of G close to 0.2 yields the smallest squared error for the controlled series. Making only the assumption then that this sample is not untypical of the behavior of the process, the value $G = 0.2$ can be adopted for the control scheme.

6.8 INSENSITIVITY TO THE CHOICE OF G

So far in this chapter we have made no assumption about the form of the disturbance model. What follows is based on the assumption that the disturbance can be represented by an IMA time series model $z_t - z_{t-1} = a_t - \theta a_{t-1}$, where the a_t values have mean zero and variance σ_a^2. If we knew the true value θ_0 of θ and we set $G = \lambda_0 = 1 - \theta_0$, then $\tilde{z}_t$ would be a *minimum mean square error* forecast of z_{t+1} and so this choice $G = 1 - \theta_0$ would provide minimum mean square error at the output. With the a_t values assumed to have mean zero, this would also be a scheme giving minimum variance $\sigma_e^2 = \sigma_a^2$ for the errors at the output. In practice, we do not know θ_0; but if we can obtain an estimate $\hat{\theta}$ from the data, because of the robustness of such an estimate, the value $\hat{G} = 1 - \hat{\theta} = \hat{\lambda}$ will produce an output variance very close to minimal.

As we said earlier, the sum of squares curve tends to be flat in the neighborhood of the minimum so that moderate departures from the theoretically best value of G are unlikely to greatly increase the mean square error. Also, inspection of actual series suggests that for many processes the interesting range of G is from zero to about 0.4. To further illustrate the robustness of this adjustment procedure, suppose that the time series describing the uncontrolled disturbance can be modeled by an IMA model for which the true value of the smoothing constant is θ_0 but a value of G different from $\lambda_0 = 1 - \theta_0$ has been used in designing the control scheme. Then we show in Appendix 6A that the variance of the output of the control scheme will be inflated by a factor

$$\frac{\sigma_e^2}{\sigma_a^2} = 1 + \frac{(G - \lambda_0)^2}{G(2 - G)}. \tag{6.12}$$

Figure 6.6 shows, for various values of λ_0, graphs of this inflation factor for the variance of the controlled process produced by using values of G not equal to λ_0.

The curve for $\lambda_0 = 0$ reaffirms the point made by Deming that total abstention from adjustment (G set equal to 0) would be best if the

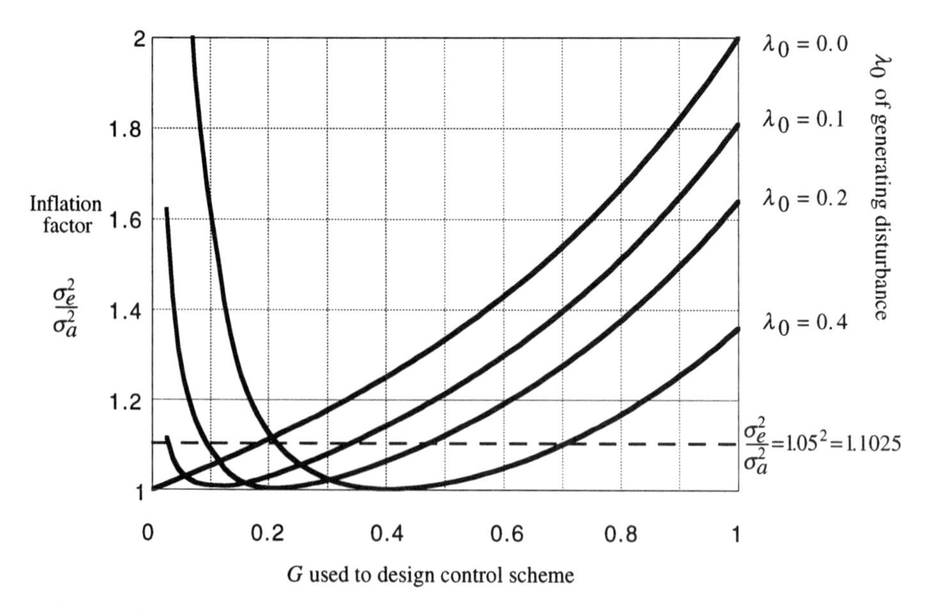

Figure 6.6 Inflation in the variance of the adjusted process arising from a choice of G different from λ_0.

process were indeed already in a state of perfect control. Also, as is evidenced by Deming's funnel experiment, the inflation factor would be 2 (the variance σ_e^2 would be double σ_a^2) if we applied full adjustment (G was set equal to 1) to a process in a perfect state of control. Note, however, that a policy of no adjustment would *not* be best if, as is likely in practice, λ_0 were somewhat greater than zero and so the process was slightly nonstationary. Then the introduction of mild control with a small value of G could greatly *reduce* the variance of the output. For illustration, we consider the relative mean square error for various values of $\lambda_0 = 1 - \theta_0$ for 100 observations from an unadjusted IMA disturbance series initiated at the target value for various values of λ_0:

λ_0	0	0.1	0.2	0.3	0.4
(MSE adjusted)/(MSE unadjusted) for a disturbance series of 100 observations	1	0.67	0.34	0.18	0.11

Thus, for example, for a disturbance series of 100 observations represented by an IMA time series model with $\lambda_0 = 0.3$, the mean square error could be reduced more than fivefold by appropriate adjustment. A justification of these results will be found in Appendix 6B.

6.9 A COMPROMISE VALUE FOR G

Inspection of Figure 6.6 also suggests the possibility of further simplification in the preparation of manual charts. In circumstances where formal estimation of G might be tedious, we can simply employ a compromise value in the neighborhood of, say, 0.2 to 0.4. The graphs show, for example, that if G was set equal to 0.2, then assuming that the disturbance was in fact represented by an IMA with $\lambda_0 = 1 - \theta_0$ having any value from 0 to 0.4, the standard deviation for the adjusted process would be, at most, only 5% larger than the minimum value achievable if λ_0 were exactly known. In particular, with G set equal to 0.2, if the process was already in a state of control ($\lambda_0 = 0$), this degree of manual adjustment would produce only a 5% increase in the value of the output standard deviation. On the other hand, if the process was not in control and λ_0 were as large as 0.4, then the standard deviation would again be only about 5% greater than it would have been for an optimal scheme with $G = 0.4$. Since, in practice, we can almost never be sure that a system is indeed in a state of perfect control, a wise policy might be to introduce mild adjustment corresponding to a value of, say, $G = 0.2$. (An alternative would be to employ a bounded adjustment scheme such as described in Chapter 8, where only an occasional change would be necessary.)

6.10 CHOOSING G TO REDUCE THE ADJUSTMENT VARIANCE σ_x^2

We now show that reducing the value of G to a "suboptimal" value less than λ_0 can considerably reduce the adjustment variance while only mildly increasing the variance of the output.

In this chapter we have confined attention to *responsive adjustment processes,* by which we mean that the effect of an adjustment is essentially fully realized at the output in one time interval. For such responsive processes we saw that for an IMA disturbance, $z_t - z_{t-1} = a_t - \theta_0 a_{t-1}$ with $G = \lambda_0 = 1 - \theta_0$, the adjustment equation $x_t = -(G/g)e_t$ with $e_t = a_t$ yields minimum mean square error σ_a^2 at the output. In the next chapter, we will consider the control of processes subject to inertia such that the full effect of a change in the compensating variable is not felt at the output by the end of one time interval. We shall see that, for such processes, control yielding minimum mean square error at the output is often undesirable, since to achieve it may require excessive manipulation at the input. Usually this is not such a serious consideration for a "responsive process," such as we are presently considering. However, here also, it is possible to considerably reduce the adjustment variance σ_x^2 at the small cost of producing output variances σ_e^2 only slightly greater than the minimum.

For simplicity, consider a scheme where the process gain is $g = 1$ and we suppose that the disturbance is represented by an IMA model with parameter $\lambda_0 = 1 - \theta_0$. Thus the minimum mean square adjustment equation is $x_t = -Ge_t$ with $G = \lambda_0 = 1 - \theta_0$, for which the output standard deviation is $\sigma_e = \sigma_a$. To achieve this, the adjustment standard deviation is $\sigma_x = G\sigma_a$. Now suppose we set G not equal to λ_0 but to some other value. Then using Equation (6.12),

$$\sigma_e = \sigma_a\sqrt{1 + \frac{(G - \lambda_0)^2}{G(2 - G)}} \quad \text{and} \quad \sigma_x = G\sigma_e.$$

For example, if $\lambda_0 = 0.2$, the smallest output standard deviation σ_a is achieved by setting $G = 0.2$ so that $\sigma_x = 0.2\sigma_a$.

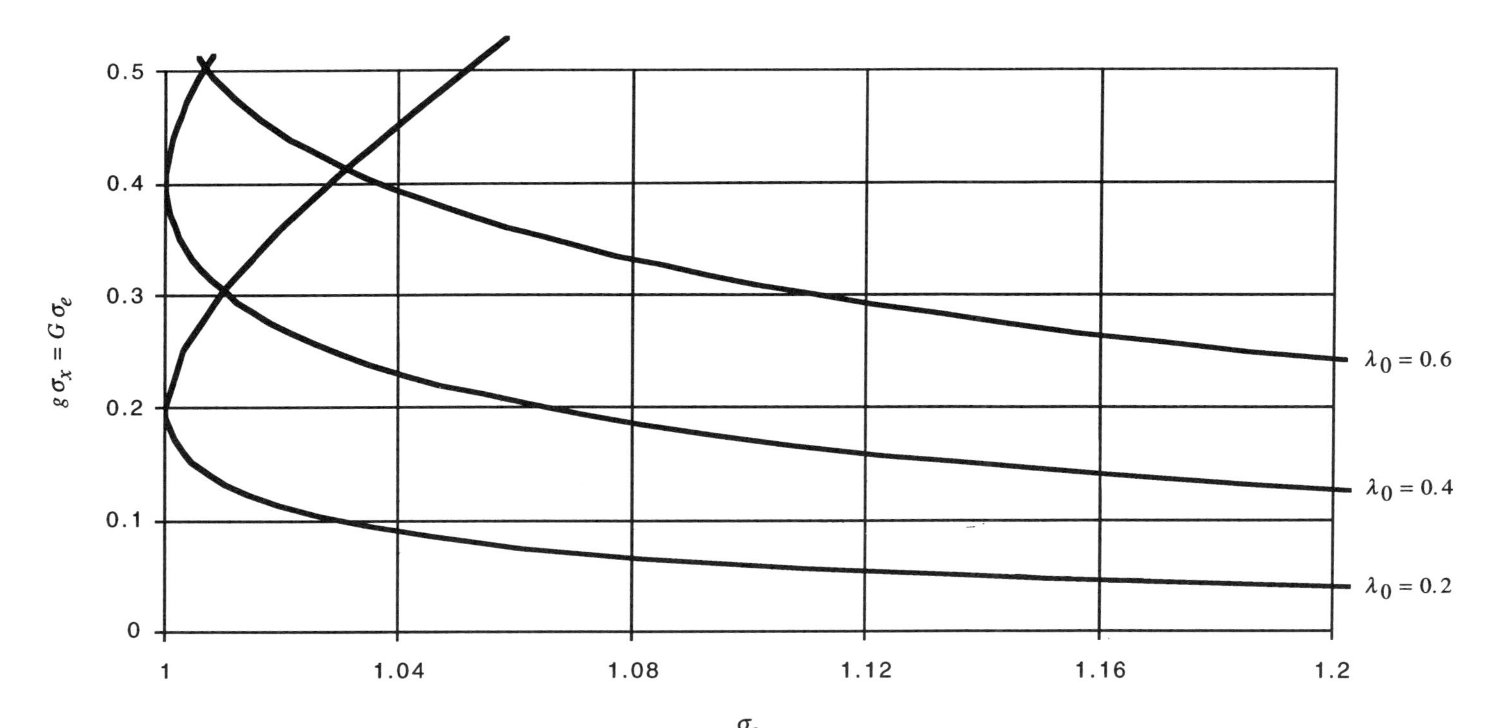

Figure 6.7 Characteristics of certain constrained adjustment schemes for an IMA disturbance with $\lambda_0 = 0.2, 0.4, 0.6$ with $g\sigma_x = G\sigma_e$ plotted against σ_e for $\sigma_a = 1.0$.

But suppose instead we set $G = 0.1$; then

$$\sigma_e = \sigma_a \sqrt{1 + \frac{(0.1)^2}{0.1 \times 1.9}} = 1.026\sigma_a$$

and, since $\sigma_x = 0.1\sigma_e$, $\sigma_x = 0.103\sigma_a$. Thus at the cost of a 2.6% increase in the output standard deviation, the *adjustment* standard deviation would be almost halved. We call these schemes, where a small increase in σ_e is tolerated in order to achieve a large reduction in σ_x, *constrained adjustment schemes.* A chart showing some of the options available is given in Figure 6.7.

The horizontal axis shows the value of σ_e relative to σ_a that is obtained with the MMSE scheme. It thus measures the "inflation" of the output standard deviation. The vertical axis shows $g\sigma_x = G\sigma_e$. The value of σ_x relative to σ_a is obtained by reading off this value on the vertical scale and dividing by g. For illustration, look again at the example we used above. Suppose $\lambda_0 = 0.2$ and we are prepared to tolerate an increase of 2.6% in the output standard deviation. Then, following the curve corresponding to $\lambda_0 = 0.2$, we see that for the value $\sigma_e = 1.026$, $g\sigma_x$ is seen to be about 0.103 and since in this case $g = 1$, $\sigma_x = 0.103$. In general, the required value of G is obtained from the relation $G = g\sigma_x/\sigma_e$, which, for this example, gives $G = 0.1$.

It will be seen that to reduce σ_x we need to *cut back* on the value of the damping factor G. The corresponding feedback chart for constrained adjustment is obtained by appropriately rescaling the vertical axis so that the scales for e_t and for x_t on the adjustment chart are as before proportional to G/g. In terms of Figure 6.4, the use of a value of G less than λ_0 implies compensating the disturbance by an EWMA $\tilde{z}_t$ with a smoothing constant larger than the "optimal value" θ_0. The use of this suboptimal value of θ slightly increases σ_e but greatly reduces σ_x.

6.11 ERRORS IN ADJUSTMENT

The process of adjustment itself may of course be subject to error. A fuller discussion of the effect of errors in adjustment is given in Section 7.6, after we consider the control of less responsive adjustment systems. However, it turns out that the effect of such errors is surprisingly small. For example, it is shown that even with the standard deviation of the errors of adjustment as large as 20% of the standard deviation of the adjustments themselves, and with G as large as 0.6, the inflation of the standard deviation at the output is less than 1%.

6.12 USE OF A DEAD BAND

There are a number of ways in which, at the cost of small losses in efficiency, manual adjustment charts may further be simplified. In particular, in Section 7.6, we briefly discuss *rounded adjustment charts* (BJR), which limit the number of adjustment options that need to be considered.

A different form of "rounding" is one where adjustments are made exactly as in Figure 6.2 except that small deviations are ignored; that is, a *dead band* is placed around the target value within which no action is taken. In Figure 6.8 the idea is illustrated using the thickness data of Figure 6.1 controlled as before by Equation (6.2) but with a dead band of width ± one standard deviation about the target value. Since the estimated standard deviation of the fully adjusted process in Figure 6.2 is about 11, this means that no action is taken if the measured thickness is within the limits 80 ± 11. The resulting increase in the output standard deviation σ_e, as a result of using this dead band, is just over 1%. Furthermore, its use ensures that about two-thirds of the time no action is required.

When, as in the example, a record of the process disturbance is available, it is easy to estimate by direct computation the effect of modifying the adjustment process in any manner whatsoever. In particular, the inflation of σ_e produced by dead bands of different widths may be studied empirically in this way and are shown below for dead bands B standard deviations in width.

B	0	0.5	1.0	1.25	1.5
Dead bands ± 11B	0	±5.5	±11	±13.75	±16.5
$\hat{\sigma}_e$	11.14	11.25	11.28	11.62	11.87
Percentage increase	0	0.9	1.2	4.3	6.5
Number of adjustments needed	99	65	35	23	16

In this particular example, a dead band of ± one standard deviation hardly increases the standard deviation of the controlled process but greatly reduces the number of adjustments necessary. For larger values of B, a more rapid rise occurs in the output standard deviation and it may then be preferable to adopt an approach using the more efficient *bounded adjustment charts* discussed in Chapters 8 and 9.

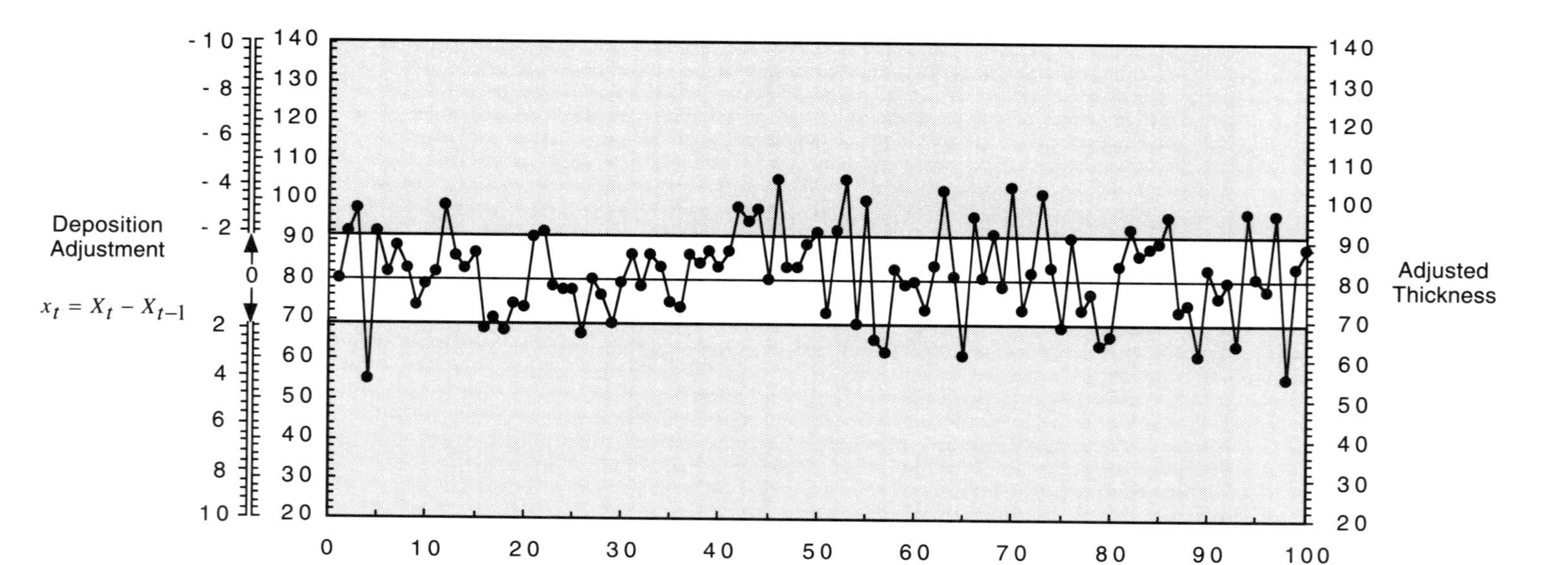

Figure 6.8 A manual adjustment chart to control metallic thickness with a dead band of one standard deviation.

Note that the reason for introducing a dead band is not to establish "statistical significance" of a deviation nor is it to avoid "tinkering"; the purpose is simply to relieve the operator from making small adjustments that would only marginally improve the control action. In some situations, where an operator must in any case regularly take data and it costs nothing to make an adjustment or take a sample, little is gained by the use of a dead band and similar schemes. Indeed, the process of regular adjustment whenever data are taken may improve morale and vigilance. However, when specific costs are incurred by making an adjustment and/or by obtaining an observation, it is important to employ the bounded adjustment methods of Chapters 8 and 9. These minimum cost schemes are developed so as to explicitly strike an economic balance between the cost factors.

6.13 FEEDBACK CONTROL OF A FIRST-ORDER AUTOREGRESSIVE DISTURBANCE

We have seen that discrete integral control with adjustment equation

$$gx_t = -Ge_t \tag{6.13}$$

yields minimum mean square error (MMSE) if the disturbance is represented by the IMA model

$$z_{t+1} - z_t = a_{t+1} - \theta a_t$$

and $G = 1 - \theta = \lambda$. To further illustrate the use of feedback control and also the robustness of discrete integral control, suppose the unadjusted disturbance is in fact generated by a stationary first-order autoregressive model

$$z_{t+1} = \phi z_t + a_{t+1}$$

and that, as before, the effect of adjustment is fully experienced at the output in one time interval. Using Equation (6.7), we could attain MMSE control for this AR(1) disturbance if we could set the level of the compensating variable such that

$$gX_t = -\hat{z}_{t+1} = -\phi z_t. \tag{6.14}$$

The controlled output series would then be such that

$$e_t = a_t \quad \text{and} \quad \sigma_e^2 = \sigma_a^2.$$

But since z_t follows an AR(1) model, its variance is

$$\sigma_z^2 = \sigma_a^2 / (1 - \phi^2).$$

Thus theoretically, at least, the effect of the feedback control would be to reduce the output variance by a factor of $1/(1 - \phi^2)$. From the following table it is seen that for highly autocorrelated series this reduction is quite large but is considerably less for moderate autocorrelation.

ϕ	Reduction Factors	
	Variance	**Standard Deviation**
0.95	10.3	3.20
0.90	5.3	2.29
0.70	2.0	1.40
0.50	1.3	1.15

Now the equation $-gX_t = \phi z_t$ of (6.14) cannot be used directly to achieve feedback control since, when the scheme is in operation, the observed output at time t is not z_t but the output error e_t. However, the adjustment equation (6.14) may alternatively be written in terms of the output e_t values as[2]

[2] Equation (6.15) may be derived as follows. If we write the AR(1) model for times t, $t - 1$, $t - 2$, . . . and multiply the second equation by ϕ the third by ϕ^2, and so on, we obtain

$$\begin{aligned} z_t &= \phi z_{t-1} + a_t, \\ \phi z_{t-1} &= \phi^2 z_{t-2} + \phi a_{t-1}, \\ \phi^2 z_{t-2} &= \phi^3 z_{t-3} + \phi^2 a_{t-2}, \\ &\vdots \end{aligned}$$

Adding n such equations where n is large enough so that $\phi^n z_{t-n}$ is negligible we get

$$z_t = a_t + \phi a_{t-1} + \cdots = \frac{\tilde{a}_t}{1 - \phi}.$$

$$gX_t = -\frac{\phi}{1-\phi}\tilde{e}_t, \tag{6.15}$$

where, following the previous notation, $\tilde{e}_t$ is an EWMA of the output deviations from target with smoothing constant ϕ. The corresponding MMSE adjustment equation is then, using Equation B of Table 4.2,

$$gx_t = -\frac{\phi}{1-\phi}(\tilde{e}_t - \tilde{e}_{t-1}) = -\phi(e_t - \tilde{e}_{t-1}). \tag{6.16}$$

Feedback control charts are then easily devised to obtain the required level X_t of the compensating variable from Equation (6.15) or the amount of adjustment $x_t = X_t - X_{t-1}$ from Equation (6.16).

However, doubts arise about the value of such schemes because of the following:

1. The potential for improvement to be gained depends heavily on the availability of a responsive system of compensation, and autocorrelated series are frequently associated with system having a high degree of inertia. Consequently, if the adjustment signal had to pass through the whole process, it might be impractical to control an autocorrelated output in this manner. However, if responsive compensation could be applied more directly near the system output, and if ϕ were fairly large, such schemes could be of potential value.
2. The AR(1) is appropriate for a system assumed to be stationary. As we have explained earlier, such an assumption is frequently suspect, particularly so if the estimate of ϕ is large. This is most important since using Equation (6.15) or (6.16) to control a nonstationary disturbance would produce a nonstationary series of output errors. This means that the variances of the output errors would increase indefinitely.
3. Over the range where the potential reduction in standard deviation is appreciable, discrete integral adjustment using Equation (6.13) produces reasonably good control even if the disturbance follows a stationary AR(1) model. Thus, for example, with G set equal to ϕ in Equation (6.13), the percentage increase in standard deviation over σ_a (which we call the ISD) would be rather small, as the following table shows:

ϕ	0.5	0.7	0.9	0.95
ISD	8.9%	7%	2.5%	1.3%

APPENDIX 6A INFLATION OF THE VARIANCE OF AN ADJUSTED PROCESS ARISING FROM A CHOICE OF G DIFFERENT FROM λ_0

Suppose a disturbance z_t is generated by an IMA model with smoothing constant θ_0, that is, with nonstationarity parameter $\lambda_0 = 1 - \theta_0$, but the process is adjusted using a chart based on some value $G \neq \lambda_0$. Then the model actually generating the disturbance is

$$z_t - z_{t-1} = a_t - \theta_0 a_{t-1},$$

where $\{a_t\}$ is a white noise sequence, whereas the sequence of observed forecast errors $\{e_t\}$ resulting from the use of $\theta = 1 - G$ instead of θ_0 is such that

$$z_t - z_{t-1} = e_t - \theta\, e_{t-1}.$$

Since the left-hand sides of both equations are the same, the e_t values must be generated by the model

$$e_t - \theta\, e_{t-1} = a_t - \theta_0 a_{t-1}.$$

This is a "first-order autoregressive–first-order moving average" time series model for the e_t values (e.g., see BJR). For these models it is known that for $|\theta| < 1$ and $|\theta_0| < 1$:

$$\frac{\sigma_e^2}{\sigma_a^2} = \frac{1 + \theta_0^2 - 2\theta\theta_0}{1 - \theta^2}.$$

Now, after substituting $G = 1 - \theta$ and $\lambda_0 = 1 - \theta_0$, we have

$$\frac{\sigma_e^2}{\sigma_a^2} = 1 + \frac{(G - \lambda_0)^2}{G(2 - G)}.$$

This is Equation (6.12) of the text and is used to draw the curves in Figure 6.6.

APPENDIX 6B AVERAGE REDUCTION IN MEAN SQUARE ERROR DUE TO ADJUSTMENT FOR OBSERVATIONS GENERATED BY AN IMA MODEL

For an IMA model with $\theta = \theta_0 = 1 - \lambda_0$, the following relation holds:

$$z_t - \hat{z}_t = a_t.$$

This implies that

$$\hat{z}_t - \hat{z}_{t-1} = \lambda_0 a_{t-1}$$

so that

$$z_m - \hat{z}_1 = a_m + \lambda_0 (a_{m-1} + \cdots + a_1).$$

Consider an unadjusted process initially on target so that $\hat{z}_1 = 0$. Then, since $\{a_t\}$ is a white noise sequence, the mean value of $z_m^2 = (y_m - T)^2$ is

$$\{1 + (m - 1)\lambda_0^2\}\sigma_a^2.$$

The mean square error for an *unadjusted* series of length n is therefore the mean value

$$\sum_{m=1}^{n} \frac{z_m^2}{n} = \left\{1 + \tfrac{1}{2}(n - 1)\lambda_0^2\right\}\sigma_a^2.$$

For the adjusted sequence of length n, the mean square error is σ_a^2. Thus

$$\frac{\text{MSE for unadjusted series}}{\text{MSE for adjusted series}} = 1 + \tfrac{1}{2}(n - 1)\lambda_0^2.$$

Problem 6.1. Consider again the metallic film series shown in Figure 6.1, which is given as Series A at the end of the book, for which Figure 6.2 shows a feedback adjustment chart with $G = 0.2$. Supposing as before that $g = 1.2$.

(a) Design a feedback control scheme for $G = 0.08$ and draw the appropriate manual control chart using the same scale for adjusted thickness as was used in Figure 6.2.

(b) Plot the points that would appear on the chart.

(c) Plot the e_t values for the scheme with $G = 0.08$ against those for the scheme with $G = 0.2$.

(d) Supposing the disturbance to be generated by an IMA model with $\lambda_0 = 0.24$, use Equation (6.12) to obtain σ_e/σ_a and σ_x/σ_a for the new scheme and compare with the corresponding values for the earlier scheme.

(e) Calculate the values for σ_a and for σ_e and σ_x for the two schemes from the observed data. How good is the agreement with the theoretical values? □

CHAPTER 7

Control of a Process with Inertia

"You must acquire and beget a temperance, that may give it smoothness."

Hamlet, WILLIAM SHAKESPEARE

In the last chapter we explored the consequences of a system of feedback control whereby a compensatory adjustment x_t was made at the process input, proportional to the last output deviation from target e_t, so that

$$gx_t = c_1 e_t. \tag{7.1}$$

The positive constant $G = -c_1$ was between zero and one and was the damping factor by which the full adjustment $x_t = -(1/g)\, e_t$ needed to be reduced. We saw that in practice desirable values for G were often between about 0.1 and 0.4.

The *adjustment equation*

$$gx_t = -Ge_t \tag{7.2}$$

could be put into effect by using a chart in which the scales for x_t and e_t were such that one unit deviation at the output corresponded to $-G/g$ units of adjustment at the input.

By summing Equation (7.1) a *control equation* was obtained,

$$gX_t = k_0 + k_I \sum_{i=1}^{t} e_i \tag{7.3}$$

with $k_I = c_1 = -G$. This equation gave the level X_t of the adjustment variable at the time t and was the discrete analog of integral control.

Finally, it was shown that such control was equivalent to setting

$$gX_t = -\tilde{z}_t, \tag{7.4}$$

where $\tilde{z}_t$ was an EWMA with smoothing constant $\theta = 1 - G$ of current and past values of the output disturbance $\{z_t\}$.

7.1 ADJUSTMENT DEPENDING ON THE LAST TWO OUTPUT ERRORS

Now consider the slightly more adventurous possibility of making the adjustment x_t depend on the last two errors e_t and e_{t-1}, so that the adjustment equation is

$$gx_t = c_1 e_t + c_2 e_{t-1}. \tag{7.5}$$

Relation to Proportional Integral Control

By summing Equation (7.5), you will find that the control equation for X_t contains a proportional action term as well as an "integral" action term. Thus

$$gX_t = k_0 + k_P e_t + k_I \sum_{i=1}^{t} e_i, \tag{7.6}$$

where $k_P = -c_2$ and $k_I = c_1 + c_2$.

Also, for simplicity, we shall in future suppose that X_t is measured from a suitable origin so that $k_0 = 0$.

A Manual Adjustment Chart for PI Control

Once more, it helps to illustrate using a manual adjustment chart, suitable for putting this PI control into effect. The manual adjustment chart of Figure 7.1 is for the PI control of a dyeing operation, where the object is to keep as small as possible the deviations $\{e_t\}$ of the color index of dyed yarn (the output variable) from the target value $T = 9.0$ by manipulation of the dye addition rate X_t (the adjustment variable).[1] For this process, one unit of change in X_t produced $g = 0.08$ units of change in the color index. For illustration, we will first suppose that the

[1] This illustration is loosely based on an example given in Box, Hunter, and Hunter (1978).

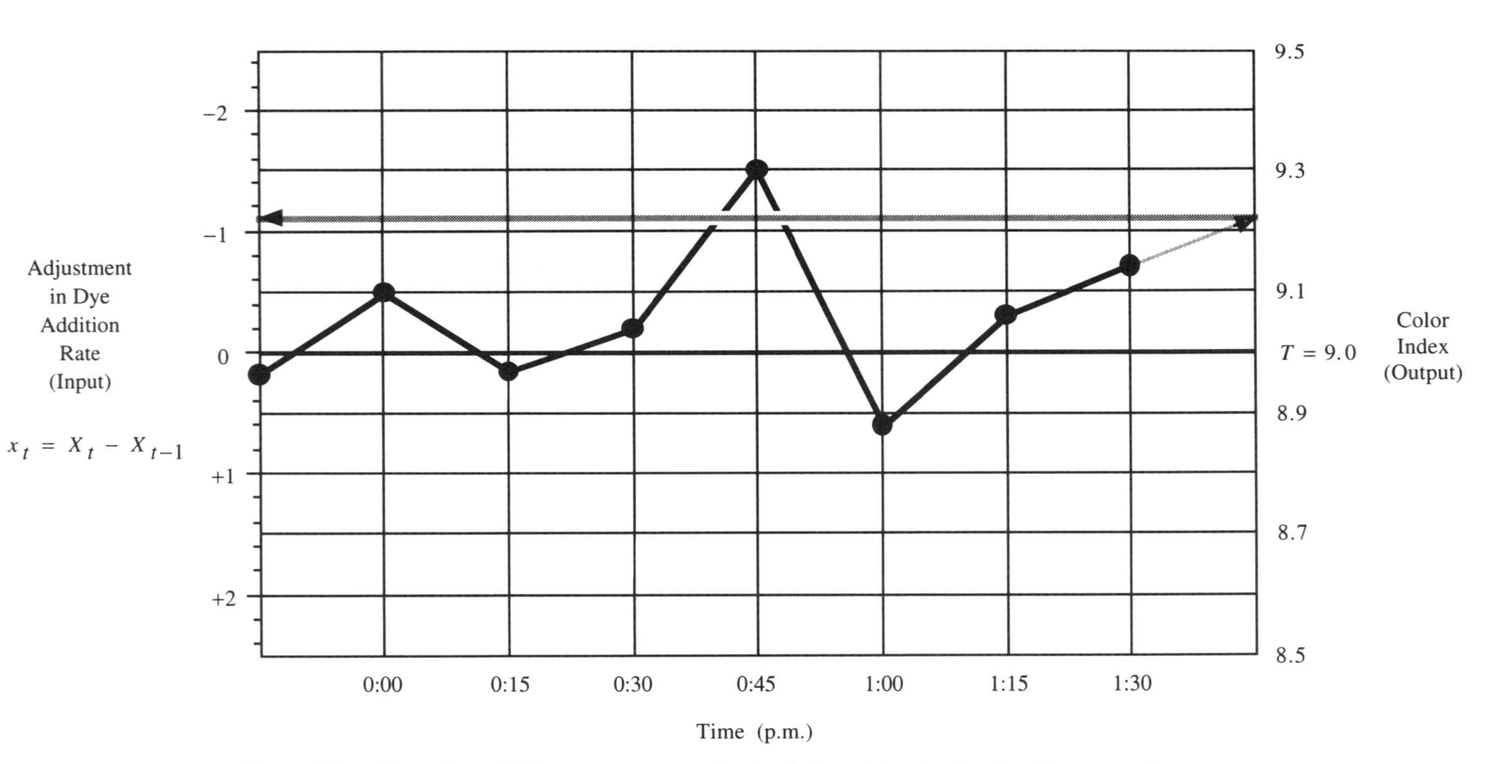

Figure 7.1 Using dye addition rate to control color index with a feedback adjustment chart.

control constants are chosen so that $c_1 = -0.8$ and $c_2 = 0.4$. Then the adjustment equation (7.5) is

$$0.08x_t = -0.8e_t + 0.4e_{t-1} \quad \text{or} \quad x_t = -10e_t + 5e_{t-1}, \qquad (7.5a)$$

and the corresponding PI control equation (7.6) is

$$0.08X_t = -0.4e_t - 0.4\sum_{i=1}^{t} e_i \quad \text{or} \quad X_t = -5e_t - 5\sum_{i=1}^{t} e_i. \qquad (7.6a)$$

The adjustment chart is constructed by writing Equation (7.5) in the form

$$gx_t = -G\{e_t + P(e_t - e_{t-1})\} \qquad (7.7)$$

with

$$G = -(c_1 + c_2) \quad \text{and} \quad P = -c_2/(c_1 + c_2) \qquad (7.8)$$

or, in terms of the PI control equation,

$$k_I = -G \quad \text{and} \quad k_P/k_I = P. \qquad (7.9)$$

Thus for the dyeing example, $G = 0.4$ and $P = 1.0$ and the adjustment equation (7.7) is

$$0.08x_t = -0.4\{e_t + 1.0(e_t - e_{t-1})\} \quad \text{or}$$

$$x_t = -5\{e_t + 1.0(e_t - e_{t-1})\}. \qquad (7.10)$$

The chart in Figure 7.1 requires, as before, two scales: that on the right shows the output variable (color index) with the target value ($T = 9.0$) at the center of the scale so that e_t is the deviation at time t from $T = 9.0$; the left-hand scale shows the adjustment x_t in the input variable (dye addition rate) centered at 0 and arranged so that one unit of deviation at the output corresponds to $-G/g = -5$ units of adjustment. To use the chart, the operator records the current output value of the color index on the chart and extrapolates (or interpolates) a line joining the last two points P time intervals. In this example $P = 1$ so that the line through the last two points is extrapolated one whole time interval and the resulting value is read off on the adjustment scale. [Had P been negative, Equation (7.7) would call for interpolation between the two points.] In Figure 7.1, at 1:30 p.m., the operator has just recorded the color index value of 9.14. The value previous to this was 9.06. The extrapolated line falls on 9.22, and reading off this value on the left-hand scale gives -1.1 so the operator would now reduce the dye addition rate by 1.1 units. As we shall see later it is not important to be very precise about this adjustment and "eyeballing" the extrapolation (or interpolation) would usually be enough.

Relation of PI Control to Exponential Smoothing

If we set $\delta = -c_2/c_1$, then $c_1 = -G/(1 - \delta)$ and Equation (7.5) becomes

$$gx_t = \frac{-G}{1 - \delta}\{e_t - \delta e_{t-1}\}. \tag{7.11}$$

It is shown in Appendix 7A that the required action is then equivalent to setting

$$g\tilde{X}_t = -\tilde{z}_t. \tag{7.12}$$

So PI control is equivalent to equating $g\tilde{X}_t$, an EWMA of current and past levels $X_t, X_{t-1}, \ldots$ having smoothing constant $\delta = -c_2/c_1$, with $\tilde{z}_t$, an EWMA of current and past values of the output disturbances z_t, $z_{t-1}, \ldots$ having smoothing constant $\theta = 1 - G$.

Note that so far we have made *no* assumption about models—in particular, about a model for the process dynamics or a model for the time series representing the disturbance. Everything discussed follows from Equation (7.5), which merely postulates that an adjustment is made depending on the last two errors at the output. The action produced by this simple adjustment equation:

1. Is the discrete analog of PI control [Equation (7.6)].
2. Can be represented by a simple adjustment chart [Equation (7.7)].
3. Is such that an exponential average of past levels of the compensatory variable just cancels a different exponential average of the disturbances [Equation (7.12)].

To learn more about the characteristics of such control schemes, we need to introduce specific models for process dynamics and for the disturbance.

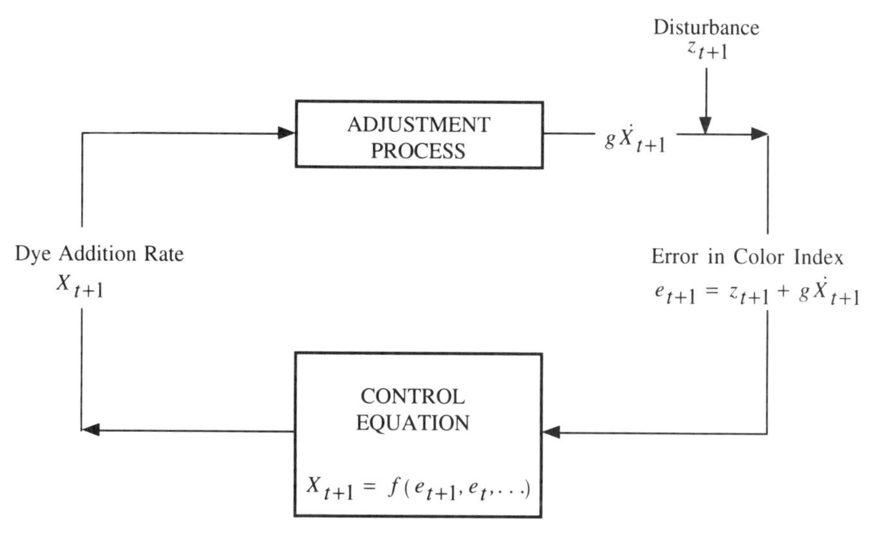

Figure 7.2 A flow diagram for a process subject to feedback control considered at some time $t + 1$.

A Feedback Loop

In Chapter 6 we discussed the control of a responsive system for which an adjustment had its full effect at the output in one time interval. We now consider the control of a process subject to inertia.

Figure 7.2 is a diagram of the feedback loop at time $t + 1$ illustrated for the dyeing example. At this time the setting of the compensatory variable, the input dye addition rate, is X_{t+1}. The total effect at the output of all past manipulations of the compensating variable is denoted by $g\dot{X}_{t+1}$ and the extent to which these manipulations have failed to compensate the disturbance z_{t+1} is denoted by e_{t+1}, which is the error of the output color index.

The diagram illustrates how the error e_{t+1} in the color index is

$$e_{t+1} = z_{t+1} + g\dot{X}_{t+1}. \tag{7.13}$$

If $\hat{z}_{t+1}$ is a forecast of z_{t+1}, then $e_{t+1} = z_{t+1} - \hat{z}_{t+1} + \hat{z}_{t+1} + g\dot{X}_{t+1}$, and if we can set $g\dot{X}_{t+1} = -\hat{z}_{t+1}$, then

$$e_{t+1} = z_{t+1} - \hat{z}_{t+1}$$

and the error of the forecast becomes the error at the output.

7.2 MINIMUM MEAN SQUARE ERROR CONTROL OF A PROCESS WITH FIRST-ORDER DYNAMICS

We now introduce the following assumptions:

1. The inertia of the adjustment process can be represented by a first-order dynamic system with parameter δ_0 so that, as in Chapter 4,

$$\dot{X}_{t+1} = \tilde{X}_t = (1 - \delta_0)(X_t + \delta_0 X_{t-1} + \delta_0^2 X_{t-2} + \cdots).$$

2. The disturbance can be represented by an IMA model

$$z_{t+1} - z_t = a_{t+1} - \theta_0 a_{t+1} \quad \text{and} \quad \lambda_0 = 1 - \theta_0.$$

Then if the control produced is

$$g\tilde{X}_t = -\tilde{z}_t,$$

where $\tilde{X}_t$ and $\tilde{z}_t$ are EWMAs with smoothing parameters δ_0 and θ_0, respectively, the error at the output is $e_t = a_t$. It can be formally shown (see BJR) that this is a MMSE scheme that minimizes the output error variance $\sigma_e^2 = \sigma_a^2$.

The constants of Equation (7.5) expressed in terms of λ_0 and δ_0 are as follows:

$$c_1 = -\frac{\lambda_0}{1 - \delta_0}, \quad c_2 = \frac{\lambda_0 \delta_0}{1 - \delta_0}. \tag{7.14}$$

Equivalently, such adjustment yields discrete PI control in the form of Equation (7.6) with

$$k_P = -\frac{\lambda_0 \delta_0}{1 - \delta_0}, \quad k_I = -\lambda_0. \tag{7.15}$$

Finally, the adjustment chart would use

$$G = \lambda_0 \quad \text{and} \quad P = \delta_0/(1 - \delta_0) \tag{7.16}$$

in Equation (7.7).

The control scheme defined in any one of these three alternative ways will yield minimum variance at the output.

TABLE 7.1 Six Alternative Control Schemes for the Dyeing Process Illustrated for the Disturbance Generated by an IMA Model with $\lambda_0 = 0.4$: Empirical Values $\hat{\sigma}_e$ and $\hat{\sigma}_x$ Are Shown with Theoretical Values σ_e and σ_x in Brackets When Available

Scheme	Adjustment Equation	σ_e Output Standard Deviation	σ_x Input Standard Deviation
(a) MMSE control	$x_t = -5\{e_t + (e_t - e_{t-1})\}$	1.00 (1.00)	11.5 (11.2)
(b) Optimal constrained control	$x_t = 0.83x_{t-1} - 0.21x_{t-2} - 1.93e_t + 0.97e_{t-1}$	1.12 (1.10)	2.55 (2.50)
(c) Optimal constrained PI control	$x_t = -2.6\{e_t - 0.25(e_t - e_{t-1})\}$	1.12 (1.10)	2.55 (2.50)
(d) Control with $G = 0.3$, $P = 0$	$x_t = -3.75e_t$	1.06 (1.04)	3.96 (3.91)
(e) Scheme (d) with rounded adjustment	Options for adjustment are limited to zero adjustment, or ± 1, or ± 2 units of change	1.06	4.11
(f) Control with Shewhart chart	$X_t = \begin{Bmatrix} -(1/0.08)\, e_t, & \text{if } \lvert e_t \rvert > 3\sigma_a \\ 0, & \text{otherwise} \end{Bmatrix}$	1.53	9.25

Suppose, for the dyeing example, that the models adequately represent the system with $\delta_0 = 0.5$ and $\theta_0 = 0.6$ ($\lambda_0 = 0.4$). Then the reader can readily confirm that the control scheme could be defined by any of the following pairs of constants: $c_1 = -0.8$, $c_2 = 0.4$ in Equation (7.14); $k_P = -0.4$, $k_I = -0.4$ in Equation (7.15); $G = 0.4$, $P = 1$ in Equation (7.16). These are precisely the values that we have used already in Equations (7.5a), (7.6a), and (7.10) to illustrate the various forms of this PI scheme. To facilitate comparison in what follows, we have set $\sigma_a = 1$. The MMSE adjustment equation, together with associated values[2] for σ_e and σ_x, is given in Table 7.1 scheme (a), and its performance is illustrated in Figure 7.3a (see also the chart of Figure 7.1). On the left of Figure 7.3 is shown a typical disturbance series z_t generated by an IMA time series model with $\lambda_0 = 0.4$, which appears as Series B at the back of this book. Figure 7.3a shows the output series e_t and the needed adjustments x_t for the MMSE control scheme. This scheme produces excellent control at the output, eliminating the changes in level of the disturbance z_t. Unfortunately, however, the corresponding x_t series shows that excessive manipulation is needed to bring this about.

[2] For the dyeing example, the actual value σ_a was estimated to be about 0.25. The associated values for σ_e and σ_x would be obtained by multiplying the tabulated values by this factor.

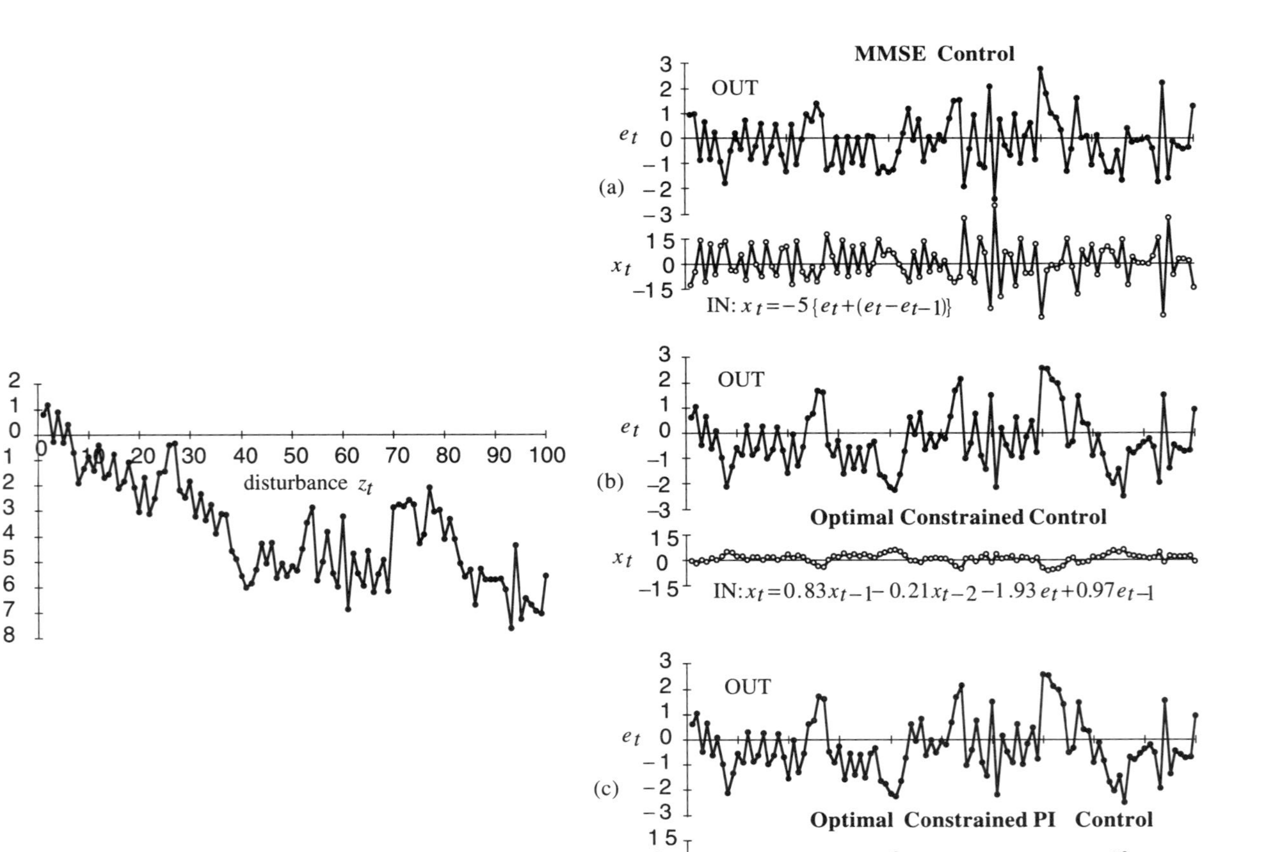

Figure 7.3 One hundred values of a disturbance generated by an IMA model with $\lambda_0 = 0.4$ and $\sigma_a = 1$. Also shown are the outputs and inputs for three of the adjustment schemes (a, b, and c) of Table 7.1.

You can understand why the MMSE scheme behaves in this way by considering the corresponding adjustment chart of Figure 7.1. Remember that P is the degree of extrapolation along a line joining the last two points, which for the MMSE scheme is $P = \delta_0/(1 - \delta_0)$. Thus, for the dyeing example with $\delta_0 = 0.5$, $P = 1$ and one whole unit of extrapolation is required. For even slower dynamics, corresponding to larger values of δ_0, even greater extrapolation would be needed, requiring even greater manipulation of the adjustment variable.

Effect of Dead Time

Sometimes adjustment processes are subject to *dead time*: that is, in response to a change, a period of pure delay occurs before *any* response is experienced at the output. In the process industries, dead time is sometimes caused by pipes and pumps that hold back the product. Alternatively, dead time may occur because, after sampling the product, some time elapses before it is possible to get the result needed to take action. If dead time can be avoided, it should be, because the efficiency of even the best feedback control scheme can greatly be reduced by dead time.

It may be shown (BJR) that, for d_0 periods of delay followed by first-order dynamics with parameter δ_0 and assuming an IMA disturbance with parameter $\lambda_0 = 1 - \theta_0$, the MMSE adjustment equation is

$$x_t = -\lambda_0(x_{t-1} + x_{t-2} + \cdots + x_{t-d_0}) - \frac{\lambda_0}{g(1 - \delta_0)}(e_t - \delta_0 e_{t-1}). \qquad (7.17)$$

However, this form of adjustment, although theoretically producing the smallest sum of squares of the output errors, is of little practical interest because the problem of excessive manipulation becomes even worse.

7.3 SCHEMES WITH CONSTRAINED ADJUSTMENT

The phenomenon of excessive manipulation required by MMSE schemes is well known and many authors (e.g., Aström, 1970: Aström and Wittenmark, 1984; Box and Jenkins, 1970; MacGregor, 1972; Tunnicliffe-Wilson, 1970; Whittle, 1963) have considered the problem of constraining the degree of manipulation needed at the input while still producing good control of the output. One way to do this is to calculate an optimal linear scheme that minimizes[3]

$$\sigma_x^2 + \alpha\sigma_e^2. \tag{7.18}$$

Where by a linear scheme we mean one in which the adjustment is some linear function of previous adjustments $x_{t-1}, x_{t-2}, \ldots$ and of the errors at the output $e_t, e_{t-1}, \ldots$

It can then be shown that different choices of α produce an "envelope" of constrained schemes; each of which, for a given increase in σ_e^2 above its minimum value, yields the smallest possible value for σ_x^2.

Remarkable reductions in σ_x can be achieved with such schemes with only very slight increases in σ_e. For instance, in the dyeing example if we use scheme (b) in Table 7.1 we see that, for an increase of only 10% in σ_e, the optimal constrained control yields a more than fourfold reduction in σ_x. The adjustment equation for this constrained scheme is

$$x_t = 0.83x_{t-1} - 0.21x_{t-2} - 1.93e_t + 0.97e_{t-1}. \tag{7.19}$$

For the same disturbance as before, Figure 7.3b illustrates how this scheme produces control of the output e_t very nearly as good as that for the MMSE scheme but with very much less adjustment. Unfortunately, such optimal constrained schemes are not particularly easy to derive and are somewhat complicated in form.

7.4 PI SCHEMES WITH CONSTRAINED ADJUSTMENT

As an alternative, Box and Luceño (1995) studied the properties of constrained PI schemes chosen to minimize the expression (7.18). Rather surprisingly, these schemes can perform very nearly as well as the more complicated general linear schemes. For example, an optimal constrained PI scheme (Figure 7.3c) that, to two-decimal accuracy, gives the same input and output variances as the more complicated scheme of Equation (7.19) is

$$x_t = -2.6\{e_t - 0.25(e_t - e_{t-1})\}. \tag{7.20}$$

[3] The quantity $\sigma_x^2 + \alpha\sigma_e^2$ can also be thought of as a measure of the total cost, with α measuring the relative costs of deviations from target and of input manipulations. Note also that minimizing the expression (7.18) is equivalent to minimizing $\beta\, \sigma_x^2 + \sigma_e^2$, where the constants α and β satisfy $\alpha\beta = 1$.

This is shown in Table 7.1 as scheme (c). Comparing scheme (c) with the MMSE scheme (a), note that, for the constrained PI scheme, the factor G/g decreases from 5.0 to 2.6, because the damping factor G is almost halved from 0.4 to 0.208. Also, the constant P, which was +1.0 for the MMSE scheme, is now negative with $P = -0.25$ for the constrained PI scheme. Thus instead of requiring extrapolation of a line joining the last two points by one whole interval, this scheme requires *interpolation* between these points. This is a much more stable procedure. This interpolation is easily achieved in practice by providing the operator with a chart like that in Figure 7.4 with lighter lines three-quarters of the way between each time line. The operator then reads off the adjustment opposite the intersection with the lighter line. For comparison, the performance of this scheme in controlling the same disturbance as before is shown in Figure 7.3c.

In Appendix 7C (see also Box and Luceño, 1995), extensive tables of optimal choices for G and P for a wide range of choices of λ_0 and δ_0 are given, with and without one whole unit of dead time ($d_0 = 0$ and $d_0 = 1$). It appears that near equality in performance between the more complicated optimal constrained schemes and the simpler constrained PI schemes holds very generally.

7.5 OPTIMAL AND NEAR-OPTIMAL CONSTRAINED PI SCHEMES: CHOICE OF *P*

Figure 7.5 shows the relation between $g\sigma_x$ and σ_e for the particular but not untypical case where the process parameters for the disturbance and for the dynamics are $\lambda_0 = 0.4$ and $\delta_0 = 0.25$ and for convenience we again set $\sigma_a = 1$. The continuous curve shows the envelope of optimal PI schemes for various values of G and P and is obtained by minimizing the expression (7.18) for different values of α. The series of short bold curves is each appropriate for the particular value of G indicated. The labeled points on all these curves correspond, from left to right, with values of P for (a) 0.33, (b) 0.25, (c) 0, (d) −0.25, (e) −0.5, and (f) −0.75. For each G, the best value of P corresponds to the point where the bold curve touches the envelope. The point on the extreme left of the diagram corresponds to the MMSE scheme with $G = 0.4$ and $P = 0.33$, which yields $\sigma_e = \sigma_a = 1.00$ and $g\sigma_x = 0.55$. However, by setting $G = 0.2$ and $P = -0.25$, for example, an alternative scheme corresponding to the point d on the bold curve for $G = 0.2$ could be used with the greatly decreased manipulation

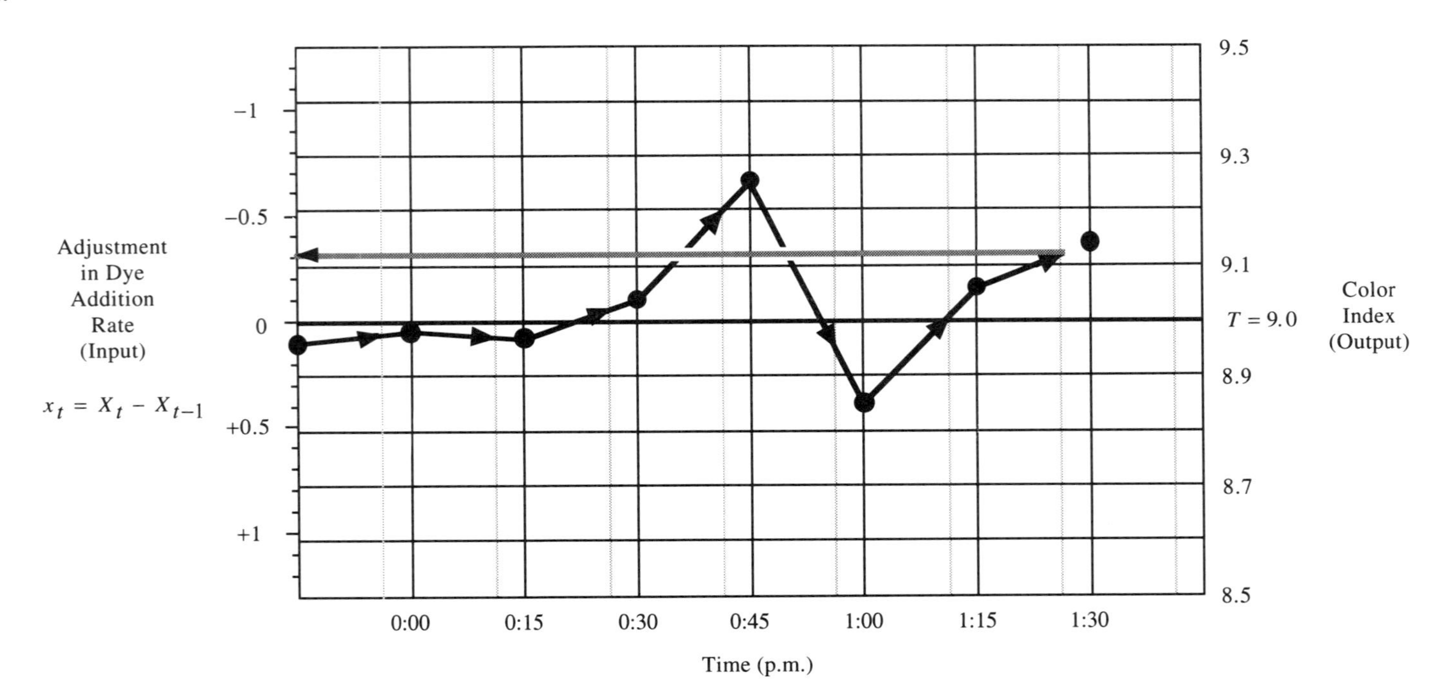

Figure 7.4 A chart for the dyeing example to put the adjustment equation $x_t = -2.6\{e_t - 0.25(e_t - e_{t-1})\}$ into effect.

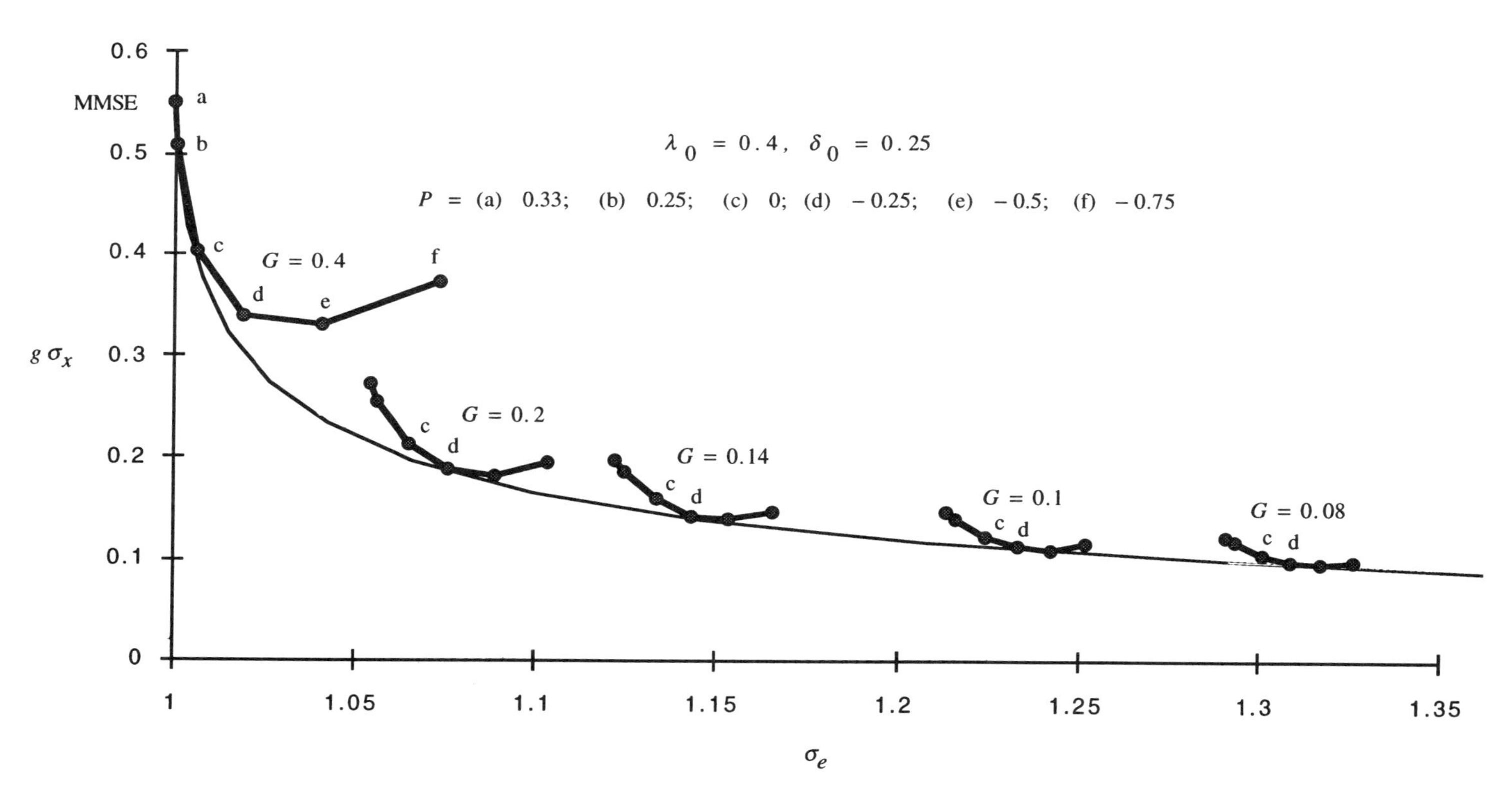

Figure 7.5 Relation between σ_x and σ_e for optimal PI schemes when $\lambda_0 = 0.4$ and $\delta_0 = 0.25$.

standard deviation of $g\sigma_x = 0.19$ and only a slight increase in the output standard deviation to $\sigma_e = 1.07$. We have made diagrams like Figure 7.5 for the ranges of most practical interest—$0 < \lambda_0 \le 0.6$ and $0 \le \delta_0 \le 0.5$—assuming both zero dead time and a dead time of one unit time interval. We have found throughout this wide field of alternatives that, apart from schemes close to the MMSE scheme which are not of much practical interest, the following findings are broadly true:

1. The best value of P remains close to $P = -0.25$, indicated by the points marked d in Figure 7.5.
2. The properties of the schemes are not sensitive to moderate changes in P.
3. The simpler choice of $P = 0$ is almost as good as $P = -0.25$ and may sometimes even be better.

In what follows, therefore, we consider in more detail only schemes for which $P = 0$ and $P = -0.25$.

7.6 CHOICE OF G FOR $P = 0$ AND FOR $P = -0.25$

Schemes with $P = 0$

When $P = 0$ we return to the simple form of control considered in the last chapter, where the adjustment is made proportional to the last error and the corresponding control equation represents discrete integral control. Thus

$$x_t = -\left(\frac{G}{g}\right)e_t \quad \text{and} \quad X_t = -\left(\frac{G}{g}\right)\sum_{i=1}^{t} e_i.$$

We saw that such schemes were equivalent to continuously setting X_t so that

$$X_t = -(1/g)\tilde{z}_t$$

with the EWMA $\tilde{z}_t$ having smoothing constant $1-G$. Also, if desired, they could be put into effect manually using a very simple feedback adjustment chart. Some choices for such schemes when the system has first-order dynamics characterized by δ_0 and an IMA disturbance characterized by $\lambda_0 = 1-\theta_0$ are shown in Figure 7.6. Comparisons are made in relation to MMSE feedback schemes for which the output standard deviation is $\sigma_e = \sigma_a = 1$ and $G = \lambda_0$. The horizontal scale shows the value of the output standard deviation σ_e and the vertical scale that of $g\sigma_x = G\sigma_e$ for constrained schemes when λ_0 is 0.2, 0.4, and 0.6. The shaded bands are for schemes with δ_0 from 0 to 0.5 with the dotted line corresponding to $\delta_0 = 0.25$. For schemes with σ_a having some value other than 1, the values read off the scales are multiplied by σ_a.

The vertical scale serves to characterize both the input standard deviation σ_x and the parameter G. Since over the range considered σ_e is only slightly greater than 1, the constant G is always slightly less than $g\sigma_x$ shown on the left scale. The lines characterized by λ_0 and δ_0 show the trade-offs that are possible for various combinations of these parameters.

How precisely we can determine the performance of a given scheme depends, of course, on how much is known about λ_0 and δ_0. However, the robustness of these schemes ensures that our knowledge need not be very precise. For example, suppose we know, or can speculate from experience, that λ_0 is about 0.4 but we know very little about δ_0; then we have the choice of schemes included in the shaded band labeled $\lambda_0 = 0.4$. Having selected a suitable pair of values for $g\sigma_x$ and σ_e within this band, the equality $G = g\sigma_x/\sigma_e$ gives the required value G to be used in the control scheme. Note that the bands defining δ_0 tend to be quite narrow in the regions of most interest. For example, with $\lambda_0 = 0.4$ and allowing for an increase in σ_e over σ_a of 10%, we see that a value $g\sigma_x$ between 0.17 and 0.21 would be obtained by setting G between $0.15 = 0.17/1.10$ and $0.19 = 0.21/1.10$. This band covers the range of δ_0 between 0 and 0.5, and the scheme is surprisingly insensitive to this choice. By adopting such a scheme, σ_x will essentially be halved since for the MMSE scheme $g\sigma_x = 0.4$.

If you know little about the parameters, you can see from Figure 7.6 that you are unlikely to go very far wrong by setting G at about 0.3. Even for the rather extreme case $\lambda_0 = 0.6$ and $\delta_0 = 0.5$, σ_e will be only about 1.16. For illustration, we again use the earlier example with $g = 0.08$, $\lambda_0 = 0.4$, and $\delta_0 = 0.5$. The adjustment equation and associated values for σ_e and σ_x are given in Table 7.1, scheme (d). Also, with the same disturbance series, Figure 7.7d shows the input and output series, which may be compared with those for schemes (a), (b), and (c) in Figure 7.3.

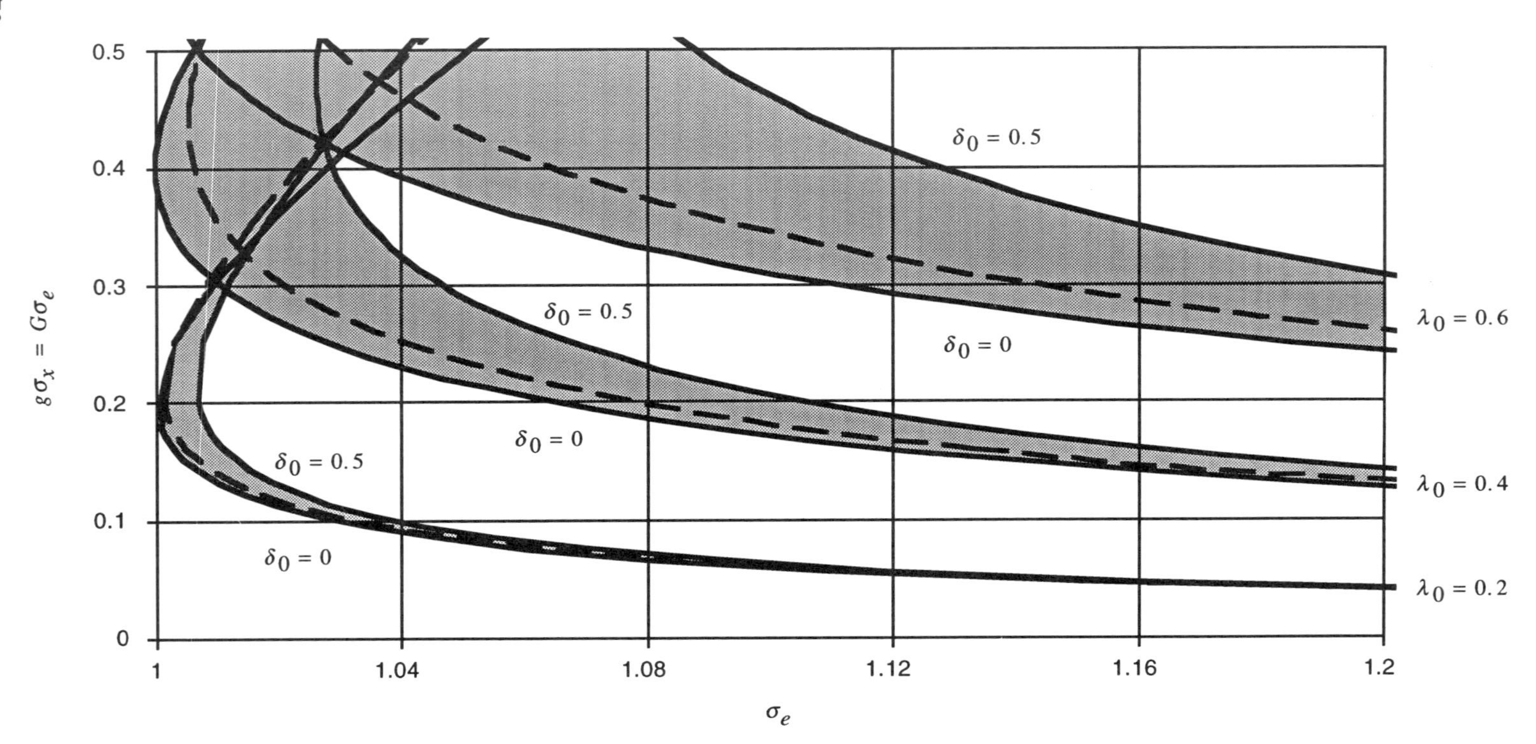

Figure 7.6 Chart for choosing G for an adjustment scheme with $P = 0$. The process has zero dead time and dynamics characterized by (g, δ_0). The disturbance is characterized by λ_0.

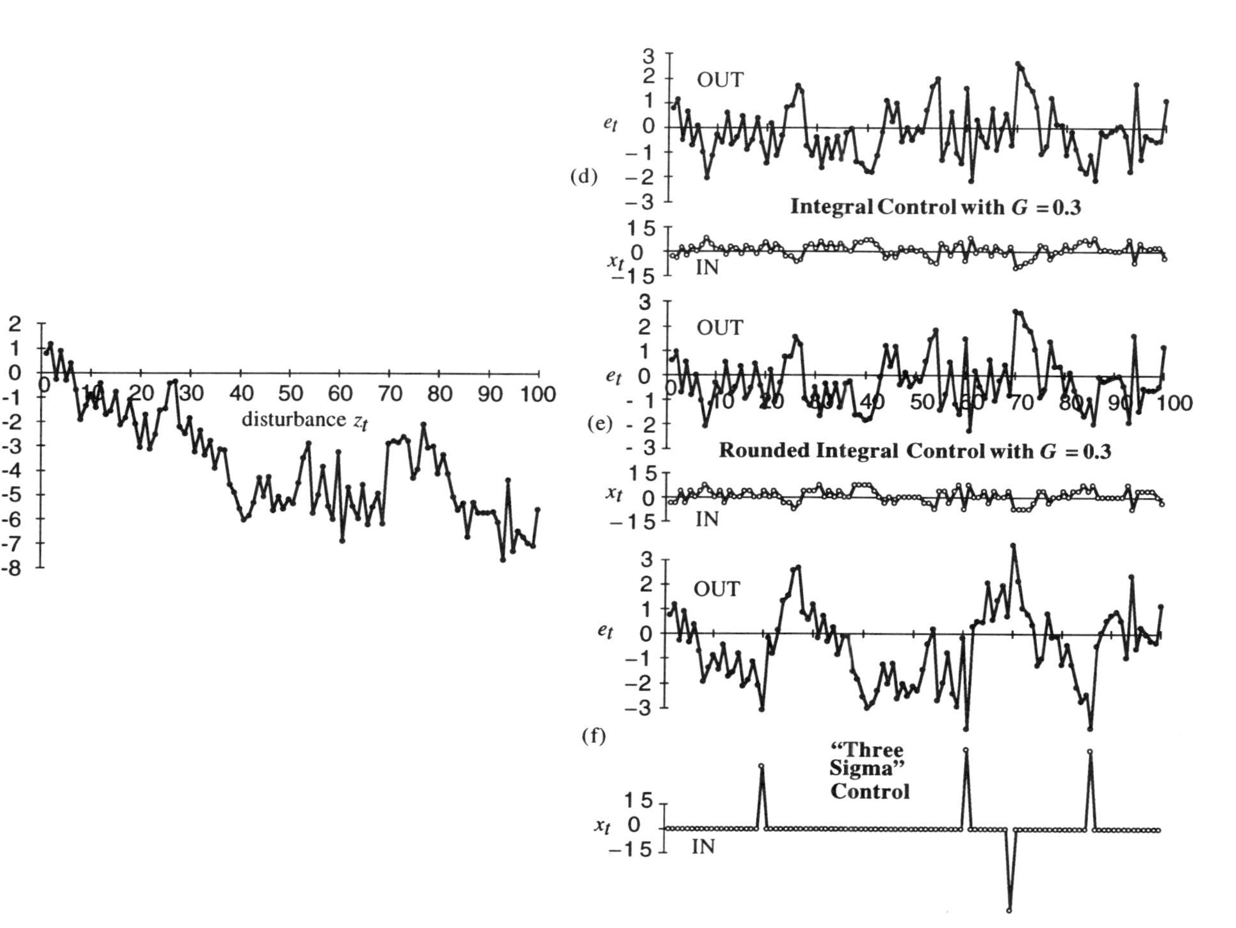

Figure 7.7 Outputs and inputs for schemes (d), (e), and (f) of Table 7.1.

Schemes with $P = -0{:}25$

Figure 7.8 shows the corresponding performance chart for the (often slightly better) choice $P = -0.25$. The difference appears to be slight and we think the slight loss of efficiency incurred by setting $P = 0$ will be compensated in practice by the added simplicity of the corresponding feedback schemes and the charts it yields.

Errors in Adjustment

The effect of errors in adjustment was discussed briefly in Section 6.11. We now consider the problem in greater detail. Suppose there is an error in making an adjustment so that, with x_t the adjustment called for, the actual adjustment made is $x_t + u_t$, where the adjustment errors $\{u_t\}$ are distributed about zero independently of each other and of the x_t values. Then a measure of the relative error of adjustment is $k = \sigma_u/\sigma_x$ or the percentage error $100k = 100\sigma_u/\sigma_x$.

Suppose that the adjustment equation

$$gx_t^{(0)} = -Ge_t^{(0)},$$

corresponding to integral control, is used with G not necessarily equal to the true value of λ, so that $\{e_t^{(0)}\}$ is the sequence of output deviations from target [and the superscript (0) attached to the e_t and x_t terms indicates that the adjustment errors (u_t) are zero].

When there is a nonnull sequence $\{u_t\}$ of errors in adjustment, the corresponding sequences of output deviations from target $\{e_t\}$ and adjustments $\{x_t\}$ satisfy

$$g(x_t + u_t) = -Ge_t,$$

and we show in Appendix 7B that, if the process dynamics can be represented by a first-order dynamic system with inertia parameter δ and gain g, then the inflation of the variance of the output deviations from target due to the adjustment errors is given by

$$\frac{\mathrm{var}(e_t)}{\mathrm{var}(e_t^{(0)})} = 1 + k^2 \frac{G}{2 - G(1 - \delta)/(1 + \delta)},$$

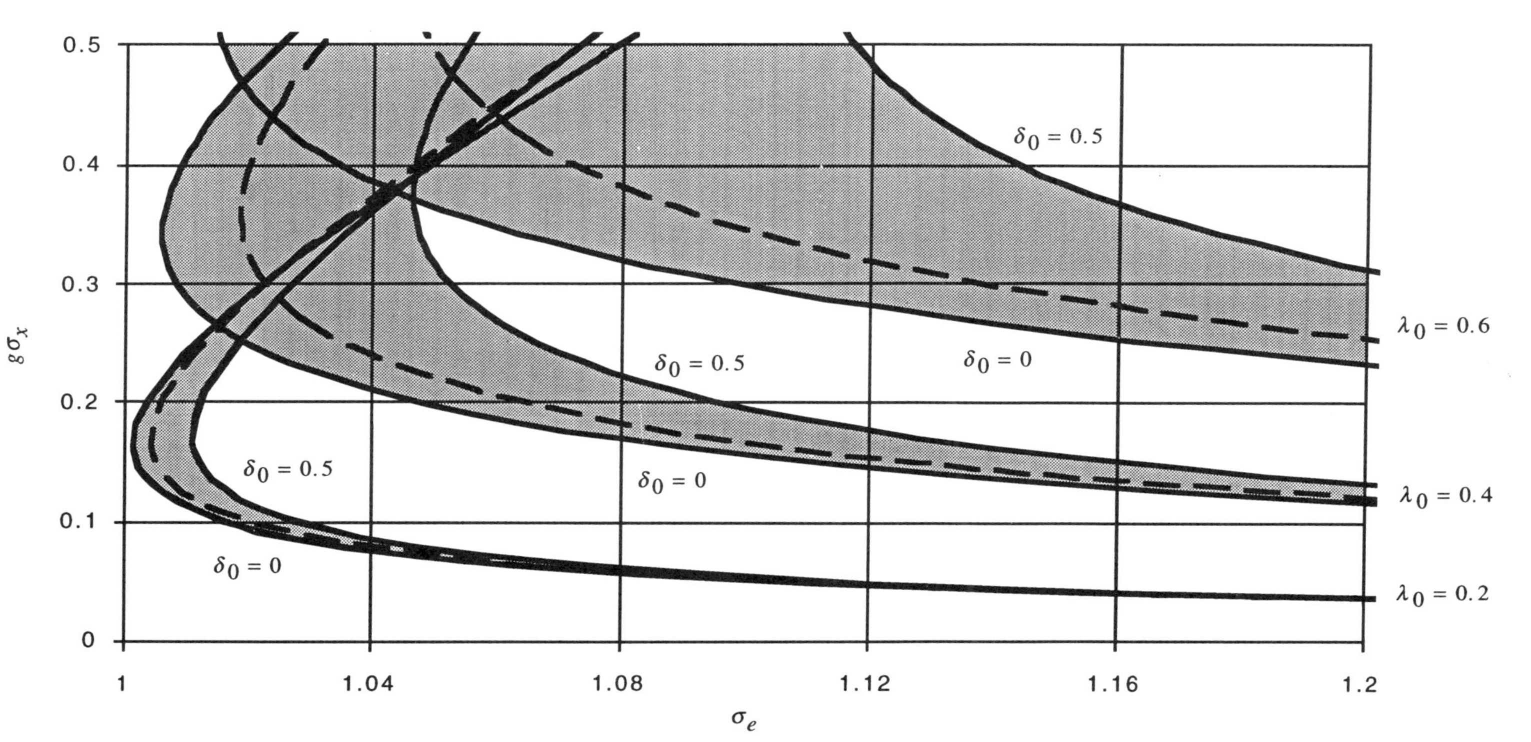

Figure 7.8 Chart for choosing G for an adjustment scheme with $P = -0.25$. The process has zero dead time and dynamics characterized by (g, δ_0). The disturbance is characterized by λ_0.

where

$$k^2 = \frac{\text{var}(u_t)}{\text{var}(x_t^{(0)})},$$

which does not depend on λ.

For $\delta = 0$ and $k = 0.10, 0.20$, or 0.30 (a 10%, 20%, or 30% "adjustment" error), the percentage increase in output standard deviation $\sqrt{\text{var}(e_t)}/\sqrt{\text{var}(e_t^{(0)})}$ is given in the following table:

	Percentage error in adjustment (100k)		
G	**10%**	**20%**	**30%**
0.1	0.03%	0.11%	0.24%
0.2	0.06%	0.22%	0.50%
0.3	0.09%	0.35%	0.79%
0.4	0.12%	0.50%	1.12%
0.5	0.17%	0.66%	1.49%
0.6	0.21%	0.85%	1.91%

Thus even with a 20% error in adjustment ($k = 0.20$) and G as high as 0.6 the increase in the error of the output is less than 1%. If δ is greater than zero, the inflation of output standard deviation would be even smaller. This might suggest that, when there is inertia, successive adjustment errors tend to partially cancel out the effect of previous adjustment errors.

Rounded Charts

Even greater simplicity, but with a further small loss of efficiency, is obtained by using rounded charts. With these charts, action is not required after each observation but only from time to time. Figure 7.9 shows rounded adjustment applied to the adjustment equation $x_t = -3.75\, e_t$ of scheme (d) in Table 7.1. Possible action is limited to five options for adjustment of the dye addition rate: no change, or ± 1, or ± 2 units of change. A justification for such charts is given in BJR, where it is suggested that the width of the bands should be about one

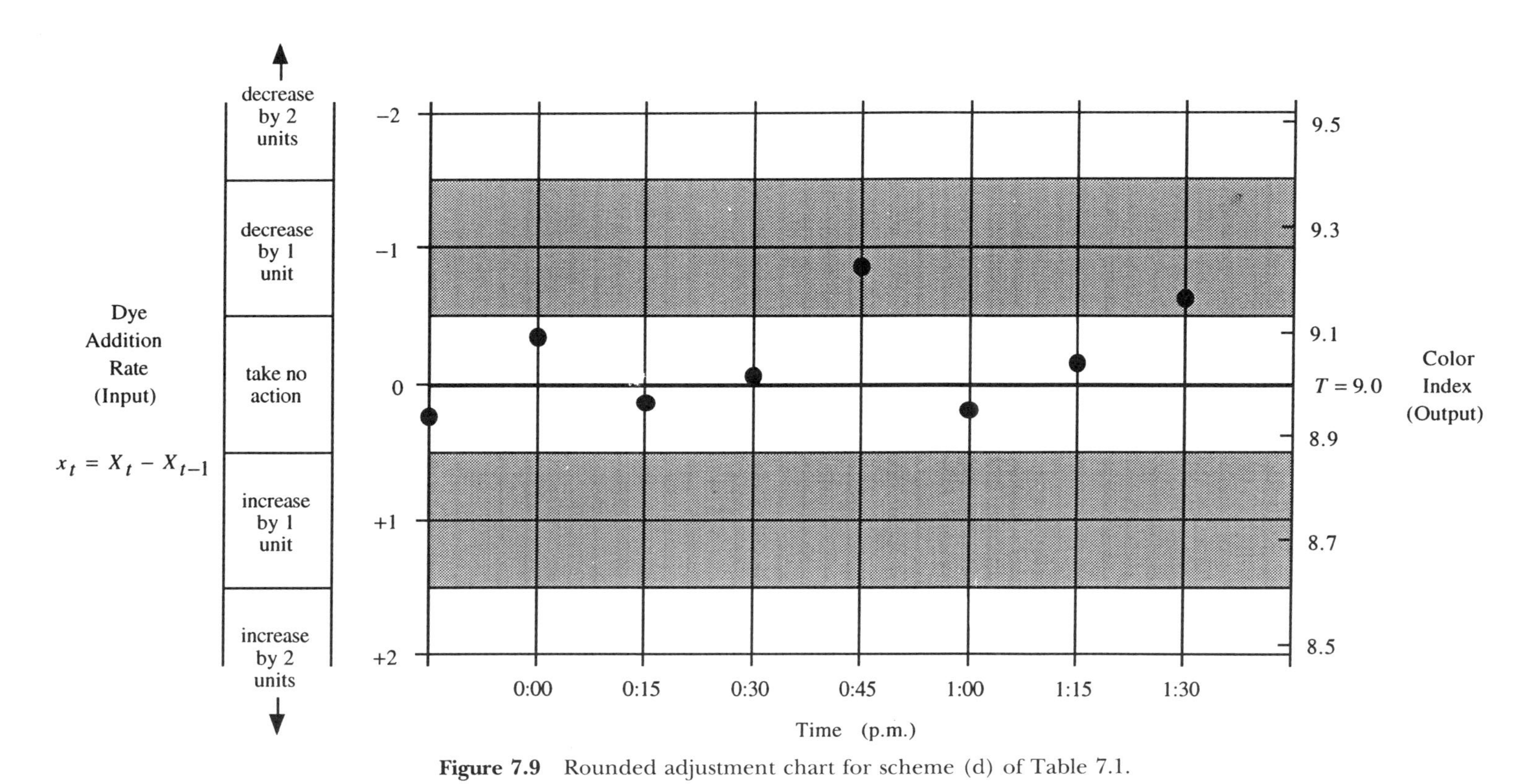

Figure 7.9 Rounded adjustment chart for scheme (d) of Table 7.1.

standard deviation of the output error or slightly less. The chart accords with this recommendation since, for the dye example, σ_a was estimated to be about 0.25. Thus σ_e for scheme (d) would be about $0.25 \times 1.04 = 0.26$. More generally, when a scheme of this kind is used, you can define "one unit of change" to approximate this rule. Figure 7.7e shows the input and output to a rounded scheme based on that in Figure 7.7d in which, as before, adjustments are computed using the empirically chosen values $G = 0.3$ and $P = 0$, and then rounded. Use of the rounded chart implies that repeated process adjustment is inconvenient or is associated with additional fixed cost. However, as we see in the next chapter, when there are costs that are additional to the cost of being off-target it may be better to use bounded adjustment charts discussed there, which take account of such costs explicitly.

7.7 PI SCHEMES FOR PROCESSES WITH DEAD TIME

A first-order dynamic model—when necessary with one or two periods of dead time—frequently provides adequate representation of system inertia. For such systems, suitably chosen integral schemes or PI schemes can provide control with rather small inflation of the output error from the theoretical MMSE value and greatly reduced need for manipulation.

Figures 7.10 and 7.11 show the performance charts corresponding to $P = 0$ and $P = -0.25$ respectively when the system has a pure delay (dead time) of one unit time interval ($d_0 = 1$) and first-order dynamics with parameter δ_0. The output standard deviation corresponding to MMSE control is then $\sigma_a\sqrt{1 + \lambda_0^2}$ rather than σ_a. This yields $1.02\sigma_a$, $1.08\sigma_a$, or $1.17\sigma_a$ for $\lambda = 0.2$, 0.4, or 0.6, respectively. Thus while the values of σ_e are now larger than when there is no dead time, the reverse may sometimes be true for the factor $\sigma_e/(\sigma_a\sqrt{1 + \lambda_0^2})$ by which the output standard deviation σ_e exceeds that obtained from the MMSE scheme.

For example, suppose that there is no dead time ($d_0 = 0$), $g = 1$, $\lambda_0 = 0.4$, $\sigma_a = 1$, and $\delta_0 = 0.25$. Then a scheme using $G = 0.2$ and $P = 0$ provides $\sigma_e = 1.065$. By contrast, if there is a dead time of one time interval ($d_0 = 1$) and, as before, $g = 1$, $\lambda_0 = 0.4$, $\sigma_a = 1$, and $\delta_0 = 0.25$, then the same scheme with $G = 0.2$ and $P = 0$ leads to $\sigma_e = 1.121$ (> 1.065), but $1.121/\sqrt{1 + 0.16} = 1.041$ is smaller than 1.065. The same calculation

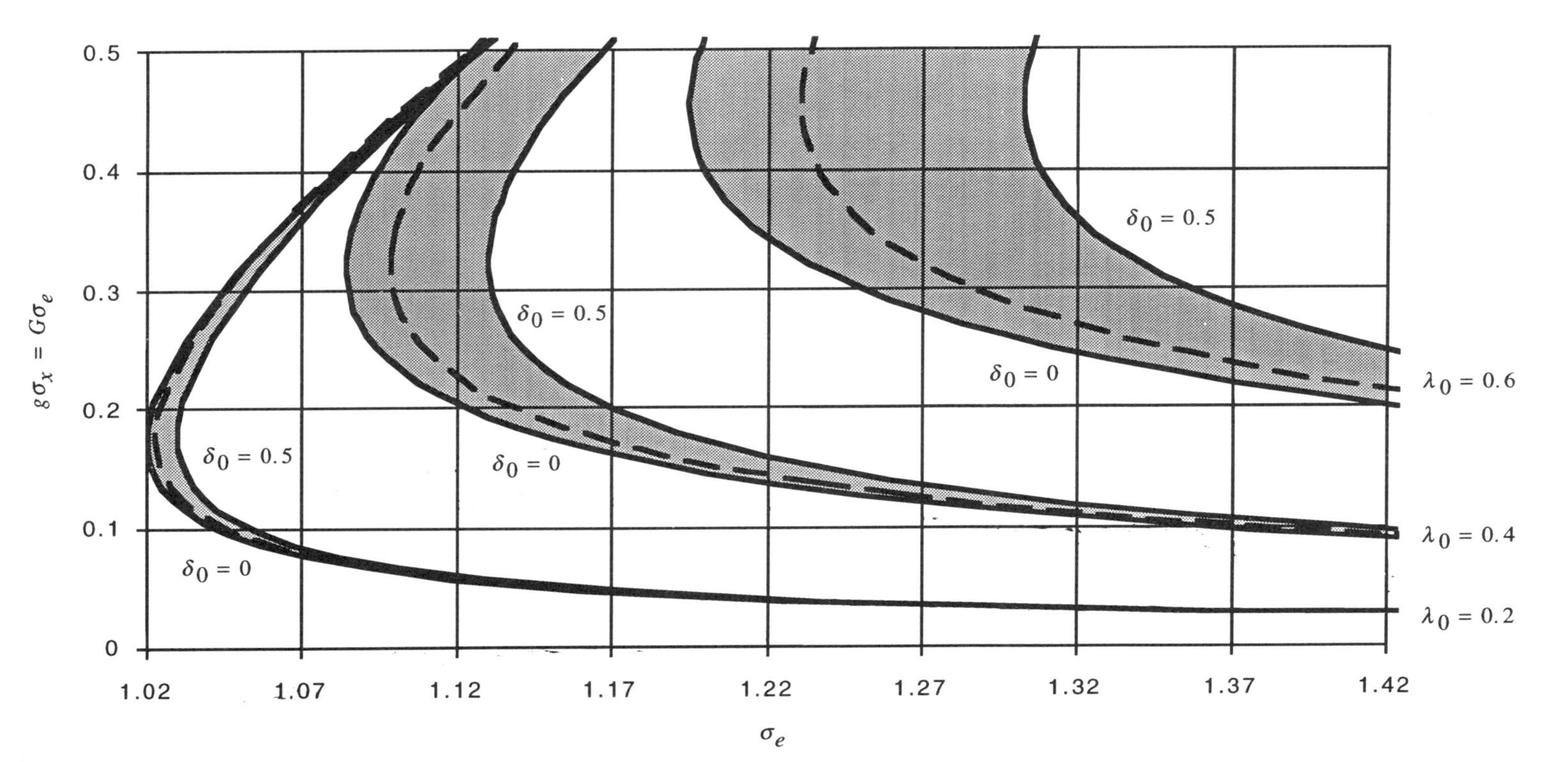

Figure 7.10 Chart for choosing G with $P = 0$ when the process has one unit of dead time.

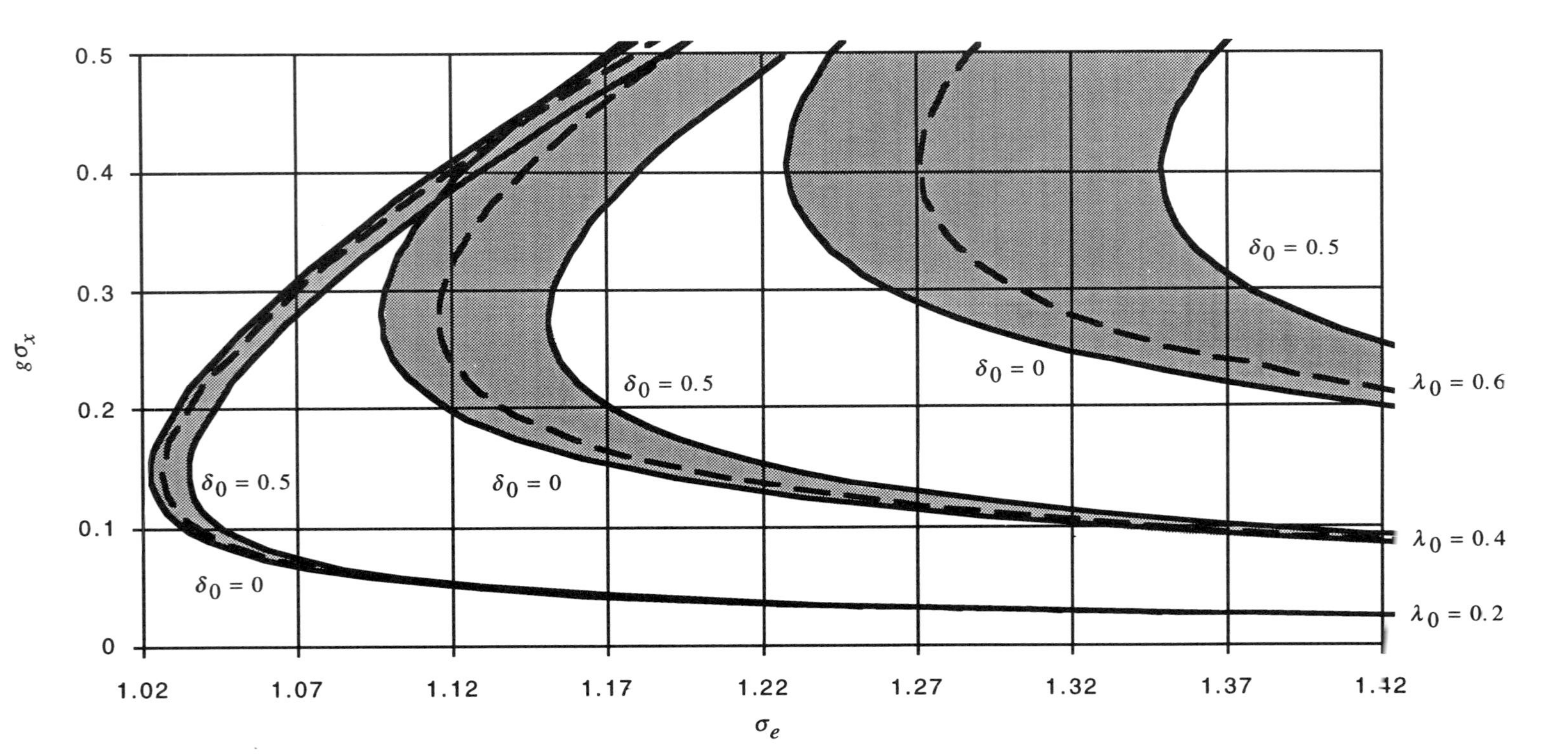

Figure 7.11 Chart for choosing G with $P = -0.25$ when the process has one unit of dead time.

using $G = 0.3$ and $P = 0$ yields $\sigma_e = 1.019$ when $d_0 = 0$ and $\sigma_e = 1.099$ (> 1.019) when $d_0 = 1$, but now $1.099/\sqrt{1 + 0.16} = 1.020$ is larger than 1.019. Therefore the relative inefficiency (compared to the MMSE scheme) caused by using an empirically chosen scheme when there is dead time may be larger or smaller than when there is no dead time.

Example of the Control of a Process with Dead Time

As an example of discrete feedback control for a process with dead time, we consider an application by Fearn and Maris (1991). Their objective was to control gluten addition in a flour mill so as to maintain the gluten content of the outgoing flour as closely as possible to a fixed target value. From extensive preliminary runs these authors showed that the disturbance (the gluten in the incoming flour) was closely represented by an IMA process with $\lambda_0 = 0.25$. Also, they found that the inertia of the system was modeled by a simple delay of one time period so that $d_0 = 1$, and $\delta_0 = 0$. They scaled the input so that $g = 1$, and they used the result of Equation (7.17) to obtain the minimum mean square error adjustment equation

$$x_t = -0.25\{x_{t-1} + e_t\}. \tag{7.21}$$

The authors found this scheme to be satisfactory. However, alternative schemes are possible that can yield considerable reduction in the manipulation standard deviation for small increases in the output standard deviation (see Appendix 7C and Box and Luceño, 1995).

For illustration, two such schemes are shown below. The inflation and reduction factors for σ_e and σ_x are shown using the MMSE scheme (7.21) for reference.

Scheme	Inflation Factor for σ_e	Reduction Factor for σ_x
MMSE [Equation (7.21)]	1.000	1.000
PI: $G = 0.2$, $P = 0.0$	1.001	0.826
PI: $G = 0.15$, $P = -0.25$	1.014	0.535

Implications of Negative *P* Values

As we have seen, optimal constrained PI schemes frequently call for negative values of P and hence for adjustments in which control action is determined by interpolation between e_t and e_{t-1} rather than by the more unstable extrapolation.

Now the adjustment equation (7.11) can be written in the form (see Appendix 7A)

$$g\tilde{x}_t = -Ge_t \tag{7.22}$$

with

$$\tilde{x}_t = (1 - \delta)\{x_t + \delta x_{t-1} + \delta^2 x_{t-2} + \cdots\} \tag{7.23}$$

an EWMA of current and previous adjustments $x_t, x_{t-1}, \ldots$. Also, since $\delta = P/(1 + P)$, when P is negative so is the smoothing parameter δ and, in particular, when $P = -0.25$ then $\delta = -0.33$. Now when an EWMA is used for exponential smoothing, the value of the smoothing parameter (δ in this case) is usually thought of as positive. This ensures that the weights

$$w_1 = 1 - \delta, \quad w_2 = (1 - \delta)\delta, \quad w_3 = (1 - \delta)\delta^2, \ldots \tag{7.24}$$

applied to current and previous values of the series are all positive.

However, you will see from Equation (7.24) that, for negative values of δ, the weights $w_1, w_2, \ldots$ will *alternate* in sign. Figure 7.12 shows a graph of P versus δ with specimen weightings of EWMAs for the specific values $\delta = 0.66$, 0.50, 0.00, and -0.33 corresponding to the values $P = 2$, 1, 0, and -0.25. For values of P between 0 and -0.25, particularly useful for constrained PI schemes, the weights w_3, w_4, $w_5, \ldots$ are small and if we ignore them the adjustment equation (7.22) is approximated by

$$g(1 - \delta)(x_t + \delta x_{t-1}) = -Ge_t,$$

or

$$x_t = -\delta x_{t-1} - \frac{G}{g(1 - \delta)} e_t. \tag{7.25}$$

Scheme (c) in Table 7.1, for example, which is the optimal constrained PI scheme with $P = -0.25$ and can be written

$$x_t = -1.95e_t - 0.65e_{t-1}, \tag{7.26}$$

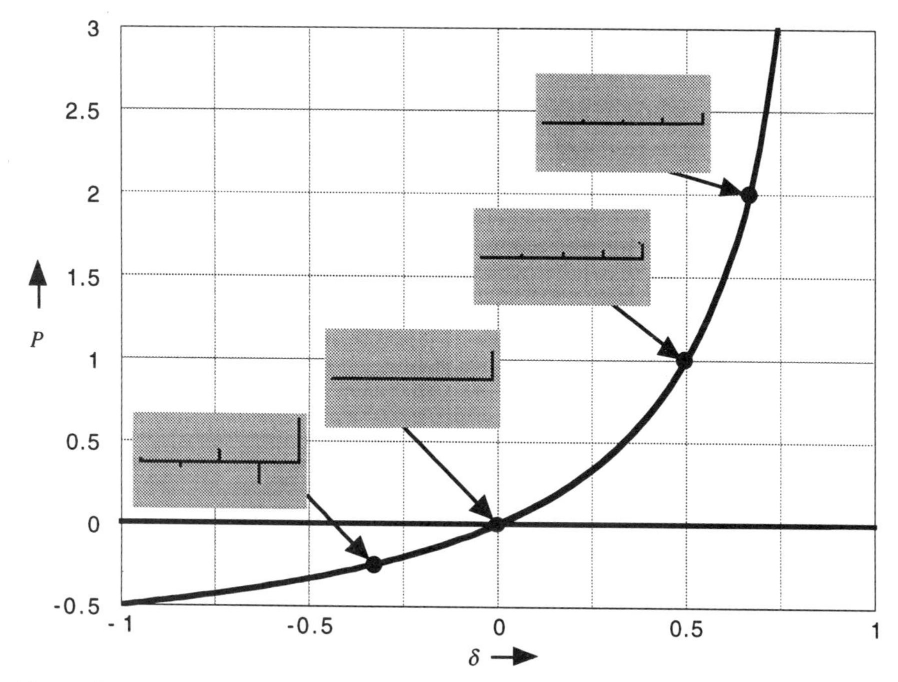

Figure 7.12 P plotted against δ with specimen weight functions for $\delta = 0.66, 0.50, 0.00, -0.33$ ($P = 2, 1, 0, -0.25$).

is closely approximated by

$$x_t = 0.33x_{t-1} - 1.95e_t. \tag{7.27}$$

Thus the implication of a negative value for P is that the current adjustment x_t, as well as depending on e_t, amplifies the adjustment x_{t-1} made previously.

7.8 PROCESS MONITORING AND PROCESS ADJUSTMENT

As mentioned earlier, problems of process adjustment have sometimes been confused with those for process monitoring. However, as was made clear both by Shewhart and by Deming, the purpose of, for example, a Shewhart chart is to signal the possibility of a special cause, which it may be possible to track down and remove once and for all. For this purpose, it makes sense to do nothing unless a deviation from the norm occurs that is so large as to have a very small probability of being due to chance.

This policy is appropriate in a context where the process is supposed to be usually in a *stable* state but from time to time a, possibly removable, problem occurs that signals its presence by the occurrence of abnormal data. By contrast, feedback adjustment is appropriate when the normal state of the process if no control were applied would be one of wandering *instability*. Such instability occurs, for example, because of the variation in naturally occurring feedstocks (oil, wood, wool, sewage, etc.) and because of uncontrolled environmental variables (e.g., ambient temperature) and for many other reasons. These causes are sometimes known but if they cannot be economically removed they must be compensated for. It is of interest to consider the consequences of (1) applying feedback adjustment to a process that is in fact in a perfect state of control, and (2) using a Shewhart chart to *adjust* a wandering unstable process.

7.9 FEEDBACK ADJUSTMENT APPLIED TO A PROCESS THAT IS IN A PERFECT STATE OF CONTROL

Suppose feedback control of the form

$$gx_t = -Ge_t$$

is applied to a responsive process that is *already in a perfect state of control* with the deviations e_t from target forming a random series exemplified by the white noise series a_t (or, what is equivalent, represented by the IMA model with the nonstationarity parameter λ_0 exactly equal to 0). Then Equation (6.12) shows that, because of this feedback control, the process standard deviation would be increased by a factor $\sqrt{2/(2-G)}$. For example, if as was recommended in Section 6.9, in the absence of special knowledge about the process, we used a feedback scheme with $G = 0.2$ on a process already in perfect control, then the output standard deviation would be increased by a factor of 1.054—that is, by only slightly more than 5%. For this modest premium we would gain the insurance that if, in fact, the system were *not* in perfect control, then it would be adjusted by the feedback scheme.

7.10 USING A SHEWHART CHART TO ADJUST AN UNSTABLE PROCESS

As we have seen, the Shewhart chart is a tool for process monitoring. It can signal the possibility of special cause, which may be assignable and possibly permanently removable. Nevertheless, this chart has sometimes mistakenly been used for process regulation—to signal the need for process adjustment. The rationale for doing this is often stated somewhat as follows: "A process should be adjusted only when there is a demonstrated need for change, signaled, for example, by a point or points falling outside the control limits." The purpose of process regulation, however, is not to minimize the probability of making an adjustment when "no real process change has occurred" but to minimize variation in the product. The policy described above can *introduce* a great deal of unnecessary variation. [See, for example, scheme(f) in Table 7.1 and Figure 7.7.] The above rationale is obviously inappropriate when no additional cost is incurred by adjustment—where, for example, an operator able to apply control manually, or special equipment able to do so automatically, must be kept available in any case. At first sight, however, it appears reasonable when there is an appreciable cost associated with making an adjustment—when, for example, to make a change it is necessary to stop the machine and replace an expensive tool. However, if cost is really the issue then it should be taken account of explicitly. This is done in the next two chapters where feedback schemes are introduced that minimize overall cost arising from (1) being off target, (2) making a process change, and (3) sampling and testing. The limits for the bounded adjustment charts developed to put these schemes into effect have nothing to do with control limits designed to give small probabilities of exceding the limits when the process is "in control." They are determined by the relative costs of (1), (2), and (3) above.

APPENDIX 7A EQUIVALENCE OF EQUATIONS 7.11 AND 7.12

Consider Equation (7.12),

$$g\tilde{X}_t = \tilde{z}_t, \qquad (7A.1)$$

where $\tilde{X}_t$ is an EWMA of $X_t, X_{t-1}, \ldots$ with smoothing constant δ and $\tilde{z}_t$ is an EWMA of $z_t, z_{t-1}, \ldots$ with smoothing constant $1 - G$.

From Equation (7A.1)

$$g(\tilde{X}_t - \tilde{X}_{t-1}) = g\tilde{x}_t = -(\tilde{z}_t - \tilde{z}_{t-1}).$$

Then Equation (4A.2) from Appendix 4A gives $g\tilde{x}_t = -Ge_t$. Hence

$$g(\tilde{x}_t - \delta\tilde{x}_{t-1}) = -G(e_t - \delta e_{t-1}). \tag{7A.2}$$

Also, the updating formula $\tilde{x}_t = (1 - \delta)x_t + \delta\tilde{x}_{t-1}$ can be written

$$\tilde{x}_t - \delta\tilde{x}_{t-1} = (1 - \delta)x_t, \tag{7A.3}$$

according to Equation (4A.1) from Appendix 4A.

Then combining Equations (7A.2) and (7A.3) gives

$$g(1 - \delta)x_t = -G(e_t - \delta e_{t-1}),$$

so that

$$gx_t = \frac{-G}{1 - \delta}(e_t - \delta e_{t-1}),$$

which is Equation (7.11) and is equivalent to discrete PI control.

APPENDIX 7B EFFECT OF ERROR IN ADJUSTMENT

Let the disturbance z_t be modeled according to the IMA process

$$z_t - z_{t-1} = a_t - \theta\, a_{t-1}, \tag{7B.1}$$

the first-order dynamic system be represented by

$$Y_t - \delta Y_{t-1} = g\,(1 - \delta)X_{t-1}, \tag{7B.2}$$

and the adjustment equation (including the adjustment errors u_t) be given by

$$g(x_t + u_t) = -Ge_t. \tag{7B.3}$$

Then, combining Equations (7B.1), (7B.2), and (7B.3), the error at the output e_t, given by

$$e_t = z_t + Y_t,$$

can be shown (after some algebra) to satisfy

$$\begin{aligned} e_t - \{1 + \delta - G(1 - \delta)\}e_{t-1} - \{-\delta\}e_{t-2} \\ = a_t - (\delta + \theta)\, a_{t-1} + \delta\theta\, a_{t-2} - g(1 - \delta)u_{t-1}. \end{aligned} \tag{7B.4}$$

In particular, when there are no adjustment errors ($u_t = 0$ at any time t), the errors $\{e_t^{(0)}\}$ at the output satisfy

$$e_t^{(0)} - \{1 + \delta - G(1 - \delta)\}\, e_{t-1}^{(0)} - \{-\delta\} e_{t-2}^{(0)}$$
$$= a_t - (\delta + \theta)\, a_{t-1} + \delta\theta\, a_{t-2}. \qquad (7B.5)$$

Equations (7B.4) and (7B.5) lead to (again after some algebra)

$$\text{var}(e_t) = \text{var}(e_t^{(0)}) + \frac{(1 + \delta) g^2\, \text{var}(u_t)}{2G(1 + \delta) - G^2(1 - \delta)}. \qquad (7B.6)$$

Now if $\text{var}(u_t) = k^2\, \text{var}\,(x_t^{(0)})$, where the $x_t^{(0)}$ terms are the adjustments that would be made if there were no adjustment errors, and satisfy $gx_t^{(0)} = -Ge_t^{(0)}$, we get

$$g^2\, \text{var}(x_t^{(0)}) = G^2\, \text{var}(e_t^{(0)})$$

and then

$$g^2\, \text{var}(u_t) = g^2 k^2\, \text{var}(x_t^{(0)}) = k^2 G^2\, \text{var}(e_t^{(0)}). \qquad (7B.7)$$

Then Equations (7B.6) and (7B.7) yield

$$\frac{\text{var}(e_t)}{\text{var}(e_t^{(0)})} = 1 + k^2 \frac{(1 + \delta) G}{2(1 + \delta) - G(1 - \delta)}.$$

APPENDIX 7C CHOICES FOR G AND P TO ATTAIN OPTIMAL CONSTRAINED PI CONTROL FOR VARIOUS VALUES OF λ_0 AND δ_0 WITH $d_0 = 0$ AND $d_0 = 1$

Tables 7C.1 and 7C.2 are taken from Box and Luceño (1995). They give the values of G, P, $\text{var}(e_t)/\sigma_a^2$, and $\text{var}(gx_t)/\sigma_a^2$ for a variety of alternative optimal constrained PI control schemes that have been obtained minimizing the expression (7.18) for different values of α. These tables may be used to choose a suitable constrained PI scheme by consideration of the resulting input and output variances. Table 7C.1 with $d_0 = 0$ is used for systems with no dead time. Table 7C.2 with $d_0 = 1$ is for systems with one period of dead time. This table also shows the input variance for the optimal unconstrained scheme $\text{Var}_U(gx_t)/\sigma_a^2$. The output variance for the optimal unconstrained scheme is given by $\text{Var}_U(e_t)/\sigma_a^2 = 1 + \lambda^2$. For intermediate values of the parameters linear interpolation is sufficient to find approximate values. The rationale for the choice of the α values is discussed in Appendix B of Box and Luceño (1995).

Problem 7.1 Suppose you wish to devise a feedback scheme to maintain the product temperature as close as possible to 70°C. Data are available every hour and analysis of past records suggests that the system disturbance can be approximated by an IMA model with $\lambda = 0.4$ and $\sigma_a = 0.6$. Also, the process engineers tell you that the inertia of the adjustment process is such that about half the eventual effect of a step change is experienced at the output in one hour and that $g = 2.8$. Stating any assumptions you make:

(a) Devise a feedback chart, with $P = 0$, that produces the smallest output standard deviation σ_e.

(b) Supposing that a ratio σ_e/σ_a of 1.10 can be tolerated, devise a scheme with $P = 0$ producing the smallest adjustment standard deviation σ_x.

(c) and **(d)** Repeat the above for charts with $P = -0.25$.

(e) Make a table showing the characteristics of the four schemes. Which would you choose?

(f) Describe precisely how you would construct the four feedback control charts for schemes (a), (b), (c), and (d). □

TABLE 7C.1. Values of G, P, $\mathrm{Var}(e_t)/\sigma_a^2$, and $\mathrm{Var}(gx_t)/\sigma_a^2$ for the Optimal PI Scheme when $d_0 = 0$ and $\delta_0 = 0.0\ (0.1)\ 0.5$, for Several Values of α and λ_0

	$\delta_0 = 0.0$						$\delta_0 = 0.1$						$\delta_0 = 0.2$					
$\lambda_0 =$	0.200	0.300	0.400	0.500	0.600	1.000	0.200	0.300	0.400	0.500	0.600	1.000	0.200	0.300	0.400	0.500	0.600	1.000
$G =$	0.200	0.300	0.400	0.500	0.600	1.000	0.200	0.300	0.400	0.500	0.600	1.000	0.200	0.300	0.400	0.500	0.600	1.000
$P =$	0.000	0.000	0.000	0.000	0.000	0.000	0.111	0.111	0.111	0.111	0.111	0.111	0.250	0.250	0.250	0.250	0.250	0.250
$\mathrm{Var}(e_t)/\sigma_a^2 =$	1.000	1.000	1.000	1.000	1.000	1.000	1.000	1.000	1.000	1.000	1.000	1.000	1.000	1.000	1.000	1.000	1.000	1.000
$\mathrm{Var}(gx_t)/\sigma_a^2 =$	0.040	0.090	0.160	0.250	0.360	1.000	0.050	0.112	0.200	0.312	0.449	1.247	0.065	0.146	0.260	0.406	0.585	1.625
	0.181	0.261	0.336	0.406	0.473	0.698	0.184	0.267	0.344	0.417	0.485	0.715	0.187	0.272	0.352	0.427	0.498	0.733
	−0.221	−0.194	−0.167	−0.140	−0.113	0.000	−0.161	−0.133	−0.104	−0.076	−0.047	0.078	−0.080	−0.050	−0.020	0.010	0.041	0.177
	1.004	1.009	1.016	1.025	1.036	1.100	1.004	1.009	1.016	1.025	1.036	1.100	1.004	1.009	1.016	1.025	1.036	1.100
	0.022	0.049	0.087	0.135	0.194	0.537	0.025	0.057	0.101	0.157	0.226	0.628	0.030	0.068	0.122	0.190	0.275	0.763
	0.170	0.240	0.305	0.364	0.418	0.592	0.174	0.247	0.314	0.375	0.431	0.608	0.178	0.254	0.324	0.387	0.446	0.626
	−0.284	−0.252	−0.219	−0.185	−0.151	0.000	−0.244	−0.209	−0.173	−0.136	−0.099	0.066	−0.187	−0.149	−0.110	−0.070	−0.029	0.150
	1.007	1.017	1.031	1.050	1.072	1.200	1.007	1.017	1.032	1.050	1.072	1.200	1.008	1.018	1.032	1.050	1.072	1.200
	0.018	0.039	0.069	1.106	0.152	0.420	0.020	0.044	0.077	0.120	0.173	0.480	0.023	0.051	0.091	0.142	0.204	0.568
	0.151	0.209	0.261	0.306	0.347	0.465	0.156	0.216	0.269	0.317	0.358	0.478	0.161	0.224	0.279	0.329	0.372	0.494
	−0.347	−0.312	−0.275	−0.235	−0.193	0.000	−0.326	−0.287	−0.246	−0.203	−0.157	0.052	−0.295	−0.252	−0.207	−0.159	−0.108	0.119
	1.015	1.035	1.063	1.099	1.144	1.400	1.015	1.035	1.063	1.099	1.144	1.400	1.015	1.035	1.063	1.100	1.144	1.400
	0.014	0.029	0.050	0.077	0.110	0.303	0.015	0.032	0.055	0.085	0.122	0.337	0.017	0.036	0.062	0.096	0.138	0.385
	0.137	0.187	0.230	0.267	0.299	0.388	0.141	0.193	0.237	0.276	0.309	0.398	0.146	0.200	0.246	0.286	0.320	0.411
	−0.380	−0.345	−0.307	−0.265	−0.219	0.000	−0.366	−0.329	−0.287	−0.241	−0.191	0.043	−0.347	−0.306	−0.260	−0.209	−0.154	0.099
	1.023	1.053	1.095	1.150	1.216	1.600	1.022	1.052	1.095	1.149	1.215	1.600	1.022	1.052	1.094	1.149	1.216	1.600
	0.011	0.023	0.040	0.061	0.087	0.240	0.012	0.025	0.043	0.066	0.095	0.262	0.013	0.028	0.048	0.074	0.106	0.293
	0.126	0.169	0.206	0.238	0.264	0.333	0.129	0.174	0.212	0.245	0.272	0.342	0.134	0.181	0.220	0.254	0.082	0.352
	−0.399	−0.366	−0.328	−0.285	−0.237	0.000	−0.391	−0.354	−0.313	−0.266	−0.215	0.037	−0.378	−0.338	−0.293	−0.242	−0.185	0.085
	1.031	1.072	1.128	1.200	1.288	1.800	1.031	1.071	1.127	1.199	1.288	1.800	1.030	1.070	1.126	1.199	1.287	1.800
	0.010	0.020	0.033	0.051	0.072	0.200	0.010	0.021	0.036	0.055	0.078	0.216	0.011	0.023	0.039	0.060	0.086	0.237
	0.116	0.155	0.187	0.214	0.237	0.293	0.119	0.159	0.193	0.220	0.243	0.300	0.123	0.165	0.199	0.228	0.252	0.308
	−0.413	−0.380	−0.343	−0.299	−0.250	0.000	−0.406	−0.372	−0.331	−0.285	−0.232	0.033	−0.398	−0.360	−0.316	−0.265	−0.207	0.075
	1.040	1.091	1.161	1.251	1.361	2.000	1.039	1.090	1.160	1.250	1.360	2.000	1.038	1.088	1.158	1.249	1.359	2.000
	0.008	0.017	0.029	0.044	0.062	0.172	0.009	0.018	0.030	0.047	0.066	0.183	0.009	0.019	0.033	0.050	0.072	0.199
	0.098	0.129	0.153	0.173	0.189	0.225	0.100	0.132	0.157	0.177	0.194	0.230	0.103	0.136	0.162	0.183	0.199	0.236
	−0.433	−0.403	−0.366	−0.323	−0.272	0.000	−0.429	−0.398	−0.360	0.313	−0.259	0.025	−0.425	−0.391	−0.350	−0.301	−0.242	0.057
	1.065	1.141	1.245	1.379	1.542	2.500	1.063	1.139	1.243	1.377	1.541	2.500	1.061	1.136	1.241	1.375	1.540	2.500
	0.006	0.012	0.021	0.032	0.046	0.127	0.006	0.013	0.022	0.034	0.048	0.134	0.007	0.014	0.024	0.036	0.052	0.142

(Continued)

TABLE 7C.1. *(Continued)*

	$\delta_0 = 0.3$						$\delta_0 = 0.4$						$\delta_0 = 0.5$					
$\lambda_0 =$	0.200	0.300	0.400	0.500	0.600	1.000	0.200	0.300	0.400	0.500	0.600	1.000	0.200	0.300	0.400	0.500	0.600	1.000
$G =$	0.200	0.300	0.400	0.500	0.600	1.000	0.200	0.300	0.400	0.500	0.600	1.000	0.200	0.300	0.400	0.500	0.600	1.000
$P =$	0.429	0.429	0.429	0.429	0.429	0.429	0.667	0.667	0.667	0.667	0.667	0.667	1.000	1.000	1.000	1.000	1.000	1.000
$\mathrm{Var}(e_t)/\sigma_a^2 =$	1.000	1.000	1.000	1.000	1.000	1.000	1.000	1.000	1.000	1.000	1.000	1.000	1.000	1.000	1.000	1.000	1.000	1.000
$\mathrm{Var}(gx_t)/\sigma_a^2 =$	0.089	0.200	0.356	0.556	0.801	2.224	0.129	0.290	0.516	0.806	1.160	3.222	0.200	0.450	0.800	1.250	1.800	5.000
	0.190	0.278	0.360	0.438	0.511	0.752	0.193	0.283	0.368	0.449	0.524	0.773	0.195	0.287	0.375	0.458	0.536	0.793
	0.035	0.065	0.096	0.127	0.160	0.308	0.202	0.230	0.260	0.292	0.325	0.486	0.450	0.474	0.502	0.532	0.566	0.742
	1.004	1.009	1.017	1.026	1.037	1.100	1.004	1.010	1.017	1.026	1.037	1.100	1.004	1.010	1.017	1.026	1.037	1.100
	0.039	0.088	0.156	0.245	0.352	0.974	0.054	0.122	0.217	0.338	0.485	1.325	0.085	0.189	0.332	0.513	0.730	1.952
	0.182	0.262	0.334	0.401	0.462	0.648	0.186	0.269	0.345	0.415	0.478	0.671	0.190	0.276	0.356	0.429	0.495	0.696
	−0.105	−0.064	−0.022	0.022	0.066	0.263	0.019	0.063	0.108	0.154	0.202	0.418	0.212	0.257	0.303	0.352	0.404	0.644
	1.008	1.019	1.033	1.051	1.073	1.200	1.009	1.020	1.034	1.053	1.075	1.201	1.009	1.020	1.035	1.054	1.076	1.201
	0.028	0.062	0.111	0.175	0.253	0.703	0.036	0.082	0.147	0.232	0.334	0.924	0.053	0.121	0.216	0.338	0.486	1.319
	0.166	0.233	0.291	0.343	0.388	0.514	0.172	0.243	0.305	0.359	0.407	0.537	0.178	0.253	0.319	0.377	0.427	0.563
	−0.249	−0.201	−0.150	−0.096	−0.040	0.209	−0.177	−0.122	−0.064	−0.004	0.059	0.335	−0.058	0.005	0.071	0.140	0.211	0.521
	1.015	1.036	1.064	1.101	1.146	1.401	1.017	1.038	1.067	1.104	1.148	1.402	1.018	1.040	1.070	1.108	1.153	1.404
	0.019	0.041	0.073	0.114	0.164	0.457	0.022	0.050	0.090	0.142	0.205	0.573	0.029	0.067	0.122	0.193	0.280	0.778
	0.152	0.209	0.257	0.299	0.335	0.427	0.159	0.219	0.271	0.315	0.352	0.447	0.166	0.231	0.287	0.333	0.373	0.472
	−0.319	−0.272	−0.220	−0.163	−0.102	0.174	−0.274	−0.220	−0.160	−0.095	−0.026	0.281	−0.199	−0.133	−0.062	0.014	0.094	0.440
	1.022	1.052	1.095	1.150	1.217	1.601	1.023	1.054	1.098	1.153	1.221	1.602	1.025	1.058	1.103	1.159	1.227	1.606
	0.015	0.031	0.055	0.085	0.122	0.339	0.017	0.037	0.065	0.102	0.147	0.411	0.021	0.046	0.083	0.131	0.191	0.536
	0.139	0.189	0.230	0.265	0.294	0.366	0.146	0.199	0.243	0.279	0.309	0.383	0.154	0.211	0.258	0.297	0.329	0.405
	−0.359	−0.315	−0.264	−0.207	−0.144	0.150	−0.330	−0.278	−0.219	−0.154	−0.082	0.242	−0.279	−0.217	−0.146	−0.069	0.014	0.381
	1.030	1.070	1.126	1.199	1.288	1.801	1.030	1.071	1.128	1.202	1.292	1.803	1.032	1.074	1.133	1.208	1.299	1.807
	0.012	0.025	0.044	0.067	0.097	0.269	0.014	0.029	0.051	0.079	0.114	0.318	0.016	0.036	0.063	0.098	0.143	0.401
	0.128	0.172	0.208	0.238	0.262	0.320	0.135	0.181	0.219	0.250	0.276	0.334	0.143	0.193	0.234	0.267	0.293	0.354
	−0.384	−0.343	−0.293	−0.237	−0.173	0.132	−0.364	−0.316	−0.259	−0.195	−0.123	0.213	−0.329	−0.271	−0.230	−0.127	−0.043	0.336
	1.038	1.087	1.158	1.249	1.360	2.001	1.038	1.088	1.159	1.251	1.363	2.003	1.039	1.090	1.163	1.257	1.370	2.008
	0.010	0.021	0.037	0.056	0.080	0.222	0.012	0.024	0.042	0.064	0.092	0.258	0.013	0.029	0.050	0.078	0.113	0.318
	0.107	0.141	0.168	0.189	0.206	0.243	0.113	0.148	0.176	0.198	0.216	0.253	0.120	0.158	0.187	0.211	0.229	0.267
	−0.419	−0.382	−0.337	−0.283	−0.219	0.101	−0.409	−0.369	−0.317	−0.256	−0.186	0.163	−0.392	−0.344	−0.285	−0.214	−0.132	0.257
	1.060	1.134	1.239	1.374	1.539	2.501	1.058	1.132	1.238	1.374	1.540	2.503	1.057	1.132	1.239	1.377	1.546	2.508
	0.007	0.015	0.026	0.039	0.056	0.155	0.008	0.017	0.029	0.044	0.063	0.174	0.009	0.019	0.033	0.051	0.073	0.204

TABLE 7C.2. Values of G, P, $\mathrm{Var}(e_t)/\sigma_a^2$, and $\mathrm{Var}(gx_t)/\sigma_a^2$ for the Optimal PI Scheme when $d_0 = 1$ and $\delta_0 = 0.0\ (0.1)\ 0.5$, for Several Values of α and λ_0*

	$\delta_0 = 0.0$						$\delta_0 = 0.1$						$\delta_0 = 0.2$					
$\lambda_0 =$	0.200	0.300	0.400	0.500	0.600	1.000	0.200	0.300	0.400	0.500	0.600	1.000	0.200	0.300	0.400	0.500	0.600	1.000
$G =$	0.166	0.228	0.278	0.319	0.351	0.420	0.165	0.227	0.276	0.316	0.347	0.412	0.165	0.225	0.274	0.312	0.342	0.404
$P =$	0.193	0.279	0.355	0.423	0.483	0.651	0.317	0.407	0.487	0.558	0.620	0.797	0.470	0.564	0.649	0.724	0.791	0.980
$\mathrm{Var}(e_t)/\sigma_a^2 =$	1.040	1.090	1.162	1.256	1.374	2.137	1.040	1.091	1.163	1.258	1.378	2.158	1.040	1.091	1.164	1.260	1.382	2.180
$\mathrm{Var}(gx_t)/\sigma_a^2 =$	0.039	0.086	0.146	0.217	0.298	0.698	0.048	0.102	0.171	0.252	0.340	0.773	0.060	0.126	0.208	0.300	0.401	0.880
	0.164	0.224	0.273	0.313	0.344	0.411	0.164	0.224	0.272	0.311	0.341	0.405	0.164	0.223	0.271	0.309	0.338	0.398
	0.115	0.208	0.293	0.369	0.436	0.625	0.228	0.327	0.417	0.497	0.568	0.766	0.369	0.475	0.571	0.657	0.732	0.944
	1.040	1.091	1.163	1.257	1.375	2.139	1.040	1.091	1.163	1.259	1.379	2.160	1.040	1.092	1.164	1.261	1.383	2.182
	0.034	0.075	0.129	0.194	0.269	0.643	0.040	0.089	0.151	0.225	0.308	0.715	0.051	0.109	0.183	0.269	0.363	0.816
	0.150	0.202	0.243	0.275	0.301	0.356	0.152	0.204	0.246	0.279	0.304	0.358	0.154	0.207	0.249	0.282	0.307	0.359
	−0.176	−0.081	0.018	0.117	0.210	0.493	−0.112	−0.006	0.104	0.211	0.311	0.608	−0.029	0.093	0.215	0.332	0.440	0.754
	1.047	1.105	1.187	1.293	1.423	2.243	1.047	1.106	1.188	1.293	1.424	2.253	1.047	1.106	1.188	1.294	1.426	2.265
	0.018	0.040	0.072	0.112	0.160	0.416	0.020	0.046	0.082	0.128	0.182	0.466	0.024	0.054	0.097	0.150	0.213	0.534
	0.103	0.133	0.156	0.173	0.186	0.210	0.108	0.140	0.164	0.182	0.195	0.221	0.114	0.147	0.172	0.191	0.205	0.232
	−0.411	−0.362	−0.298	−0.219	−0.128	0.254	−0.398	−0.342	−0.271	−0.184	−0.086	0.314	−0.379	−0.315	−0.234	−0.137	−0.030	0.395
	1.084	1.190	1.340	1.534	1.770	3.147	1.080	1.183	1.329	1.517	1.745	3.076	1.078	1.178	1.321	1.504	1.726	3.016
	0.007	0.015	0.026	0.039	0.056	0.156	0.008	0.017	0.028	0.044	0.063	0.174	0.009	0.019	0.032	0.049	0.071	0.177
	0.052	0.064	0.072	0.077	0.081	0.087	0.056	0.069	0.078	0.084	0.088	0.094	0.060	0.075	0.084	0.091	0.095	0.103
	−0.467	−0.440	−0.400	−0.347	−0.279	0.094	−0.464	−0.435	−0.394	−0.338	−0.267	0.117	−0.460	−0.430	−0.385	−0.326	−0.251	0.147
	1.248	1.521	1.898	2.386	2.988	6.519	1.227	1.479	1.830	2.284	2.844	6.122	1.206	1.439	1.765	2.187	2.707	5.746
	0.002	0.005	0.008	0.013	0.018	0.050	0.003	0.005	0.009	0.014	0.020	0.056	0.003	0.006	0.010	0.016	0.023	0.063
$\mathrm{Var_U}(gx_t)/\sigma_a^2 =$	0.040	0.090	0.160	0.250	0.360	1.000	0.050	0.112	0.200	0.312	0.449	1.247	0.065	0.146	0.260	0.406	0.585	1.625

TABLE 7C.2. (*Continued*)

	$\delta_0 = 0.3$						$\delta_0 = 0.4$						$\delta_0 = 0.5$					
$\lambda_0 =$	0.200	0.300	0.400	0.500	0.600	1.000	0.200	0.300	0.400	0.500	0.600	1.000	0.200	0.300	0.400	0.500	0.600	1.000
$G =$	0.165	0.224	0.272	0.309	0.338	0.396	0.164	0.223	0.270	0.306	0.334	0.388	0.164	0.222	0.268	0.303	0.329	0.380
$P =$	0.664	0.764	0.855	0.937	1.010	1.217	0.920	1.029	1.129	1.221	1.303	1.534	1.275	1.398	1.512	1.617	1.712	1.980
$\text{Var}(e_t)/\sigma_a^2 =$	1.040	1.091	1.165	1.262	1.387	2.203	1.040	1.092	1.166	1.265	1.392	2.227	1.040	1.092	1.167	1.268	1.397	2.252
$\text{Var}(gx_t)/\sigma_a^2 =$	0.078	0.161	0.261	0.372	0.490	1.039	0.108	0.217	0.346	0.484	0.628	1.285	0.158	0.312	0.487	0.672	0.861	1.696
	0.164	0.223	0.270	0.306	0.335	0.392	0.163	0.222	0.268	0.304	0.331	0.385	0.163	0.221	0.266	0.301	0.327	0.378
	0.550	0.665	0.768	0.861	0.944	1.174	0.791	0.917	1.031	1.135	1.227	1.483	1.128	1.270	1.399	1.517	1.623	1.918
	1.041	1.092	1.166	1.263	1.388	2.205	1.041	1.092	1.167	1.266	1.393	2.229	1.041	1.093	1.168	1.269	1.398	2.254
	0.066	0.140	0.231	0.334	0.445	0.965	0.091	0.189	0.306	0.435	0.572	1.195	0.134	0.273	0.434	0.607	0.785	1.580
	0.155	0.209	0.251	0.284	0.309	0.359	0.157	0.211	0.253	0.286	0.310	0.359	0.158	0.212	0.255	0.287	0.311	0.358
	0.085	0.225	0.362	0.490	0.608	0.947	0.246	0.407	0.562	0.706	0.837	1.208	0.483	0.672	0.850	1.014	1.162	1.581
	1.048	1.107	1.189	1.296	1.429	2.281	1.049	1.108	1.190	1.298	1.432	2.298	1.049	1.109	1.192	1.300	1.435	2.317
	0.029	0.067	0.119	0.185	0.260	0.633	0.038	0.088	0.157	0.240	0.333	0.785	0.056	0.128	0.223	0.335	0.459	1.038
	0.120	0.155	0.181	0.201	0.215	0.243	0.126	0.163	0.191	0.211	0.226	0.255	0.132	0.172	0.200	0.221	0.237	0.266
	−0.352	−0.277	−0.187	−0.071	0.049	0.505	−0.311	−0.218	−0.104	0.025	0.163	0.661	−0.244	−0.124	0.019	0.175	0.336	0.893
	1.076	1.175	1.316	1.495	1.712	2.968	1.076	1.176	1.315	1.493	1.706	2.933	1.078	1.180	1.320	1.496	1.705	2.912
	0.010	0.021	0.036	0.057	0.082	0.229	0.011	0.024	0.043	0.068	0.099	0.275	0.014	0.030	0.054	0.086	0.126	0.350
	0.065	0.081	0.092	0.099	0.104	0.113	0.072	0.089	0.101	0.109	0.115	0.124	0.080	0.099	0.113	0.122	0.128	0.138
	−0.455	−0.422	−0.373	−0.309	−0.228	0.191	−0.448	−0.410	−0.355	−0.283	−0.193	0.257	−0.437	−0.391	−0.326	−0.242	−0.138	0.361
	1.186	1.402	1.704	2.097	2.582	5.398	1.167	1.368	1.651	2.019	2.471	5.085	1.152	1.340	1.608	1.956	2.383	4.822
	0.003	0.007	0.012	0.018	0.026	0.072	0.004	0.008	0.014	0.021	0.030	0.084	0.005	0.010	0.017	0.025	0.037	0.102
$\text{Var}_U(gx_t)/\sigma_a^2 =$	0.089	0.200	0.356	0.556	0.801	2.224	0.129	0.290	0.516	0.806	1.160	3.222	0.200	0.450	0.800	1.250	1.800	5.000

*The table also shows the input variance for the optimal unconstrained scheme $\text{Var}_U(gx_t)$. The output variance for the optimal unconstrained scheme is given by $\text{Var}_U(e_t)/\sigma_a^2 = 1 + \lambda^2 d_0$.

CHAPTER 8

Feedback Control When There Are Adjustment Costs

"Change is not made without inconvenience, even from worse to better."
RICHARD HOOKER quoted in preface to Samuel Johnson's Dictionary

8.1 INTRODUCTION

We mentioned that engineering process control (EPC) had been developed largely in the process industries, particularly the chemical industry where continuous compensation is necessary to cancel the effects of uncontrollable disturbances. By contrast, the development of statistical process control (SPC) had been more closely associated with the parts industries, where production of a nearly uniform product was often possible by standardization of procedures and raw materials and by setting up a system of repeated monitoring for detection and removal of sources of trouble.

While the two approaches reflect characteristics of their different origins, in almost any industry some knowledge of both SPC *and* EPC is necessary to achieve the best manufacturing practice. This is true more than ever now that products and processes previously unknown are continually being developed. For example, the electronics industry frequently employs hybrid methods of manufacture that at some stages resemble those of the chemical industry and at other stages those of the parts industries.

When continuous feedback control is used in the process industries usually the tacit assumption is that the only cost is the cost of variation about the target. However, particularly in the parts industries, additional costs must often be taken into account. For example, to make an adjustment it may be necessary to incur the cost of stopping a machine and/or of changing a tool. This does not, however, imply that *all* feedback control would be useless, only that feedback control of a different kind must be employed which takes such additional costs into account.

Types of control should not be dichotomized in terms of the industry in which they were developed but by the characteristics of the process to be controlled and, in particular, by the types and magnitudes of the costs involved. Thus substantial costs of making a change and in some cases of taking and testing a sample may occur in any industry and will affect the nature of the feedback control. We begin by considering feedback control schemes that are appropriate when there is a *fixed cost* associated with an adjustment.

8.2 BOUNDED FEEDBACK ADJUSTMENT

Chapter 6 explored the commonsense idea that an adjustment should be made proportional to the last error at the output. In particular, this yielded the discrete analog of integral control:

$$gX_t = k_I \sum_{i=1}^{t} e_i. \tag{8.1}$$

Suppose now that for a responsive system there is a fixed cost involved in making an adjustment, so that you would prefer to make adjustments less frequently, but without greatly increasing the error at the output. A policy that we later justify is still to use the control equation (8.1) but not to take action at each time period. Suppose that you last made an adjustment at some time r when the level of the compensating variable stood at X_r. Then to follow this policy you should not adjust again until at some time t the absolute difference[1] between X_t and X_r reached a predetermined fixed value, say, k.

Thus the control rule would be:

$$\text{Adjust as soon as } |X_t - X_r| \geq k.$$

In Chapter 6 it was shown that the control achieved by Equation (8.1) is equivalent to setting X_t so as to cancel an EWMA estimate

[1] The *absolute* value of $X_t - X_r$ is the difference without regard to sign. The vertical lines in $|X_t - X_r|$ mean "the absolute value of." The sign $\geq$ means "greater than or equal to."

$\hat{z}_{t+1}$ of the next value of z_{t+1} of the disturbance so that

$$X_t = -\left(\frac{1}{g}\right)\hat{z}_{t+1} \tag{8.2}$$

and

$$X_t - X_r = -\frac{1}{g}(\hat{z}_{t+1} - \hat{z}_{r+1}). \tag{8.3}$$

It follows that action should be taken as soon as

$$|\hat{z}_{t+1} - \hat{z}_{r+1}| \geq L, \tag{8.4}$$

where $L = kg$. We earlier defined the disturbance by $z_t = y_t - T$ so that Equation (8.4) may equally well be written in terms of the original data y_t:

$$|\hat{y}_{t+1} - \hat{y}_{r+1}| \geq L. \tag{8.5}$$

A *bounded* adjustment chart to put such a policy into effect is shown in Figure 8.1 (Box and Jenkins, 1963; Box, 1991b; Luceño 1993; BJR). For illustration, we show how the chart might be used to control the thickness series plotted in Figure 6.1. We suppose, as before, that the process is controlled by adjusting the deposition rate, that $g = 1.2$, $T = 80$, and that the EWMA used in the calculations has smoothing constant $\theta = 0.8$. The open circles shown on the chart are the values y'_t. The addition of a prime y' to y means that it is the actual measured thickness that would be observed taking into account all the previous *adjustments that had been applied*. The EWMAs $\tilde{y}'_t = \hat{y}'_{t+1}$ are shown by solid circles. For illustration, the value of L has here been set equal to 8 so that the boundaries are at 80 ± 8, that is, at 88 and 72. Looking at the chart, suppose initially the deposition rate is at some value X_0. This remains unchanged until time $t = 13$ when the forecasted value $\hat{y}'_{14} = 88.5$ falls outside the upper limit and the chart signals that a change is needed in the deposition rate that will reduce the thickness by $(88.5 - 80.0) = 8.5$. Since $g = 1.2$, the adjustment $X_{13} - X_0$ can be read directly off the left-hand scale, which is such that, for example, -10 units of adjustment corresponds to $+12$ units on the disturbance scale.

The updated EWMA is best calculated using the formula

$$\hat{y}'_{t+1} = \lambda y'_t + \theta \hat{y}'_t. \tag{8.6}$$

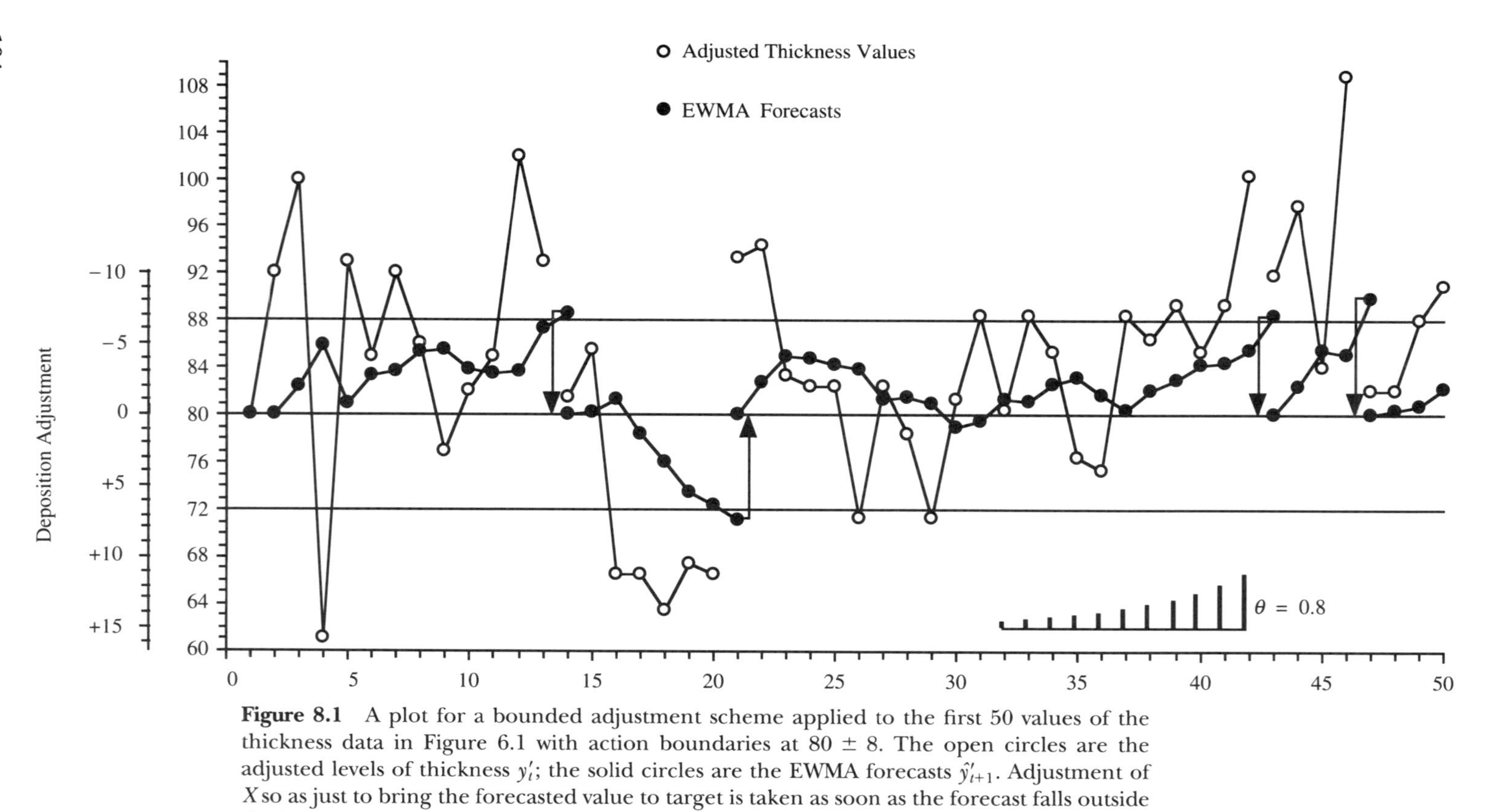

Figure 8.1 A plot for a bounded adjustment scheme applied to the first 50 values of the thickness data in Figure 6.1 with action boundaries at 80 ± 8. The open circles are the adjusted levels of thickness y_t'; the solid circles are the EWMA forecasts $\hat{y}_{t+1}'$. Adjustment of X so as just to bring the forecasted value to target is taken as soon as the forecast falls outside the boundary lines.

In this particular case, therefore, we obtain $\hat{y}'_{t+1}$ from $\hat{y}'_{t+1} = 0.2\hat{y}'_t + 0.8\hat{y}'_t$. This corresponds to an interpolation two-tenths of the way between the previous forecast $\hat{y}'_t$ and the present value y'_t. This interpolation can often be made to a sufficient approximation by eye. Alternatively, a graphical device introduced later may be used.

8.3 NUMERICAL CALCULATIONS FOR BOUNDED ADJUSTMENT

Although the process of adjustment can be carried out entirely graphically, for completeness, the corresponding calculations for the data of Figure 8.1 are set out in Table 8.1. The mechanics of the calculations are, initially at least, identical to those for the EWMA monitoring chart of Chapter 3. However, the immediate objective here is not to monitor the process in a search for removable aberrations, but to signal the need for adjustment to *compensate* irremovable disturbances.

The second column of Table 8.1 shows the unadjusted thickness y_t. The third column shows the computed value for the thickness y'_t after adjustments have been made. The fourth column shows the value $\hat{y}'_t$: the EWMA $\tilde{y}'_{t-1}$ calculated one step previously. The square brackets indicate that an adjustment has been made.

The fifth column of the table shows the effects $g(X_t - X_r)$ of the adjustments $X_t - X_r$ in the deposition rate called for by this scheme. Note that the first 13 values in this column are zero because no adjustment is made until after the 13th observation. In column six, the accumulated effects ν_t of these adjustments are shown (where ν is the greek letter *nu*). Then y'_t, the calculated thickness after adjustment, is given by

$$y'_t = y_t + \nu_t.$$

The final column shows the values of $e_t = y'_t - \hat{y}'_t$. The purpose of this column is explained later.

In this illustration the calculation has been initiated[2] by setting $\hat{y}'_1 = y'_1 = 80$. We see that at time $t = 13$ the forecasted value $\hat{y}'_{14}$ is 88.5 and is the first such value to fall outside the boundaries. As in Figure 8.1, therefore, immediately after observation 13, the mean

[2] A better starting value could have been obtained by the "back forecasting" method described in BJR.

Table 8.1 Bounded Adjustment Calculations for the First 50 Values of the Metallic Film Data Used in Figure 8.1

1	2	3	4	5	6	7
t	Unadjusted Series, y_t	Adjusted Series, y'_t	EWMA for Adjusted Series, $\hat{y}'_t$	Effects of Adjustments, $g(X_t - X_r)$	Accumulated Effects of Adjustments, ν_t	$e_t = y'_t - \hat{y}'_t$
1	80	80.0	80.0	0.0	0.0	0.0
2	92	92.0	80.0	0.0	0.0	12.0
3	100	100.0	82.4	0.0	0.0	17.6
4	61	61.0	85.9	0.0	0.0	−24.9
5	93	93.0	80.9	0.0	0.0	12.1
6	85	85.0	83.3	0.0	0.0	1.7
7	92	92.0	83.7	0.0	0.0	8.3
8	86	86.0	85.3	0.0	0.0	0.7
9	77	77.0	85.5	0.0	0.0	−8.5
10	82	82.0	83.8	0.0	0.0	−1.8
11	85	85.0	83.4	0.0	0.0	1.6
12	102	102.0	83.7	0.0	0.0	18.3
13	93	93.0	87.4	0.0	0.0	5.6
14			[88.5]			
14	90	81.5	80.0	−8.5	−8.5	1.5
15	94	85.5	80.3	0.0	−8.5	5.2
16	75	66.5	81.3	0.0	−8.5	−14.8
17	75	66.5	78.4	0.0	−8.5	−11.9
18	72	63.5	76.0	0.0	−8.5	−12.5
19	76	67.5	73.5	0.0	−8.5	−6.0
20	75	66.5	72.3	0.0	−8.5	−5.8
21			[71.1]			
21	93	93.4	80.0	8.9	0.4	13.4
22	94	94.4	82.7	0.0	0.4	11.7
23	83	83.4	85.0	0.0	0.4	−1.7
24	82	82.4	84.7	0.0	0.4	−2.3
25	82	82.4	84.2	0.0	0.4	−1.9
26	71	71.4	83.8	0.0	0.4	−12.5
27	82	82.4	81.3	0.0	0.4	1.0
28	78	78.4	81.5	0.0	0.4	−3.2
29	71	71.4	80.9	0.0	0.4	−9.6
30	81	81.4	79.0	0.0	0.4	2.4
31	88	88.4	79.5	0.0	0.4	8.9
32	80	80.4	81.2	0.0	0.4	−0.9
33	88	88.4	81.1	0.0	0.4	7.3
34	85	85.4	82.5	0.0	0.4	2.8
35	76	76.4	83.1	0.0	0.4	−6.7
36	75	75.4	81.7	0.0	0.4	−6.4
37	88	88.4	80.5	0.0	0.4	7.9
38	86	86.4	82.0	0.0	0.4	4.3
39	89	89.4	82.9	0.0	0.4	6.4

Table 8.1 *(Continued)*

1	2	3	4	5	6	7
t	Unadjusted Series, y_t	Adjusted Series, y'_t	EWMA for Adjusted Series, $\hat{y}'_t$	Effects of Adjustments, $g(X_t - X_r)$	Accumulated Effects of Adjustments, ν_t	$e_t = y'_t - \hat{y}'_t$
40	85	85.4	84.2	0.0	0.4	1.2
41	89	89.4	84.4	0.0	0.4	4.9
42	100	100.4	85.4	0.0	0.4	14.9
43			[88.4]			
43	100	92.0	80.0	−8.4	−8.0	12.0
44	106	98.0	82.4	0.0	−8.0	15.6
45	92	84.0	85.5	0.0	−8.0	−1.5
46	117	109.0	85.2	0.0	−8.0	23.8
47			[89.9]			
47	100	82.0	80.0	−9.9	−18.0	2.0
48	100	82.0	80.4	0.0	−18.0	1.6
49	106	88.0	80.7	0.0	−18.0	7.3
50	109	91.0	82.2	0.0	−18.0	8.8

thickness is reduced by an amount $-(88.5 - 80.0) = -8.5$ by making the adjustment

$$X_{13} - X_0 = \frac{1}{1.2}(-8.5) = -7.1.$$

At $t = 14$, $\nu_t = -8.5$ and for the next several time periods we have calculated the value of the adjusted series y'_t from $y'_t = y_t + \nu_t = y_t - 8.5$; for example, $y'_{14} = 90 - 8.5 = 81.5$. Adjustments of the kind made after the observation at $t = 13$ do not upset the calculation of subsequent EWMAs because adjustment affects the thickness measurement y'_t and its EWMA estimate $\hat{y}'_t$ equally. So that, for example,

$$\hat{y}'_{15} = (0.2 \times 81.5) + (0.8 \times 80.0) = 80.3,$$

where 80 is the appropriate value for $\hat{y}'_{14}$ *after* the adjustment has been made.

The next change is called for at $t = 20$, when the estimated value $\hat{y}'_{21}$ is equal to 71.1, which falls below the lower boundary of 72. This calls for a change in level of thickness of $-(71.1 - 80) = +8.9$. This change can be made as soon as y'_{20} is available, by making an adjustment

$$X_{20} - X_{13} = \frac{1}{1.2} \times 8.9.$$

The cumulative effect on thickness of these adjustments would now be $\nu_{21} = -8.5 + 8.9 = 0.4$ and for the next several time periods the adjusted level of the series would be given by

$$y'_t = y_t + 0.4.$$

8.4 A SIMPLE DEVICE FOR FACILITATING BOUNDED ADJUSTMENT

It is sometimes[3] convenient to construct an adjustment device, such as is shown in Figure 8.2, to update the forecast and to indicate when, and by how much, the process should be adjusted. We again illustrate for the thickness data with the boundaries at 72 and 88.

Suppose we are currently at some time t. Then the scale to the left labeled *previous forecast* would record the value $\hat{y}'_t$, which would be the forecast of y'_t made previously at time $t - 1$, while that to the right labeled *current observation* would record the value for y'_t actually observed at time t. The third intermediate scale labeled *updated forecast* is arranged so as to divide the horizontal distance between the two outer scales in the ratio λ/θ; this allows appropriate linear interpolation between $\hat{y}'_t$ and y'_t to obtain the updated forecast $\hat{y}'_{t+1}$. For this example $\lambda = 0.2$, so that the third scale is two-tenths of the way between the two outer scales.

On the far left, an adjustment scale for the deposition rate is shown. This is blank between the boundary values 72 and 88 within which no action is called for. Outside these limits, it is calibrated to show the adjustment in the deposition rate X required by Equation (8.3) and takes account of the fact that $g = 1.2$. The situation is depicted in Figure 8.2 for some specific time t. The previous forecast, which was made at time $t - 1$, is $\hat{y}'_t = 86$ and the actual value observed at the present time t is $y'_t = 66$. By joining the two points with a straight line, the updated forecast $\hat{y}'_{t+1} = 82$ is read off. Since this forecasted value $\hat{y}'_{t+1}$ does not lie out-side either boundary, it now becomes the previous forecast and is transferred to the left scale to await availability of the new current observation y'_{t+1} at time $t + 1$, and so on.

[3] Such a device will, of course, not be needed if a suitable process computer or programmed calculator is available.

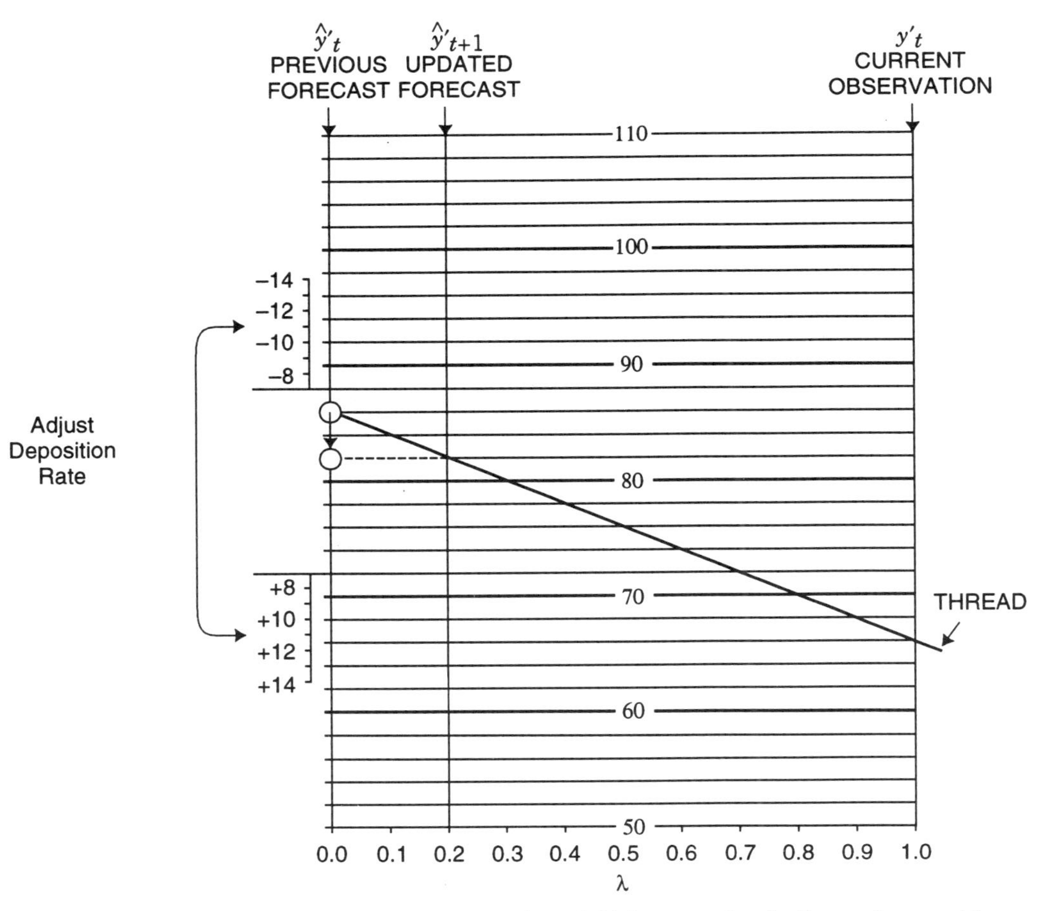

Figure 8.2 A device to update the forecast value of thickness and to indicate when, and by how much, the deposition rate should be adjusted.

Had the updated forecast $\hat{y}'_{t+1}$ fallen outside one or the other boundary, a change of deposition rate would have been made corresponding to the value read off on the scale labeled *adjust deposition rate.* Because this action is calculated to bring the forecasted thickness back on target, the "previous forecast" becomes at this point the target value of 80 to be combined with the value y'_{t+1} when this becomes available. For example, if the updated forecast had been $\hat{y}'_{t+1} = 92$, then this point would exceed the upper boundary and fall in the action zone. The chart would then call for a reduction of 10 units in the deposition rate.

A means of putting the above actions into effect (see Figure 8.2) uses a piece of black thread attached to a pushpin located on the left-hand scale to make the interpolation. Illustrating with the numbers used earlier, just before time t, the location of the pushpin on the previous forecast scale would be at $\hat{y}'_t = 86$ with the thread hanging down from the pin. As soon as the current observation $y'_t = 66$ became available, the thread would be pulled tightly to join the point 66 on the right-hand scale. The updated forecast $\hat{y}'_{t+1} = 82$ would then be read off on the intermediate scale. This value lies within the boundaries, so the pushpin would be moved down to become the new "previous forecast" with the thread hanging loose again until the next "current observation" y'_{t+1} became available to produce a new updated forecast. If the updated forecast $\hat{y}'_{t+1}$ had fallen outside either boundary, the appropriate adjustment in deposition rate would have been made, and the pushpin would then be *placed on the target value* ready for the next interpolation.

If we set $L = 0$ in a bounded adjustment chart, the two limit lines merge on the target value and the action called for becomes that for continual discrete integral control discussed in Chapter 6. The chart makes it easy to see why this is so, for, in that case, the pushpin would always have been placed on the target value. At any time t, the updated forecast of the adjusted series would then be such that

$$\hat{y}'_{t+1} - T = \lambda(y'_t - T)$$

and, consequently, the required adjustment would be

$$X_t - X_{t-1} = -\frac{1}{g}(\hat{y}'_{t+1} - T) = -\frac{\lambda}{g}(y'_t - T) = -\frac{\lambda}{g}e_t.$$

The interpolation chart also makes it easy to appreciate what are the precise functions of the quantities λ and L in a bounded adjustment chart.

Table 8.2 Expirical Illustration for the Thickness Disturbance Series of the Effect of the Choice of the Boundary Lines at $T \pm L$ for Various Values of L

L	$L/\hat{\sigma}_a$	Number of Adjustments	AAI	ISD
0	0.00	99	1.0	0.0
4	0.36	20	5.0	2.4
8	0.72	7	14.1	7.0
12	1.08	3	33.0	9.2

The value λ is concerned with separating the signal from the noise. It supplies a damping factor that extracts an estimate of the real changes that are occurring by smoothing out the noise. Increasing the value of λ will result in less smoothing, so how large a value of λ we should employ will depend on how strong the signal is in relation to the noise. A sum of squares plot like that in Figure 6.5 helps to decide this.

The value of L determines how much signal we are prepared to allow to "leak" into the adjusted series in exchange for less frequent adjustment. It is convenient to measure this "leakage" by the percentage increase of the standard deviation (ISD) over that obtained when $L = 0$ (i.e., over the integral scheme requiring repeated adjustment).

For the complete set of 100 thickness observations plotted in Figure 6.1, it turns out that, when L is set equal to 8, a total of only 7 adjustments would be required. Thus the empirical average adjustment inverval (AAI) is $99/7 = 14.1$. The corresponding empirical value for the *inflation*[4] of the standard deviation (ISD) is 7.0%. Calculations for this series with $L = 0, 4, 8$, and 12 are set out in Table 8.2. In general, we will tabulate the boundary lines in terms of L/σ_a and, for this example, we use $\hat{\sigma}_a = 11.1$—the standard deviation for the adjusted series with $L = 0$—as an estimate of σ_a. Shown in Table 8.2 for various values of L and hence $L/\hat{\sigma}_a$ are the percentage inflation of the standard deviation calculated from the sum of squares S for the adjusted series. Also shown are the numbers of adjustments that were actually required. The empirical AAI is then given by 99 divided by the number of adjustments.

These values for the AAI and for the percentage inflation of the standard deviation in Table 8.2 are for a particular run of 100 observations from this particular industrial process. Although empirical calculations of this kind are useful, it must be borne in mind that they are subject to quite large sampling errors: that is, if a second run was taken from the same process the values might differ considerably.

[4] Because of sampling variation and/or departures from the IMA model, occasionally for small values of λ a slight reduction in the standard deviation rather than an increase may sometimes be found.

8.5 BOUNDED ADJUSTMENT SEEN AS A PROCESS OF TRACKING

Bounded adjustment is equivalent to tracking a process disturbance with a series of step changes. Each of these is initiated when its change in level, as estimated by the appropriate EWMA, exceeds $\pm L$. Figure 8.3 illustrates the idea with the thickness measurement series. The unadjusted series together with the signals extracted by the process of

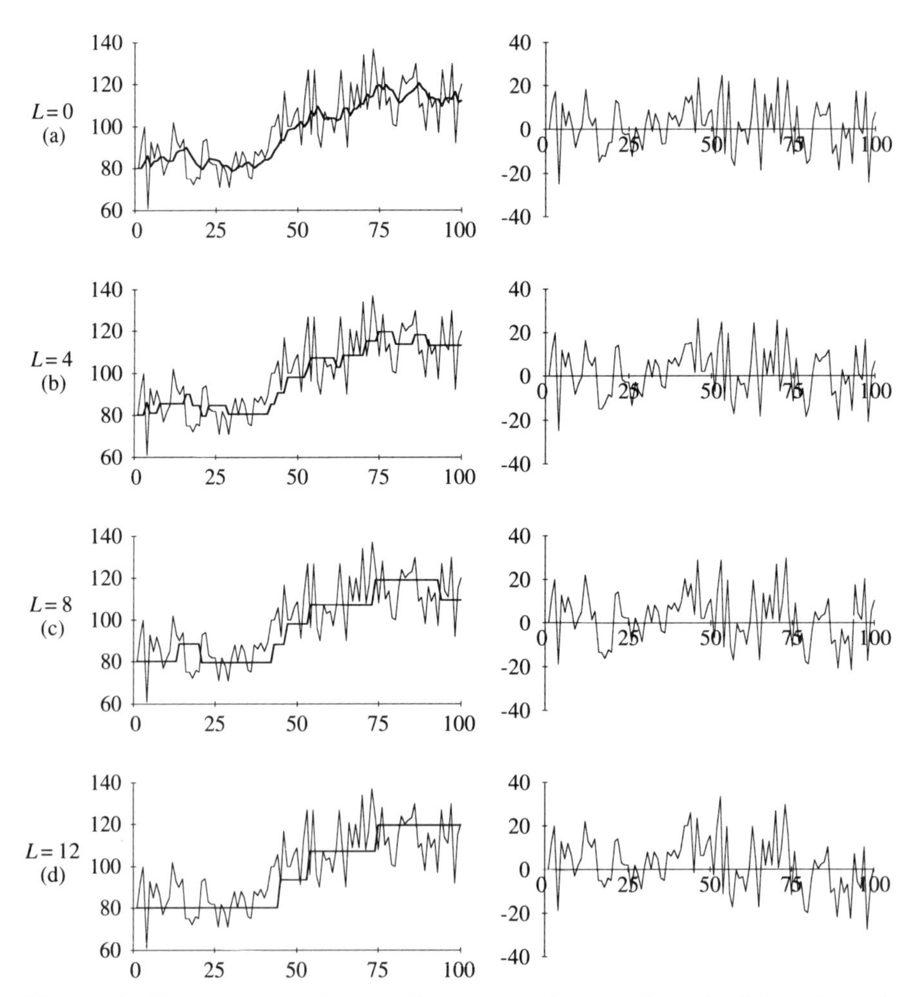

Figure 8.3 Thickness data. The unadjusted series is shown in each of the left hand diagrams. Superimposed are and signals extracted by bounded adjustment schemes with $L = 0, 4, 8, 12$. The extracted signals occur as a series of steps each slightly greater in magnitude than $|L|$. The right-hand charts show the residual deviations.

adjustment with $L = 0$, 4, 8, and 12 are shown on the left-hand diagrams. The right-hand diagrams show the residual deviations that constitute the adjusted thickness series for each case.

8.6 THEORETICAL AVERAGE ADJUSTMENT INTERVAL AND INFLATION OF MEAN SQUARE ERROR FOR AN IMA DISTURBANCE

So far we have made no assumptions about the model. It is now supposed that we have a responsive adjustment system and that the uncontrolled disturbance is adequately represented by an IMA model with parameters λ and σ_a. Then σ_a becomes the standard deviation of the MMSE scheme and the results of Box and Jenkins (1963)[5] may be used to calculate the theoretical average adjustment intervals and percentage increases in the mean square error for various values of λ and L. Table 8.3 shows the theoretical values for the AAI and ISD for various values of λ and L/σ_a. Suppose, for example, that $\lambda = 0.3$ and that the desired ISD is 10; that is, we are prepared to allow an increase of 10% in the standard deviation of the adjusted process above the minimum value obtained when action is taken after each observation. Then using the table we see that the desired scheme requires that $L/\sigma_a = 1$ and that the process would need adjustment on the average about every 16 sampling intervals. The last two columns of the table are discussed later.

Any initial scheme obtained from Table 8.3 may be checked and, if necessary, refined empirically in the light of actual experience by making suitable computer runs on the reconstructed disturbance, as in Table 8.2.

8.7 RELATION BETWEEN BOUNDED ADJUSTMENT CHARTS AND EWMA MONITORING CHARTS

As already noted, although bounded adjustment charts superficially resemble EWMA process monitoring charts, their purpose, rationale, and construction are very different.

[5] These authors showed that (assuming an IMA disturbance, a quadratic off-target loss function, and a fixed cost for each adjustment) schemes giving minimum overall cost were obtained by plotting an appropriate EWMA between *parallel* limit lines at $T \pm L$. They also provided a table from which L could be found.

Table 8.3 Average Adjustment Interval (AAI) and Percentage Increase in the Standard Deviation (ISD) for Various Values of λ and L Assuming an IMA Model[a]

λ	L/σ_a	AAI	ISD	C_A/C_T	R_A
0.5	0	1	0	0	0
0.5	0.5	2.8	2.4	0.3	1.17
0.5	1	6.9	10	1.9	7.7
0.5	1.5	13.1	21	7.2	28.8
0.5	2	21.3	34	19	77
0.5	2.5	31.4	49	42	167
0.4	0	1	0	0	0
0.4	0.5	3.6	2.5	0.3	2.03
0.4	1	9.8	10	2.5	15.5
0.4	1.5	19.0	21	10	61
0.4	2	31.4	34	27	166
0.4	2.5	47	48	59	370
0.3	0	1	0	0	0
0.3	0.25	2.3	0.5	0.1	0.8
0.3	0.5	5.3	2.6	0.4	4.3
0.3	0.75	9.8	5.8	1.4	16
0.3	1	15.6	10	3.6	40.5
0.3	1.5	31.4	20	15	166
0.3	2	53	33	41	460
0.3	2.5	80	47	95	1050
0.2	0	1	0	0	0
0.2	0.25	3.6	0.6	0.1	2
0.2	0.5	9.8	2.6	0.6	15.4
0.2	0.75	19.0	5.6	2.4	61
0.2	1	31.4	9	6.6	166
0.2	1.25	47	14	15	367
0.2	1.5	66	19	28	710
0.2	1.75	87	25	50	1255
0.2	2	112	32	82	2060
0.2	2.5	171	46	192	4800
0.1	0	1	0	0	0
0.1	0.25	9.8	0.7	0.2	15
0.1	0.5	31.4	2.4	1.6	164
0.1	0.75	66	5.2	7.1	708
0.1	1	112	9	21	2060
0.1	1.25	171	13	48	4805
0.1	1.5	243	18	97	9680
0.1	1.75	327	24	176	17600
0.1	2	424	31	296	29600
0.1	2.5	658	44	711	71050

[a] The use of the last two columns of the table is discussed later. These columns allow the bounded adjustment schemes to be chosen to minimize overall cost.

The EWMA charts discussed in Chapter 3 were used in the same way as Shewhart charts, fulfilling the function of *process monitoring* and dedicated to the sensible proposition "let's not go to the trouble of fixing it when it ain't broke." To do this:

1. The discount factor θ was chosen so that the chart would perform its monitoring function adequately.
2. The limit lines were based on a standard deviation of $\tilde{y}_t$ calculated assuming that the output from the process, when in a state of control, could be represented by white noise.
3. Aberrant behavior signaled the need to search for an assignable cause whose influence might be eliminated.

By contrast, EWMAs are used for *feedback adjustment* when, because of uncontrollable disturbances, the process—if left to itself—would wander off target. For this purpose:

1. The discount factor θ, and hence $\lambda = 1 - \theta$, is chosen to represent the nonstationary disturbance that has to be compensated.
2. The position of the limit lines are arrived at as a compromise between, on the one hand, the need to keep the standard deviation small and, on the other hand, the need to reduce the frequency of an adjustment as much as possible.
3. The crossing of the boundary lines signals the need to make an adjustment, the amount of the adjustment being based on the best estimate of how far the process has drifted from target.

Thus, on the one hand, *process adjustment* uses *statistical estimation* to determine the level of the disturbance that is to be fed back and canceled; on the other hand, *process monitoring* parallels a procedure for *statistical significance testing* with the purpose of detecting possibly assignable and removable causes. (See also Baxley, 1991, 1994.)

8.8 COMBINATION OF ADJUSTMENT AND MONITORING

To say that we must be clear headed about the fundamental differences between systems of monitoring and of adjustment is not to say that these procedures cannot be carried out simultaneously. A process that must be adjusted because of uncontrollable drifts may also exhibit occasional large superimposed deviations arising from specific assignable causes. We can expect to see these as outlying points on a monitoring chart for the residual error series $e_t = y'_t - \hat{y}'_t$ shown for the thickness series in the last column of Table 8.1. In particular, a Shewhart chart may be run on these residual errors (sometimes after appropriate filtering described in Chapter 11) to detect and to eliminate "special" causes.

Indeed, the whole idea of common and special causes, which was discussed in Chapter 2, can be generalized to include processes that require adjustment. In such cases, it is the time series model representing the disturbance that constitutes the common cause system. Outliers from this model may correspond to special causes. The standard Shewhart model can thus be thought of as a special case in which the time series representing the disturbance consists of random noise varying about a fixed mean μ. This topic is discussed in much greater detail in Chapters 10 and 11, where it is shown that, when changes of a specific nature in the more broadly defined common cause system are feared, sensitive detection methods can be applied by sequential plotting of an appropriate function of the data called a "Cuscore." As one example, a Cuscore statistic can be designed to provide the earliest possible warning of a change in the value of λ used in a control scheme or in a forecasting scheme.

8.9 BEHAVIOR OF THE BOUNDED ADJUSTMENT CHART FOR SERIES NOT NECESSARILY GENERATED BY AN IMA MODEL

Bounded EWMA adjustment yields the smallest mean square error for any given average adjustment interval on the assumption that the disturbance can be represented by an IMA time series model. However, the procedure appears to be quite robust to wide departures from this model. As was shown in Figure 8.3 it seemed to behave well in adjusting the industrial series of thickness measurements. We now consider its performance in controlling two artificially generated series. The first of these is a disturbance series generated by the Barnard model discussed earlier in Chapter 5.

Adjustment of a Series of 200 Observations Generated by the Barnard Model

The "signal" μ_t shown on the left of Figure 8.4a is such that the periods between jumps are random drawings from a Poisson distribution with mean 16, and the sizes of the jumps are random normal deviates with

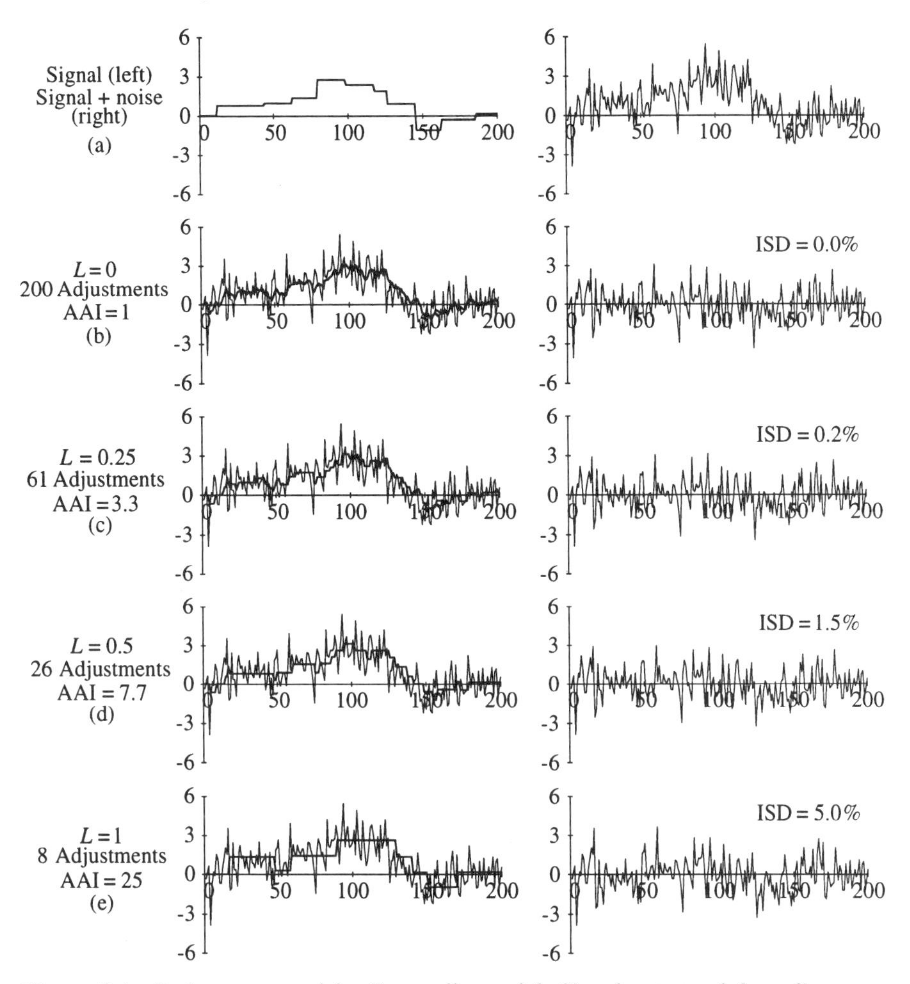

Figure 8.4 Series generated by Barnard's model. Signal extracted for adjustment using various values of L with corresponding adjusted series.

mean zero and standard deviation unity. On the right of Figure 8.4a is shown a generated disturbance y_t obtained by adding to this signal normally distributed random noise ϵ_t with mean zero and standard deviation unity. Thus the unadjusted disturbance series is $y_t = \mu_t + \epsilon_t$. The step functions shown in the left-hand Figures 8.4b, 8.4c, 8.4d and 8.4e represent the signals that would be extracted from this disturbance by bounded charts with[6] $\lambda = 0.2$, for limit lines with $L = 0.00$, 0.25, 0.50, and 1.00. Also shown are the number of adjustments made

[6]The maximum likelihood estimate of λ (see Appendix 10.A), computed as if the model were an IMA with the routine in Luceño (1995b), is 0.18. We use 0.20 as an adequate approximation.

Table 8.4 Some Comparisons of Empirical and "Theoretical" Values[a] of the AAI and ISD for the Barnard Model and the Random Slope Model

	Barnard ($\lambda = 0.2$)		Random Slope ($\lambda = 0.23$)	
L/σ_a	AAI	ISD	AAI	ISD
0.00	1 (1)	0 (0)	1 (1)	0 (0)
0.25	3.3 (3.6)	0.2 (0.6)	2.7 (3.1)	1.8 (0.6)
0.50	7.7 (9.8)	1.5 (2.6)	6.7 (7.9)	3.2 (2.6)
1.00	25.0 (31.4)	5.0 (9.4)	22.2 (24.6)	6.4 (9.6)

[a] Theoretical values, which appear in parentheses, are obtained from Table 8.3 (and Figure 8A.1, discussed later).

per 200 observations and the corresponding empirical AAIs for the various examples. The corresponding diagrams to the right show the adjusted outputs that would be produced for the various choices of L. Also shown are the percentage increases in the standard deviations (empirical ISDs) obtained by comparing the bounded adjusted schemes with the repeatedly adjusted scheme of Figure 8.4b.

Table 8.4 shows that these empirical AAIs and ISDs agree reasonably well with the corresponding "theoretical" values (shown in parentheses) obtained from the fitted IMA models.

Adjustment of a Series of 200 Observations Generated by the Random Slope Model

Figure 8.5 shows results for a model that deviates in a different way from the IMA formulation. In this model it is the slope rather than the level of the signal μ_t that is subjected to random jumps. As before, the periods between jumps are random drawings from a Poisson distribution with mean 16; however, the changes themselves consist of random alterations in slope. In Figure 8.5 the slopes (defined by the increases per unit period) are random drawings from a normal distribution with mean increase zero and standard deviation 1/16. Figure 8.5a shows on the left the course of the signal μ_t generated in this way, and on the right is shown the generated disturbance series $y_t = \mu_t + \epsilon_t$ obtained by adding to this signal random noise ϵ_t that is normally distributed with mean zero and standard deviation unity.

The maximum likelihood estimate[7] of λ for this series is 0.23 and this is the value used in the EWMA filter. As before, for various values of L, the step function adjustments are shown on the left-hand side of Figures 8.5b, 8.5c, 8.5d, and 8.5e, together with the number of adjustments made per 200 observations and the corresponding empirical

[7] Computed with the routine in Luceño (1995b) as if the model were an IMA.

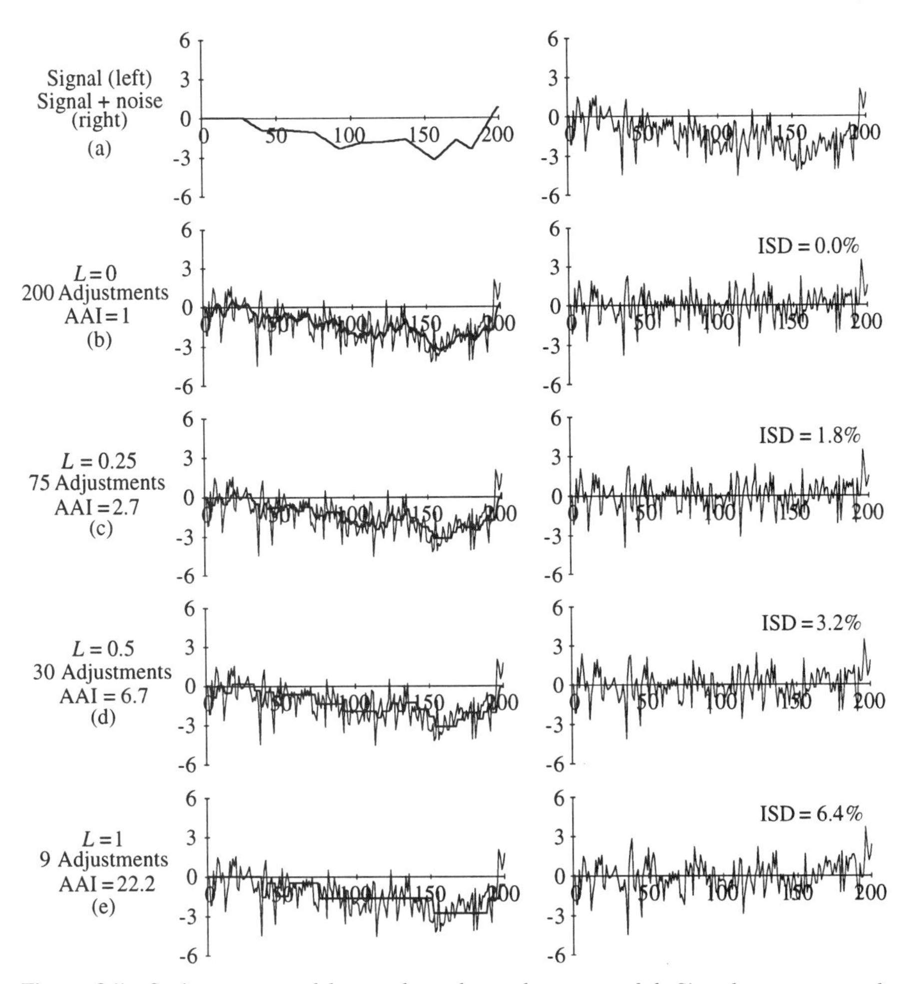

Figure 8.5 Series generated by random slope change model. Signal was extracted from adjustment using various values of L with corresponding adjusted series.

AAIs. On the right-hand side are the corresponding adjusted processes with empirical percentage changes in standard deviations. Again Table 8.4 shows reasonably good agreement between the empirical and "theoretical" values. The procedures have also been shown to be satisfactorily robust to various kinds of nonnormality in the a_t's. (Box and Luceño, 1994; Srivastava, 1995.)

8.10 EXPLICIT CONSIDERATION OF MONETARY COSTS

In the schemes described above in some sense, we are balancing the *value* of less frequent adjustment against the *value* of an increase in variation. These schemes were originally derived (Box and Jenkins,

1963) to minimize overall cost explicitly. However, it is often not an easy matter to assess value in terms of money. For example: Are plant operators less or more likely to be alert with a scheme that calls for frequent adjustment, as opposed to one calling for less frequent adjustment? If such an effect exists, what is its monetary worth?

It is because of such questions that it may be impractical to try to work directly with monetary costs. It is usually better to use Table 8.3 to provide estimates of the frequency of adjustment and of the corresponding standard deviation for a few reasonable alternatives. The choice between them should be made by someone, such as the supervising engineer, who is familiar with the human as well as the technical circumstances that apply to the particular facility.

These circumstances can be very different. For example, the making of an adjustment might simply involve the turning of a valve by an operator who must be available anyway. In that case, the adjustment cost would be essentially zero. In other circumstances, adjustment might involve the stopping of a machine and/or the replacement of a tool at considerable cost. Also, an increase of, say, 20% in the standard deviation to reduce the sampling frequency might be, in some circumstances, acceptable and, in others, out of the question.

For example, suppose that study of a section of reconstructed disturbance for a particular process produced an estimated value for λ of 0.25 and a value of $\hat{\sigma}_a$ such that with continual adjustment the process could be maintained on target with a C_p index of 3; that is, a S_σ (spec/sigma) ratio of 18 (see Chapter 3). Study of Table 8.3 with $\lambda = 0.25$ implies that allowing the standard deviation to increase by 20% (thus reducing the C_p index to about 2.5) could reduce the frequency of needed adjustment to about one adjustment for every 40 or so observations. Thus, in this instance, the adjustment frequency could be reduced by a factor of 40 while maintaining a more than adequate spec/sigma ratio of 15.

By contrast, if even with continual adjustment the C_p index was close to unity, such an option would not be attractive and some other solution might be sought—possibly a major modification of the process, or a rethinking of the control strategy.

However, costs can sometimes be arrived at, at least approximately, and in that case a scheme that approximately minimizes overall costs can be devised. The cost of making an adjustment to the process, which we call C_A, is often fairly easy to estimate. A more difficult problem is how to assess the value of increasing or reducing the output mean square error. To do this we need to know the cost of being off target. The curve that describes this is called the cost function or the loss

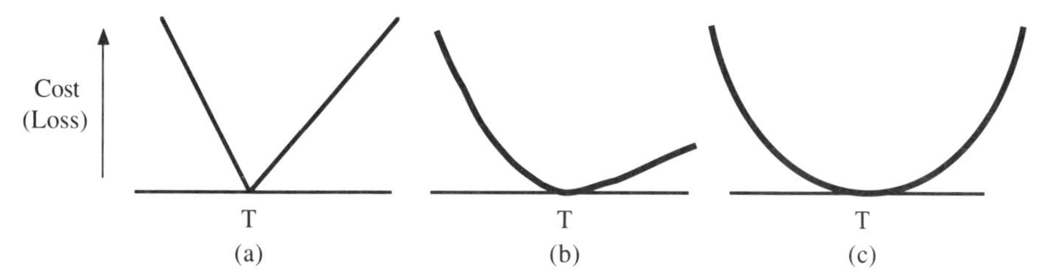

Figure 8.6 Possible cost/loss functions measuring the cost of deviating from the target value.

function. A few possibilities for such a function are shown in Figure 8.6. On the horizontal axis is the value of the adjusted quality characteristic y', and on the vertical axis is the cost of failing to produce material exactly on the target T. In each case the cost functions are drawn so that the cost is zero when $y' = T$. (Thus one point on the cost curve is known.)

The idea of a cost or loss function is a very old one and many different kinds of loss functions can be imagined. In very rare cases, the cost of being off target in either direction may be exactly known, but most frequently it is not. It seems reasonable, however, that the *incremental* increase in cost should increase as the deviation increases so that for most purposes the curved cost functions (b) and (c) in Figure 8.6 would seem more reasonable than (a). Also, the cost function may not be symmetric about T. For example, as in (b), the cost of exceeding the target value by a given amount might be much less than the cost of falling short of it by that amount. See for example Harris (1992).

The simplest cost function to manipulate mathematically is the quadratic function

$$c = k_T(y' - T)^2, \tag{8.7}$$

where c is the cost of being off target and k_T is a constant. This assumes symmetry about the target value and also that the cost is proportional to the square of the deviation from target. For this quadratic cost function, minimizing cost is equivalent to minimizing mean square error about the target value. To the extent that minimizing mean square error about the target yields sensible results, it encourages us to think that in some circumstances, anyway, a quadratic cost function may be a useful working assumption.

It will be seen that the constant k_T that appears in Equation (8.7) is the cost of the adjusted process being one unit off target for one period of time. In general, it is more convenient to work, not with k_T itself, but with a standardized constant $C_T = k_T\sigma_a^2$. Then the cost of being off target can be rewritten as

$$c = C_T\left(\frac{y'_t - T}{\sigma_a}\right)^2 \tag{8.8}$$

and C_T is the average cost of being off target for one period by an amount σ_a. It will be recalled that, on the assumption of an IMA disturbance, σ_a is the standard deviation of the output with $L = 0$ and can be estimated by making a sum of squares plot for the appropriate reconstructed disturbance.

If the cost function was quadratic with a minimum at T and assuming σ_a known, all that we would need to determine C_T would be one further point on the cost curve. Taguchi (1987) argues that such an additional point may be obtained if we know that a deviation in level of the adjusted process from target by more than $\pm(y'_0 - T)$ would result in rejection of all the material manufactured during the whole sampling interval, at a cost, say, of c_0. On that argument

$$c_0 = C_T\left(\frac{y'_0 - T}{\sigma_a}\right)^2,$$

and hence C_T is

$$C_T = \frac{c_0\sigma_a^2}{(y'_0 - T)^2}.$$

For example, it will be recalled that for the thickness data $\hat{\sigma}_a = 11.1$. Now suppose that material whose thickness deviated from the target value by more than ± 40 would be rejected at a cost of \$500 per time period. Then with $c_0 = \$500$, $y'_0 - T = 40$, and $\hat{\sigma}_a = 11.1$,

$$C_T \approx \frac{500 \times 11.1^2}{1600} = \$38.5.$$

By substituting mean values in Equation (8.8) we see that C_T is the unit cost (the implied average cost incurred in one time interval) of process variation of a magnitude measured by σ_a. That is the cost incurred when $L = 0$, and we have the continuous adjustment scheme of Chapter 6.

To design a bounded adjustment chart, we need the value of L so that the boundary lines $T \pm L$ will produce the smallest overall cost. Box and Jenkins (1963) showed that, for a quadratic loss function and with C_A the cost of making an adjustment, minimum cost schemes could be tabulated in terms of $R_A = (C_A/C_T)/\lambda^2$ and $L/\lambda\sigma_a$ and they gave a short table.

Using the simulation results of Kramer (1989) and the numerical method in Luceño, González, and Puig-Pey (1996), a more extensive tabulation is shown in Table 8.5, where values of $L/\lambda\sigma_a$, AAI, and q for various values of R_A are given. The mean square error of the adjusted process is then $\sigma_a^2\{1 + \lambda^2 q\}$. [Appendix 8A gives more details on the functions h and q, and Appendix 8B explains why minimum cost schemes can be expressed in terms of $L/\lambda\sigma_a$ and R_A.]

Table 8.5 Table for Arriving at Bounded Adjustment Schemes Yielding Minimum Overall Cost[a]

R_A	$\dfrac{L}{\lambda\sigma_a}$	$\text{AAI} = h\left(\dfrac{L}{\lambda\sigma_a}\right)$	$q\left(\dfrac{L}{\lambda\sigma_a}\right)$
1	0.93	2.6	0.16
2	1.23	3.5	0.31
3	1.43	4.3	0.43
4	1.59	4.9	0.54
5	1.72	5.5	0.63
10	2.17	7.8	1.0
20	2.70	11.0	1.5
30	3.05	13.5	1.9
40	3.33	15.5	2.3
50	3.55	17.3	2.6
60	3.75	19.0	2.9
70	3.92	20.6	3.1
80	4.07	21.9	3.3
90	4.21	23.2	3.6
100	4.34	24.5	3.8
200	5.28	34.7	5.5
300	5.91	42.4	6.8
400	6.40	49.0	7.9
500	6.81	55	8.8
600	7.14	60	9.7
700	7.45	65	10.5
800	7.72	69	11.2
900	7.98	74	12.0
1000	8.21	77	12.6
2000	9.88	110	18.0
3000	11.02	135	22.1
4000	11.85	155	25.5

[a] The table gives values of $L/\lambda\sigma_a$, AAI $= h(L/\lambda\sigma_a)$, and $q(L/\lambda\sigma_a)$ for various values of $R_A = (C_A/C_T)/\lambda^2$. The standard deviation of the scheme is given by $\sigma_a\sqrt{1 + \lambda^2 q}$. See Appendices 8A and 8B for details.

For illustration, suppose that for the thickness data example the cost of making a change is $C_A = \$100$ and as before that $C_T = \$38.5$. Then $C_A/C_T = 2.60$ and

$$R_A = \frac{C_A/C_T}{\lambda^2} = \frac{2.60}{0.04} = 65.0$$

From Table 8.5 we see that this corresponds to a value of about $L/\lambda\sigma_a = 3.8$, so $L = 3.8 \times 0.2 \times 11.1 = 8.4$ and the boundary lines should be at 88.4 and 71.6. Also, you will see from the table that the theoretical AAI is about 20 and q is about 3.0. So the inflation factor for the standard deviation is $(1 + \lambda^2 q)^{1/2} = (1 + 0.04 \times 3.0)^{1/2} = 1.06$. On average then, you will need to adjust the process about once in every twenty intervals and the increase in standard deviation over a repeated adjustment scheme is about 6%.

This scheme with $L = 8.4$ happens to be quite close to that portrayed in Figure 8.1 (see also Figure 8.3c), where $L = 8.0$. The values of the AAI = 20.0 and the ISD = 6% for this theoretical scheme are not dramatically different from the empirical values shown in Table 8.2 with an AAI equal to 14.1 and an ISD of 7%.

Relationship of the Two Approaches

The schemes discussed in this chapter can either be chosen by direct consideration of costs or indirectly in terms of the AAIs and ISDs they produce. The relationships involved can be appreciated from Figure 8.7. For various values of λ and various choices of C_A/C_T, the percentage increase in standard deviation (ISD) and the corresponding average adjustment interval (AAI) can be read off. In addition, when costs can be estimated, the curved lines of constant C_A/C_T provide optimal schemes associated with different values of λ. Thus in the example above (with $\lambda = 0.2$), we need a minimum cost scheme with $C_A/C_T = 2.6$; the diagram agrees, as of course it should, with the scheme obtained previously giving $L/\lambda\sigma_a = 3.8$ with about a 6% increase in the standard deviation and with AAI near 20. In practice, there is no point in being overly exact in these calculations. They cannot be expected to provide more than reasonable approximations.

APPENDIX 8A FUNCTIONS $h(L/\lambda\sigma_a)$ AND $q(L/\lambda\sigma_a)$ IN TABLE 8.5

The functions h and q are important because they provide the average adjustment interval (AAI) and mean square error (or mean square deviation—MSD) as a function of the quotient $L/\lambda\sigma_a$.

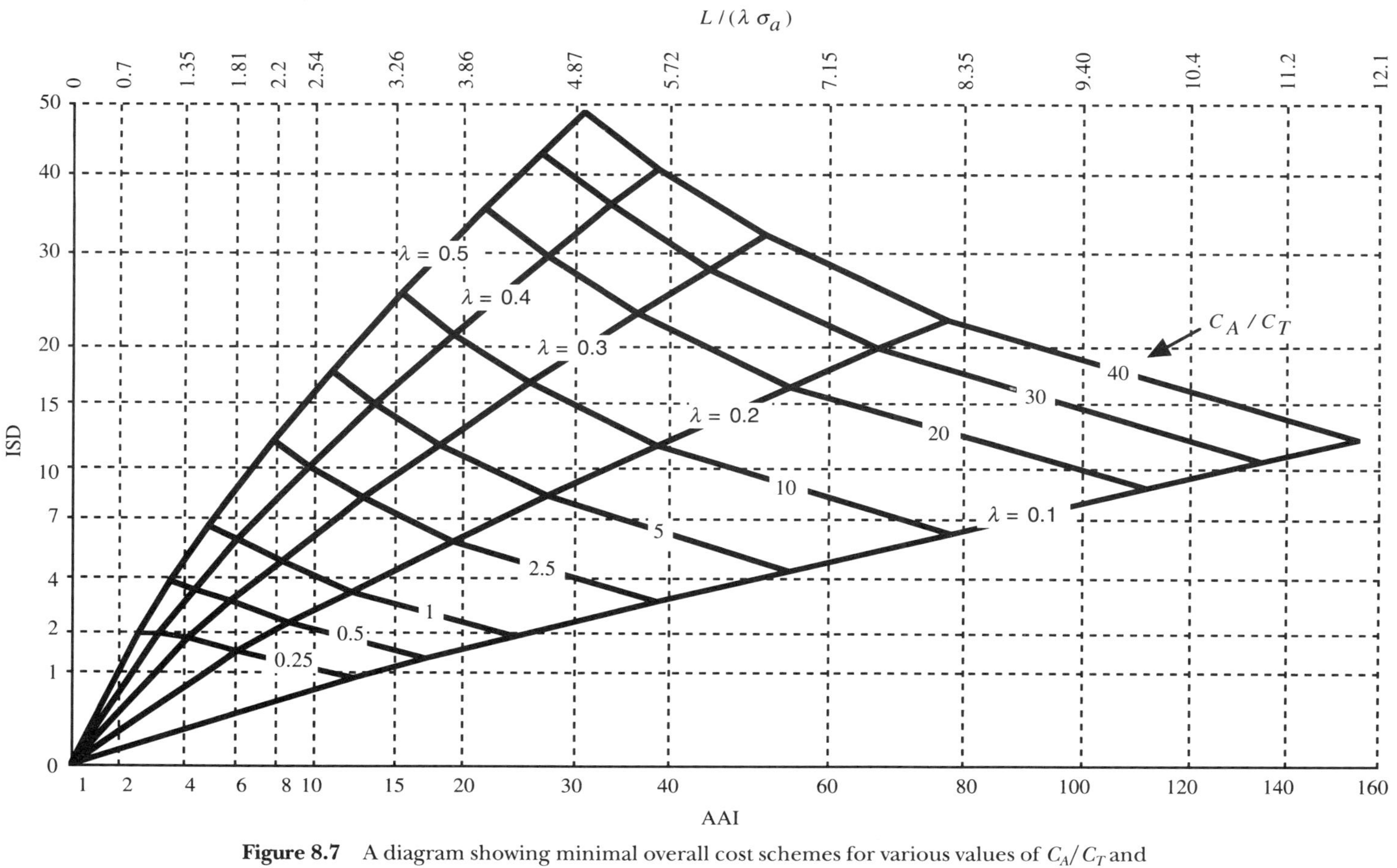

Figure 8.7 A diagram showing minimal overall cost schemes for various values of C_A/C_T and λ in relation to $L/\lambda\sigma_a$, the AAI, and the ISD. Calculations are made on the assumption that the disturbance is adequately represented by an IMA model.

Thus

$$\text{AAI} = h\left(\frac{L}{\lambda\sigma_a}\right), \tag{8A.1}$$

$$\text{MSD} = \sigma_a^2\left\{1 + \lambda^2 q\left(\frac{L}{\lambda\sigma_a}\right)\right\}. \tag{8A.2}$$

These functions can be characterized by the integral equations in Box and Luceño (1994) or their generalizations in Luceño et al. (1996). Computer programs for evaluating these functions and finding minimum cost schemes are given in the latter reference. It is interesting that, in the particular case studied in this chapter, the square roots $\sqrt{h(L/\lambda\sigma_a)}$ and $\sqrt{q(L/\lambda\sigma_a)}$ can be very well approximated by linear functions of $L/\lambda\sigma_a$ as the upper part of Figure 8A.1 shows. The linear approximation is a little worse only when $L/\lambda\sigma_a$ is close to 0, as the lower part of Figure 8A.1 shows.

APPENDIX 8B MINIMUM COST SCHEMES

The cost function is defined as

$$c = \frac{C_A}{\text{AAI}} + C_T\frac{\text{MSD}}{\sigma_a^2}, \tag{8B.1}$$

so that introducing Equations (8A.1) and (8A.2) in Equation (8B.1) gives

$$c = \frac{C_A}{h\left(\dfrac{L}{\lambda\sigma_a}\right)} + C_T\left\{1 + \lambda^2 q\left(\frac{L}{\lambda\sigma_a}\right)\right\}. \tag{8B.2}$$

Now Equation (8B.2) can be rearranged as

$$c = C_T + C_A\left[h^{-1}\left(\frac{L}{\lambda\sigma_a}\right) + \left(\frac{C_A}{C_T\lambda^2}\right)^{-1} q\left(\frac{L}{\lambda\sigma_a}\right)\right],$$

or equivalently,

$$\frac{c - C_T}{C_A} = h^{-1}\left(\frac{L}{\lambda\sigma_a}\right) + \frac{1}{R_A}q\left(\frac{L}{\lambda\sigma_a}\right),$$

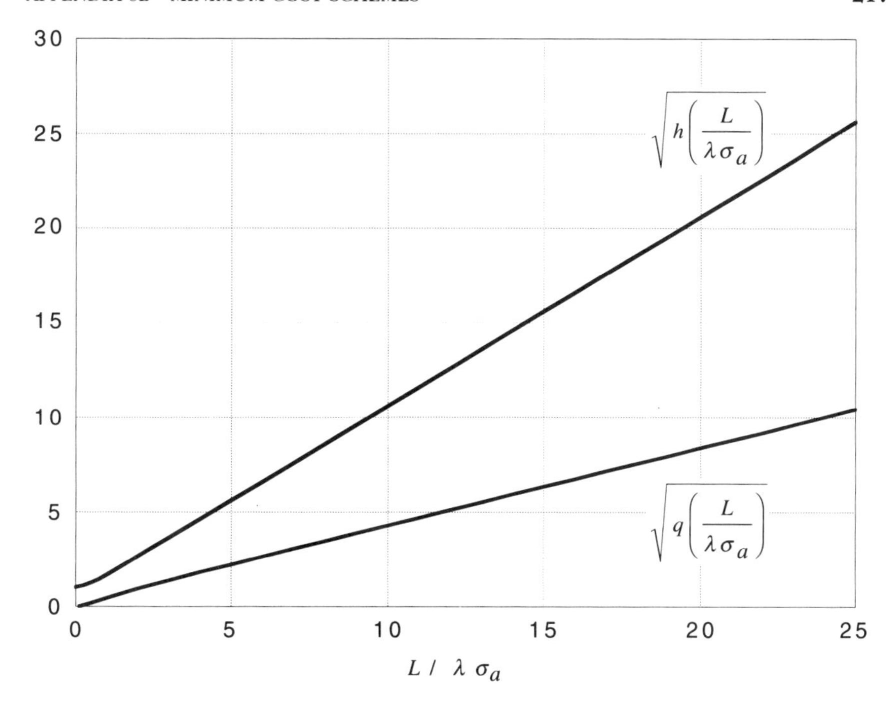

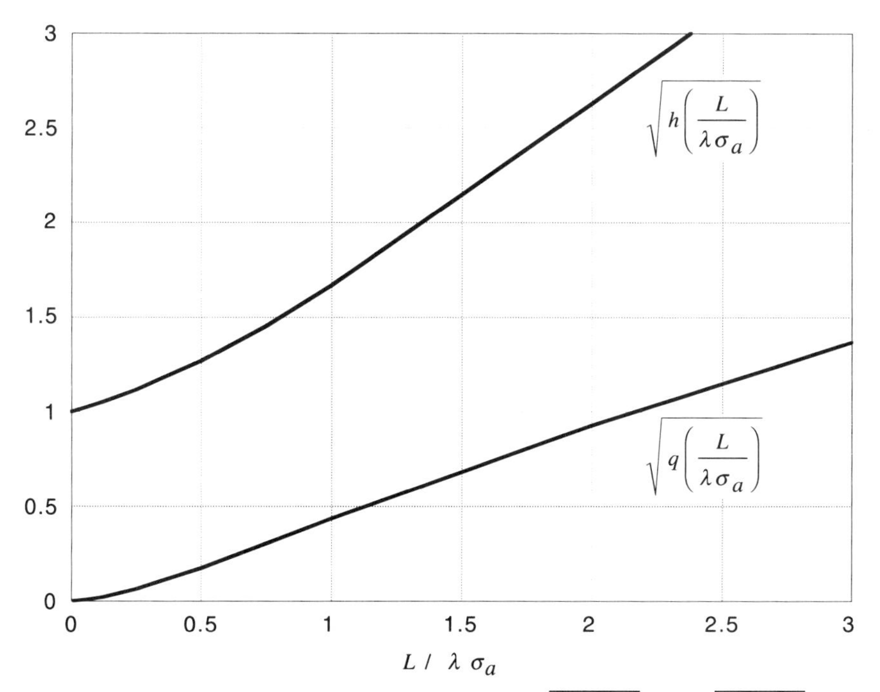

Figure 8A.1 The upper part shows the functions $\sqrt{h(L/\lambda\sigma_a)}$ and $\sqrt{q(L/\lambda\sigma_a)}$ for $0 \leq L/\lambda\sigma_a \leq 25$. The lower part shows a detail of the same functions for $0 \leq L/\lambda\sigma_a \leq 3$.

so that the (standardized) cost $(c - C_T)/C_A$ is a function of $L/\lambda\sigma_a$ and R_A exclusively. Hence the value of $L/\lambda\sigma_a$ providing minimum cost only depends on R_A.

Problem 8.1 A bounded adjustment chart is designed to control viscosity. It uses an EWMA with $\theta = 0.7$, has limit lines at 12.6 and 11.4, and is found to produce an average adjustment interval of about 9. Stating any assumption you make:

(a) Give approximate values for σ_a, the ISD, and the output standard deviation σ_e for the scheme.

(b) For an alternative scheme with an ISD of 20, what would be the values of L and of the AAI?

(c) Assuming the off-target cost constant is $C_T = \$10$, what are the implied values of C_A for the two schemes? □

CHAPTER 9

Including the Cost of Surveillance: How Often Should You Sample?

"Singing roll or bowl a ball, a penny a pitch."
Song: I've Got a Lovely Bunch of Coconuts, FRED HEATHERTON

9.1 INTRODUCTION

A further important question involves how often you should *observe* the process. The answer depends on the cost of doing this in relation to other costs (see also Abraham and Box, 1979; MacGregor, 1976). In this chapter the expressions "cost of sampling" and "surveillance cost" are used more or less interchangeably. What will be meant is the *whole* cost of getting a numerical reading. It may include, for example, the cost of actually getting a representative sample of the product as well as the cost of its physical and chemical analysis. When this cost is high, you will be especially anxious not to take observations more often than you need and will perhaps ask the question, "Can we employ a longer sampling period?"

Once again, a purely empirical strategy can sometimes be helpful. If you have the necessary knowledge about the process, a record of past performance, and the adjustments that have been made, you can reconstruct a specimen of the disturbance series. Then various trial schemes of control can be run on this series. Such an exercise can give a fairly good idea of what might be possible. Alternatively, with the assumptions made previously, we can compute the characteristics of various possible schemes.

Suppose that the process you are studying is presently observed with a specific sampling interval, which from now on will be called *the unit interval,* and you wish to explore the possibility of observing the process less frequently. The size of the unit interval could be very different depending on the process: for a wastewater treatment plant it might be a day; for a particular industrial process perhaps half an hour; and for an atomic reactor 10 seconds. In any case we will suppose, as in the previous chapter, that the adjustment system is responsive, that the disturbance is approximately modeled by an IMA time series, that the cost of being off target—characterized by C_T—is proportional to the output mean square error, and that there is a fixed cost C_A of making an adjustment. A third cost is now introduced—the total cost C_S of taking a sample. This is called the cost of surveillance.

Taking account of this additional cost, Box and Kramer (1992) showed that minimizing total cost again required bounded adjustment schemes like that of Figure 8.1 but with the additional feature that the sampling interval had to be decided. They provided charts that,—given the values of λ, C_T, C_A, and C_S—gave the best value for L/σ_a and also the best sampling interval regarded as a multiple S of the unit interval.

Because it is sometimes difficult to determine the three cost coefficients C_T, C_A, and C_S, these minimum cost schemes were reparameterized by Box and Luceño (1994) in terms of the following three alternative criteria:

1. The sampling interval (S), expressed in terms of the unit interval.
2. The average adjustment interval (AAI), also *expressed in terms of the unit interval.*
3. The resulting percentage increase in the standard deviation (ISD), expressed in terms of σ_a.

To aid in the choice of suitable schemes, these authors provided charts for $\lambda = 0.1$, 0.2, 0.3, 0.4, 0.5, and 0.6, which are reproduced in Figure 9.1. For each value of λ the chart shows the ISD and the corresponding average adjustment interval AAI for various values of S and L/σ_a. The charts included here cover the regions of most interest where there are only small or moderate percentage increases in the output standard deviations.

Suppose a time series, modeled by an IMA with nonstationarity parameter $\lambda = \lambda_1$ and generated by independent normal "random shocks" $\{a_t\}$ with standard deviation $\sigma_a = \sigma_{a1}$, is sampled every S unit intervals. Then it can be shown (BJR) that the sampled series can also be modeled by an IMA with parameter λ_S also generated by independent random shocks with standard deviation σ_{aS}, and that the relationships between the parameters of the original series and the sampled series are

$$\frac{S\lambda_1^2}{\theta_1} = \frac{\lambda_S^2}{\theta_S} \quad \text{and} \quad \sigma_{aS} = \sigma_{a1}\sqrt{\frac{\theta_1}{\theta_S}}.$$

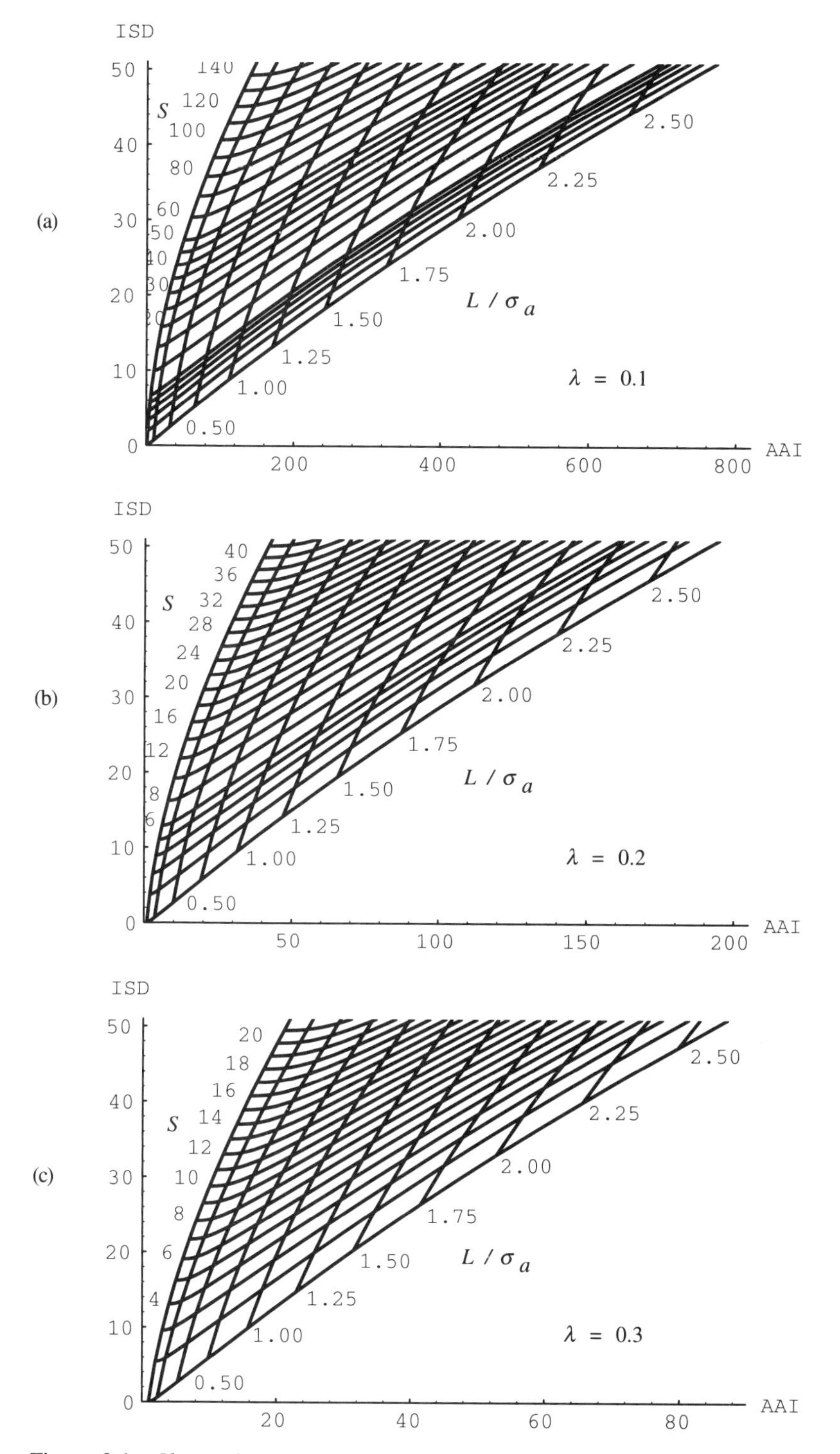

Figure 9.1 Charts showing the average adjustment interval and percentage increase in standard deviation for $\lambda = 0.1, 0.2, 0.3, 0.4, 0.5,$ and 0.6; $L/\sigma_a = 0.0, 0.25, 0.50, 0.75, 1.00, 1.25, 1.50, 1.75, 2.00, 2.25,$ and 2.50; and several values of the sampling interval S from 1 to 140. Note that the average adjustment interval (AAI) is in terms of the *original unit intervals.*

(d)

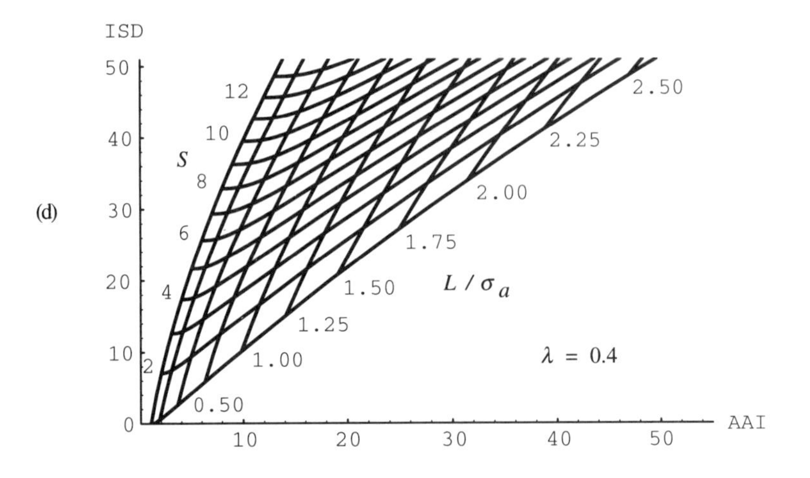

(e)

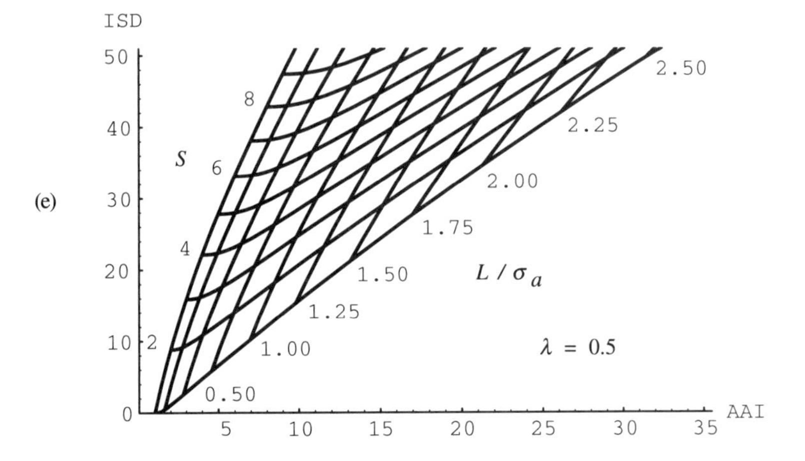

(f)

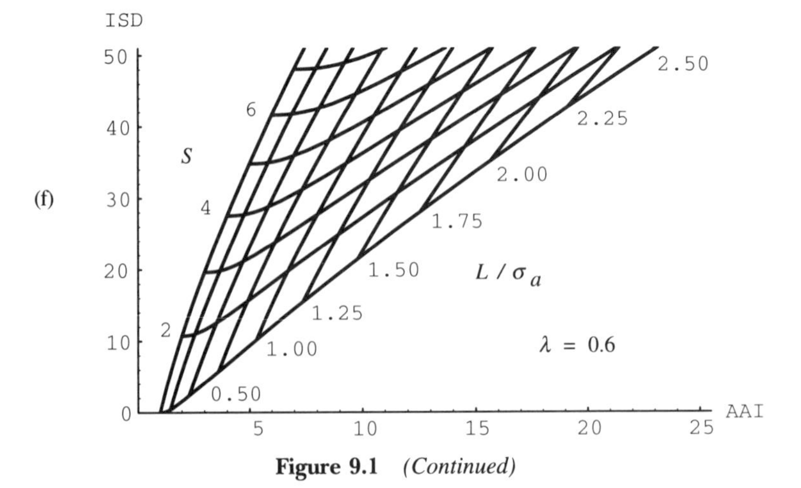

Figure 9.1 *(Continued)*

A diagram constructed using the first of these expressions from which you can read off the appropriate value of λ_S, knowing $\lambda = \lambda_1$ and S, is shown in Figure 9.2. For example, suppose $\lambda = \lambda_1 = 0.2$. If you decide to double the interval ($S = 2$), then for the new bounded adjustment chart you will need to use $\lambda_2 = 0.27$ in the updating formula (8.6). Also, the standard deviation of the random shocks for the IMA representing the sampled series will be $\sigma_{a2} = \sigma_{a1}\sqrt{0.80/0.73} = 1.05\sigma_{a1}$. If you were to increase the sampling interval tenfold ($S = 10$), you would need to use $\lambda_{10} = 0.50$ and $\sigma_{a10} = \sigma_{a1}\sqrt{0.8/0.5} = 1.26\sigma_{a1}$.

To illustrate the use of these charts, consider once more the bounded adjustment scheme of Figure 8.1. The time between observations for this "reference" scheme will be regarded as the unit interval. So $S = 1$ and we suppose the estimated parameters are $\hat{\lambda} = 0.2$ and $\hat{\sigma}_a = 11.1$ as before. The limit lines at 80 ± 8, used earlier for illustration, gave a theoretical AAI of 17.8 intervals with a 5.2% increase over the minimum value σ_a obtained by continuous adjustment. By using the estimate $\hat{\sigma}_a = 11.1$, this scheme corresponded to the choice $L/\hat{\sigma}_a = 8/11.1 = 0.72$. In Figure 9.3, which is a duplicate of Figure 9.1b somewhat enlarged, this reference scheme is indicated by an open circle (like an O) on the line for $S = 1$. Also shown are other choices A, B, C, D, E, F, G, H, I, in the immediate neighborhood of O with $S = 1, 2$, and 5. The values of their AAIs and ISDs are shown in the accompanying table and those for the reference scheme are shown above this table. The choice depends on circumstances—of how important it is to change the sampling rate (increase S), to control the process variation (the ISD), and to modify the time between needed adjustments (the AAI). Depending on the situation, any one of the schemes characterized above might be favored.

As an example, suppose your process was currently using the "reference" feedback control scheme O, but that you wanted to reduce the sampling rate because sampling and testing were expensive. Suppose also that, while it was important not to increase variation about the target, the cost of making an adjustment was sufficiently low that a moderate increase in the frequency of adjustment would be acceptable. Then you might prefer to switch to scheme E. With that scheme, the sampling rate is halved ($S = 2$), the standard deviation is almost the same (the ISD = 6%), but somewhat more frequent adjustment is necessary (the AAI is reduced from 18 to 12 unit intervals). Note that you pay for the reduction in surveillance by having to adjust the process more frequently. (While there *is* no such thing as a free lunch, you can at least choose the menu.)

A graphical comparison of the two schemes O and E in controlling the metallic disturbance series is shown in Figure 9.4. The performance

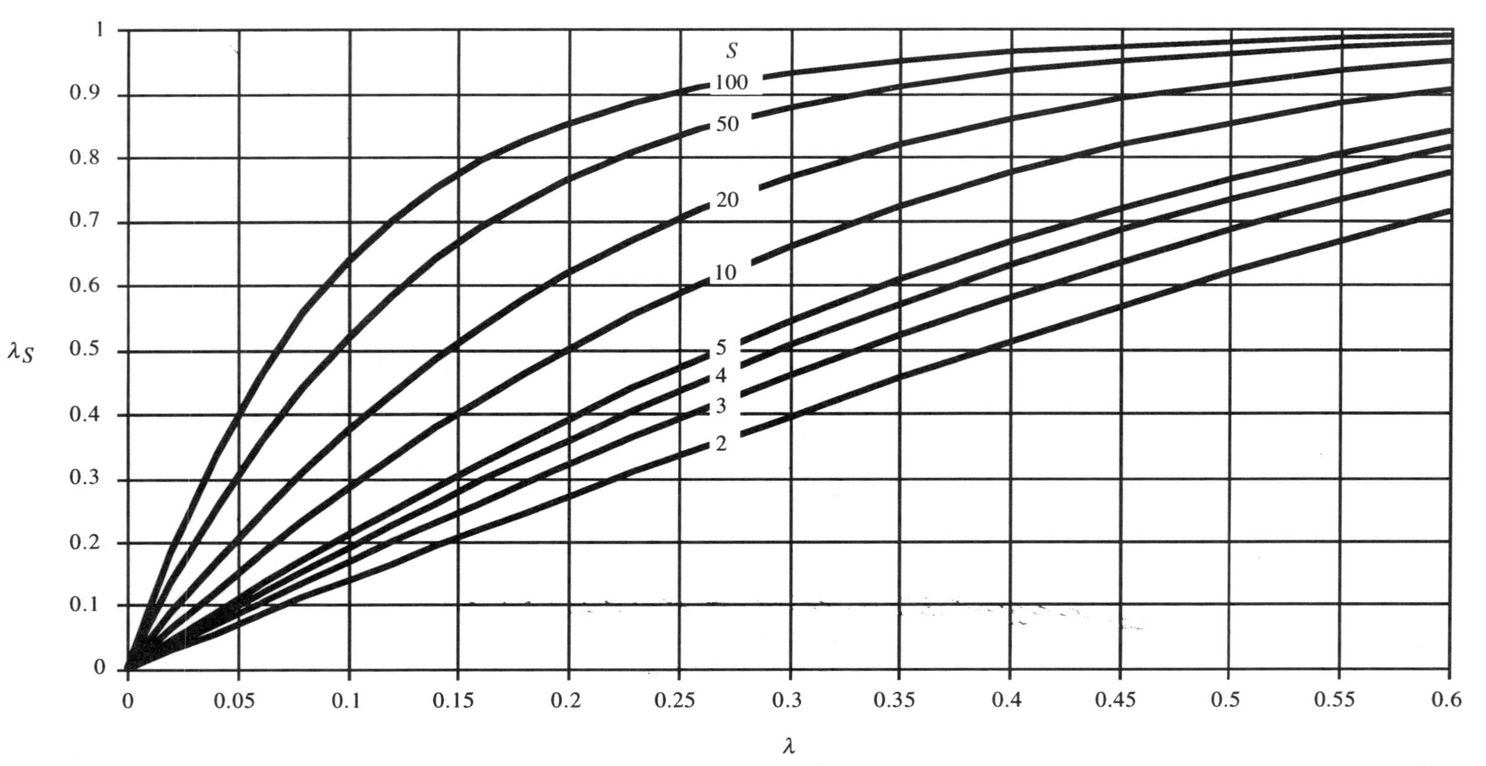

Figure 9.2 Values of λ_S for different values of $\lambda = \lambda_1$ and S.

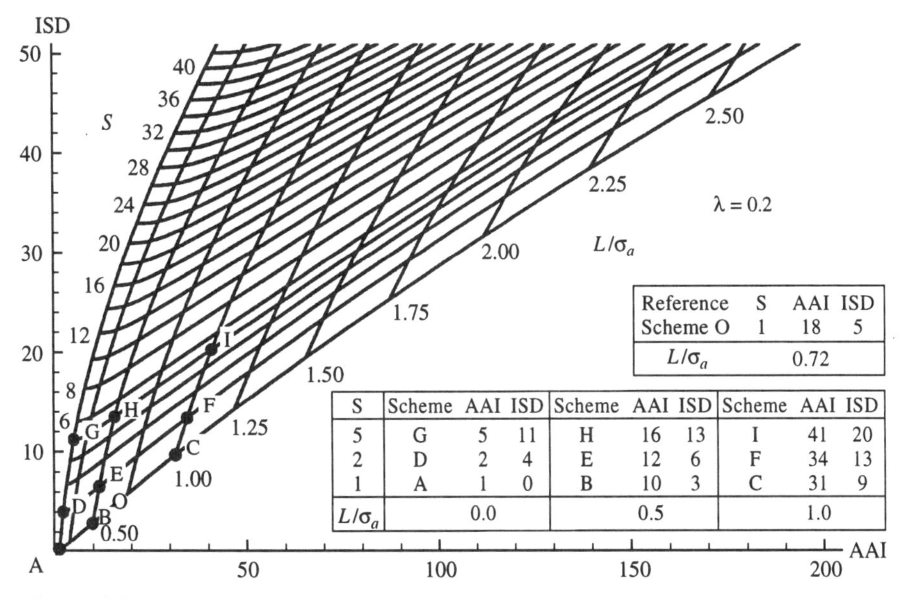

Figure 9.3 A duplicate of Figure 9.1b showing the reference scheme O and the alternative schemes A, B, C, D, E, F, G, H, and I with corresponding values for S, L/σ_a, AAI, and ISD.

of the reference scheme O is shown in Figure 9.4a. This figure is a duplicate of Figure 8.1. By using as data only the odd values of the same disturbance series, Figure 9.4b illustrates the behavior the scheme E with $S = 2$. Note for this scheme that $L/\sigma_a = 0.5$, so with $\hat{\sigma}_a = 11.1$, $L = 5.55$ and the limit lines are 74.45 and 85.55—somewhat closer together than before. Also, while the forecasts of scheme O are updated with $\lambda = 0.2$ (i.e., with smoothing constant $\theta = 0.8$), those for scheme E are updated using $\lambda_2 = 0.27$ with smoothing constant $\theta_2 = 0.73$. For comparison, the weight functions for the two EWMAs are also shown in Figure 9.4.

For any given λ, S, and L/σ_a, the charts discussed above provide the AAI and ISD on the assumptions made in Chapter 8 about the disturbance (IMA model), the dynamics (the adjustment system is responsive), and the cost function (quadratic). Thus, for given λ you can read off the values of S and L/σ_a required to produce any combination of values of the ISD and AAI. Looking at Figure 9.1, one can see that, for given ISD, the AAI decreases as S increases (and L/σ_a decreases) so that the maximum AAI is always attained with $S = 1$.

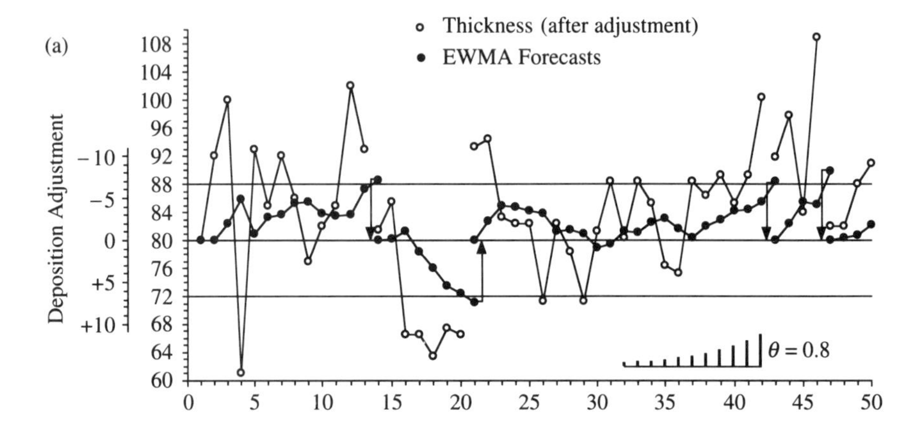

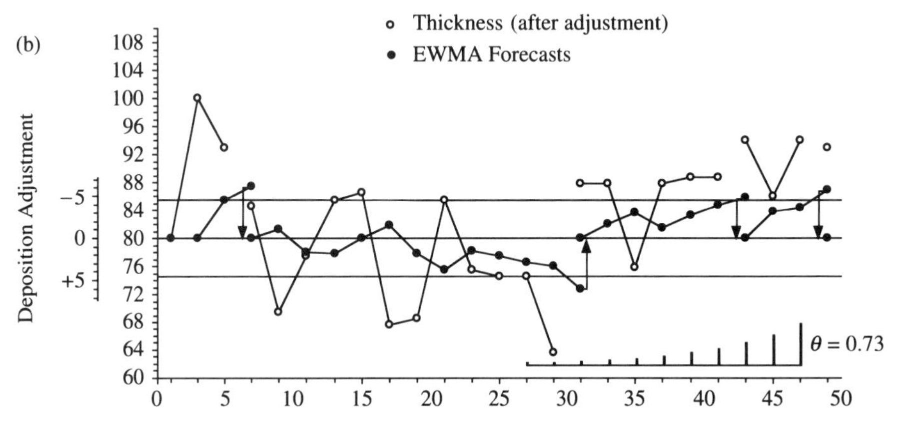

Figure 9.4 (a) Scheme O applied to the first 50 values of the thickness data with action boundaries at 80 ± 8 and unit sampling interval ($S = 1$). (b) Scheme E applied with $S = 2$ using only the odd values of the data and boundaries at 80 ± 5.55. In both cases the open circles are the adjusted levels of thickness, the solid circles are the EWMA forecasts.

9.2 CHOOSING THE FEEDBACK SCHEMES DIRECTLY IN TERMS OF COSTS

When costs can be estimated, the feedback adjustment schemes can be characterized directly in terms of the cost coefficients C_T, C_A, and C_S. On the previous assumptions—it was shown by Box and Kramer (1992) that these schemes minimize the total cost of control. The charts they developed are shown in Figures 9.5a and 9.5b. For purposes of tabulation, it is convenient to characterize these schemes in terms of the three quantities

$$R_A = \frac{C_A/C_T}{\lambda^2}, \quad R_S = \frac{C_S/C_T}{\lambda^2}, \quad \text{and} \quad \lambda.$$

For given values of R_A and R_S, Chart A in Figure 9.5a provides the best value of the standardized action limit ℓ, from which knowing σ_a and λ, the corresponding value for $L = \ell\lambda\sigma_a$ is obtained and hence the limits $T \pm L$ for the bounded adjustment chart. In addition, Chart B (Figure 9.5b) gives the best value of the sampling interval S in terms of R_S and λ.

Example

In Section 8.10 a minimum cost scheme for controlling the thickness series was given, with the sampling interval supposed fixed ($S = 1$). For that scheme

$$\hat{\lambda} = 0.2, \quad \hat{\sigma}_a = 11.1, \quad C_A = \$100,$$

$$C_T = \$38.5, \quad R_A = \frac{C_A/C_T}{\lambda^2} = 65.0$$

and, using Table 8.5, a minimum cost scheme was obtained with $L = 8.4$. From Figure 9.1b (for $\lambda = 0.2$) with $S = 1$ and $L/\sigma_a = 8.4/11.1 = 0.76$ we find as before that this scheme provides an AAI close to 20 and an ISD just under 6%. For that scheme it was tacitly assumed that it cost nothing to obtain a measurement.

Suppose now that it costs \$9 to obtain a measurement, so that $C_S = \$9$ and $R_S = (C_S/C_T)/\lambda^2 = 5.8$. Then, using Chart A (in Figure 9.5a) with $R_S = 5.8$ and $R_A = 64.9$ we see that approximately $\ell = 3.6$ so $L = \ell\lambda\sigma_a = 3.6 \times 0.2 \times 11.1 = 8.0$. Also, from Chart B (in Figure 9.5b) with $\lambda = \lambda_1 = 0.2$ and $R_S = 5.8$, we see that an approximately minimum cost scheme is obtained when $S = 2$. As before a value $\lambda_2 = 0.27$ is used for updating the forecast. Figure 9.1b for $\lambda = 0.2$, $S = 2$ and $L/\sigma_a = 8/11.1 = 0.72$ gives approximately AAI = 20 and ISD = 9%.

Thus as a result of the added cost of surveillence (\$9 to get a measurement) the minimum cost scheme requires that we halve the sampling frequency using about the same limits as before. The result will be that the AAI is about the same but the ISD is increased by about 3%.

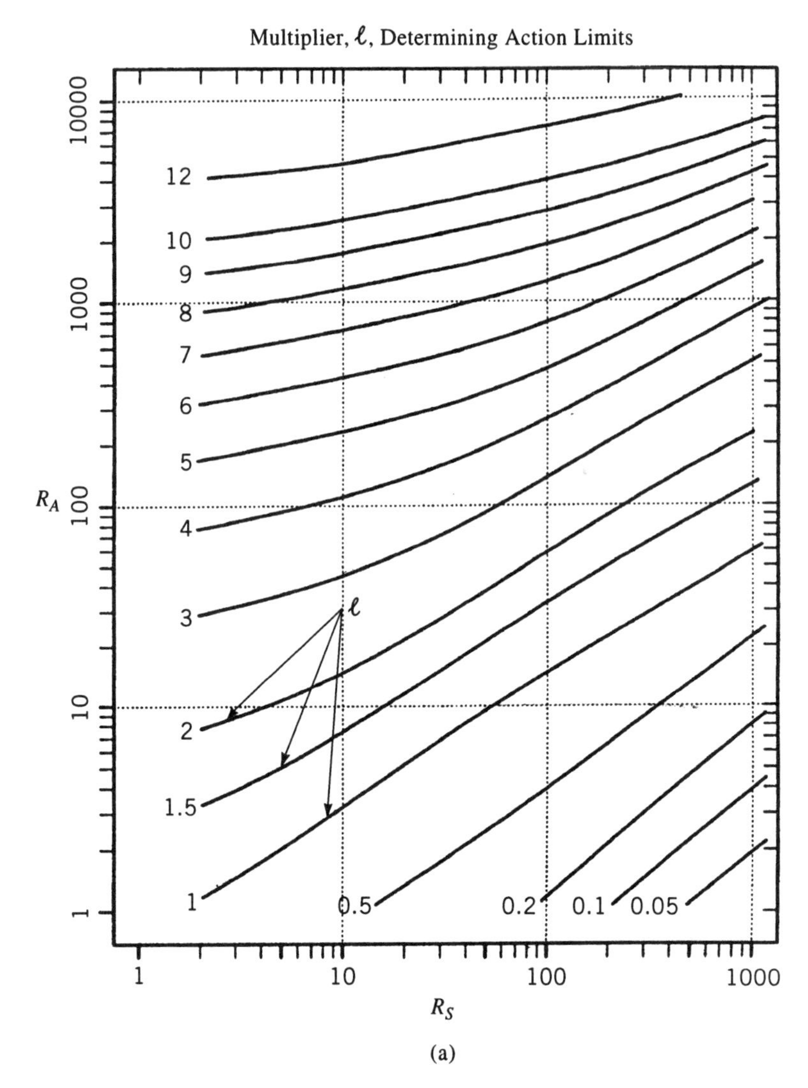

Figure 9.5 (a) Chart A shows the contours of the standardized action limit ℓ for various values of R_S and R_A. (b) Chart B shows contours of the monitoring interval S for various values of λ and R_S.

A Computer Program

The charts, while usually supplying adequate approximations, are not particularly easy to read. Alternatively, the quantities L and S providing minimum expected cost may be evaluated using the computer program in Luceño et al. (1996) (This program uses a numerical method

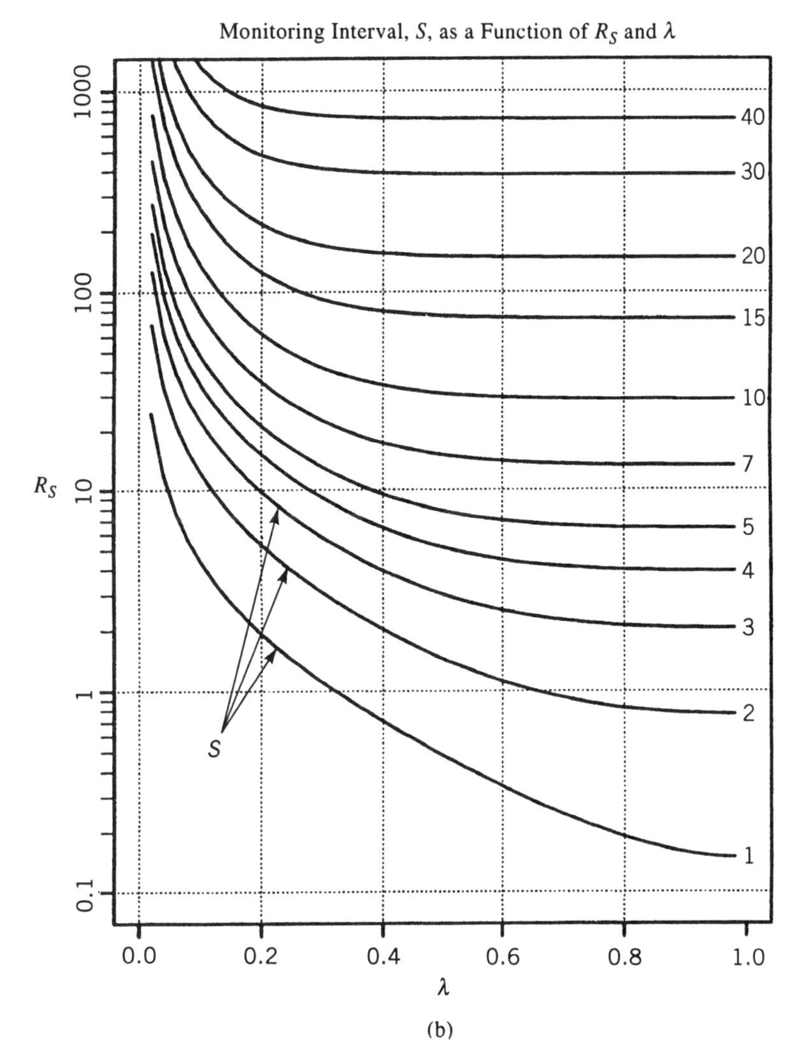

(b)

Figure 9.5 *(Continued)*

based on the characterization of h and q referred to in Appendix 8A.). The cost function is now defined by adding the term C_S/S to the right-hand side of Equation (8B.1), where the MSD is given by $(1 + \text{ISD}/100)^2\ \sigma_a^2$. For the above scheme in Section 8.10 (with $C_S = 0$), the program shows that $L = 8.5$ yields a minimum cost of \$48.16 per unit interval and leads to an AAI = 19.8 and an ISD = 5.8%. For the scheme with $C_S = \$9$, the computer program shows that a minimum expected cost of \$55.14 per unit interval can be attained by using $S = 2$ and $L = 7.9$, which corresponds to an AAI = 19.8 and a ISD = 8.8%; the value of $\lambda_2 = 0.27$ is also given. The program is useful also for exploring alternative options. It shows for example that the expected cost is only slightly larger (\$55.51 rather than \$55.14) for the scheme using $S = 3$, $L = 7.5$, and $\lambda_3 = 0.32$ with corresponding AAI = 19.9 and ISD = 11.0%.

Problem 9.1 Design a scheme appropriate to the following situation. In a process used in a pharmaceutical manufacture, an important quality characteristic was subject to a disturbance, which it had not been possible to remove. Using past hourly records, estimates of the disturbance parameters were $\hat{\lambda} = 0.3$ and $\hat{\sigma}_a = 3$. The target value was 340 and the *specification limits* were 325 and 355. The operating procedure was to take one sample every hour and to make a very expensive determination costing $C_S = \$200$ for each sample. Material that fell outside these specification limits needed to be reprocessed at cost of \$1350 per hour's worth of material. The cost of each adjustment to the process was $C_A = \$600$. □

CHAPTER 10

Directed Process Monitoring and Cuscore Charts: Looking for Signals in Noise

"Detection is, or ought to be, an exact science."
The Sign of Four, SIR ARTHUR CONAN DOYLE

10.1 INTRODUCTION

This book has been about process *monitoring*, of which there is a preliminary discussion in Chapters 2 and 3, and about process *adjustment*, considered in Chapters 6, 7, 8, and 9. The present chapter returns to a closer and more sophisticated discussion of process monitoring. We said earlier that a process is monitored by sequentially checking the "common cause" system—that is, by checking whether or not the process continues to stay in a stable state usually represented as stationary *noise* varying about a target value. The objective is to seek for "special causes" that are suggested by significant patterns. Such patterns, in turn, can suggest the existence of systematic *signals*. Information as to the nature, time of occurrence, and size of such potential signals can lead to identification and elimination of the signaling factor(s). Another way of saying this is that we are seeking to improve quality by trying to detect *signals* hidden in *noise.*

For the purpose of process monitoring, the basic charts devised by Walter Shewhart have the great advantage that they are plots of the actual data. Much emphasis has been given in the literature to their technical aspects such as the exact location of limit lines and the associated probabilities of exceeding those limits. But perhaps their most important virtue is that they are run charts that provide an excuse for exhibiting to those concerned with its operation a continuous picture of what the process is actually doing. Because they are not specialized, these charts are potentially capable of drawing attention to unusual kinds of behavior and hence to possible signals and causes previously unsuspected. That every fifth value seems unusually high, or that a low value is almost always followed by a high value, or that it looks as if there is a regular reduction of variation during a particular shift could each provide[1] a vital clue for process improvement. Note that in such situations the chart performs an *inductive* role; it can suggest *unexpected* hypotheses to be considered further. In the detective work of quality improvement, as in any process of learning, *hypothesis generation* is of paramount importance; for we certainly cannot test a hypothesis we don't yet have.

Situations occur, however, where special kinds of process deviations are feared a priori because they are known to be characteristic of, or peculiar to, a particular system. It is then possible to devise procedures that are especially sensitive to deviations of that type. This chapter considers an efficient and very general means of doing this using the concept, due to Fisher (1925), of *efficient score statistics.* This concept produces what we will call *Cuscore* statistics and hence Cuscore charts (Box and Ramírez, 1992; Box and Jenkins, 1966; Bagshaw and Johnson, 1977). See also Lai, 1995, and Crowder, Hawkins Reynolds and Yashkin, 1997, for more general discussions of sequential change-point detection. In particular, given appropriate assumptions about the noise and about the signal we want to detect, each of the charts we have already discussed—the Shewhart chart, the Cusum chart, and the EWMA chart—is an example of a Cuscore chart. One question of incidental interest is precisely what these assumptions need to be. More generally, however, Cuscore statistics can be devised for almost any kind of signal hidden in almost any kind of noise, allowing you to look into possibilities *not* covered by the standard charts.

These powerful but "blinkered" specialized Cuscore procedures should be used to augment rather than to replace the basic data plot provided by a run chart or by a Shewhart chart. The situation has been likened (Box, 1980) to that faced by a small country that fears aggression from the air. As illustrated in Figure 10.1, powerful radars might be beamed in specific likely directions of attack, but it would be wise also to have a global radar that scanned the full horizon. The highly directional radars are analogs for the Cuscore charts discussed in this chapter; the global radar is an analog for a direct plot of the data using a run chart or a Shewhart chart.

We illustrated in Chapter 3, for the "case of the delinquent calibrator," how the informal use of the Cusum chart provided a tool for the

[1] Unless such effects are reasonably large compared with the noise level, it may not be easy to detect them, however

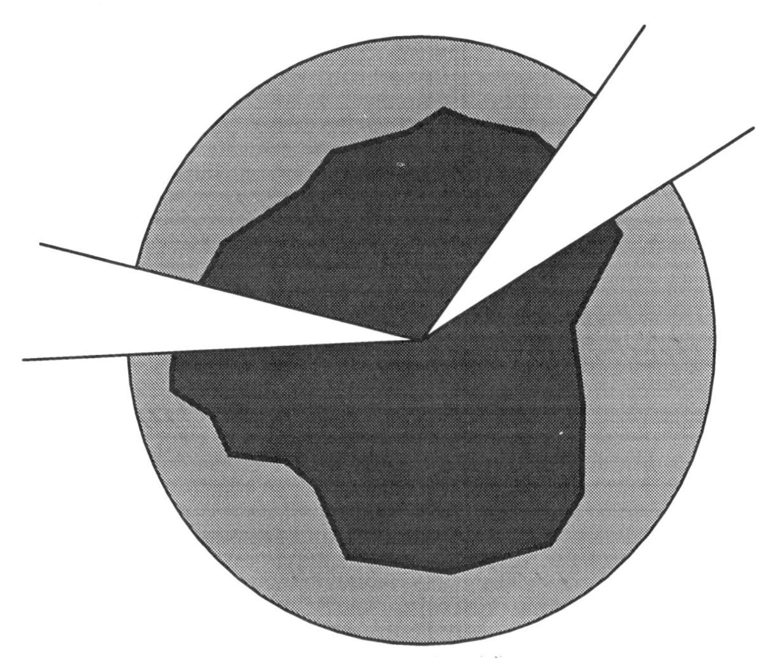

Figure 10.1 Roles of the global Shewhart charts and more focused Cusum and Cuscore charts likened to those of possible radar defenses.

elimination of sources of trouble. Used in this way, it had indicated approximately *when* and by *how much* small "step changes" in the *mean* had occurred. Thus it helped to track down, identify, and eliminate the causes of these changes. In this chapter we will use the more general Cuscore charts in a similar informal way.

Now Cusum statistics can be derived on the basis that we are looking for a possible *step change* in the mean of the process, which is buried in *white noise.* It will be remembered that by white noise is meant a sequence, denoted by $\{a_t\}$, of independent errors with standard deviation σ_a that are approximately normally distributed. However, not all "feared" signals are step changes and not all noise is white noise. To illustrate Cuscore statistics, we begin by considering the detection of a somewhat unusual signal. This, however, gives the general flavor of what is to follow. We first describe a technical problem where the technique was needed, then illustrate the solution with some data and finally provide a justification with appropriate references to more basic material.

Detecting an Intermittently Occurring Sine Wave

Consider the following problem. A process normally operates in an apparent state of control roughly modeled by white noise varying about a target value T. Occasionally, however, harmonic cycling about the target occurs. The source of this cycling is known. It is caused by a rotary system upstream that can get out of adjustment. The signal we are looking for is thus a sine wave of known period and phase but unknown amplitude. Although action has been taken that seems to have solved the problem, it is feared that it might recur and a procedure is needed that will give the earliest possible warning of its reappearance. So we need a "directional radar" specifically designed to detect a sine wave in white noise. This is provided by the Cuscore statistic. For this example, this turns out to be

$$Q = \sum (y_t - T) \sin x_t. \tag{10.1}$$

To see what this expression means, remember that the *Cusum* statistic $Q = \Sigma(y_t - T)$ is the sum of the residual deviations from the process target. You will see that this Cuscore statistic requires each of the residuals $y_t - T$ to be first multiplied by the appropriate value of $\sin x_t$, and then summed.

For illustration, consider Figure 10.2, which (a) shows the "signal"—a sine wave[2] beginning at the 48th observation and ceasing at the 96th with period 12 and amplitude one-half of the standard deviation; (b) shows this signal *plus* the noise consisting of 144 random normal deviates with zero mean and standard deviation $\sigma = 1$; and (c) shows the behavior of the appropriate Cuscore statistic. Although the sine wave is invisible in the noise in Figure 10.2b, its appearance and disappearance are clearly signaled by the Cuscore chart in Figure 10.2c.

Detecting a Change in a Rate of Increase

As a second example, suppose a procedure is needed to give early warning of a change in a rate of increase—for example, in the rate of increase in the wear of a machine tool. Suppose for a normal tool the graph of wear versus time will be a straight line with known slope β (Greek *beta*). The feared discrepancy is that for a defective tool the rate of wear will at some unknown point increase more rapidly than this.

[2] Specifically, at time t, $x_t = 360t/p$ is the angle measured in degrees and p is the period of the cycle. In the example, the period of 12 means that it takes $p = 12$ time intervals to complete one cycle, so the x_t values will be a series starting at some *known* point in the sequence of angles 30°, 60°, 90°, 120°, ...

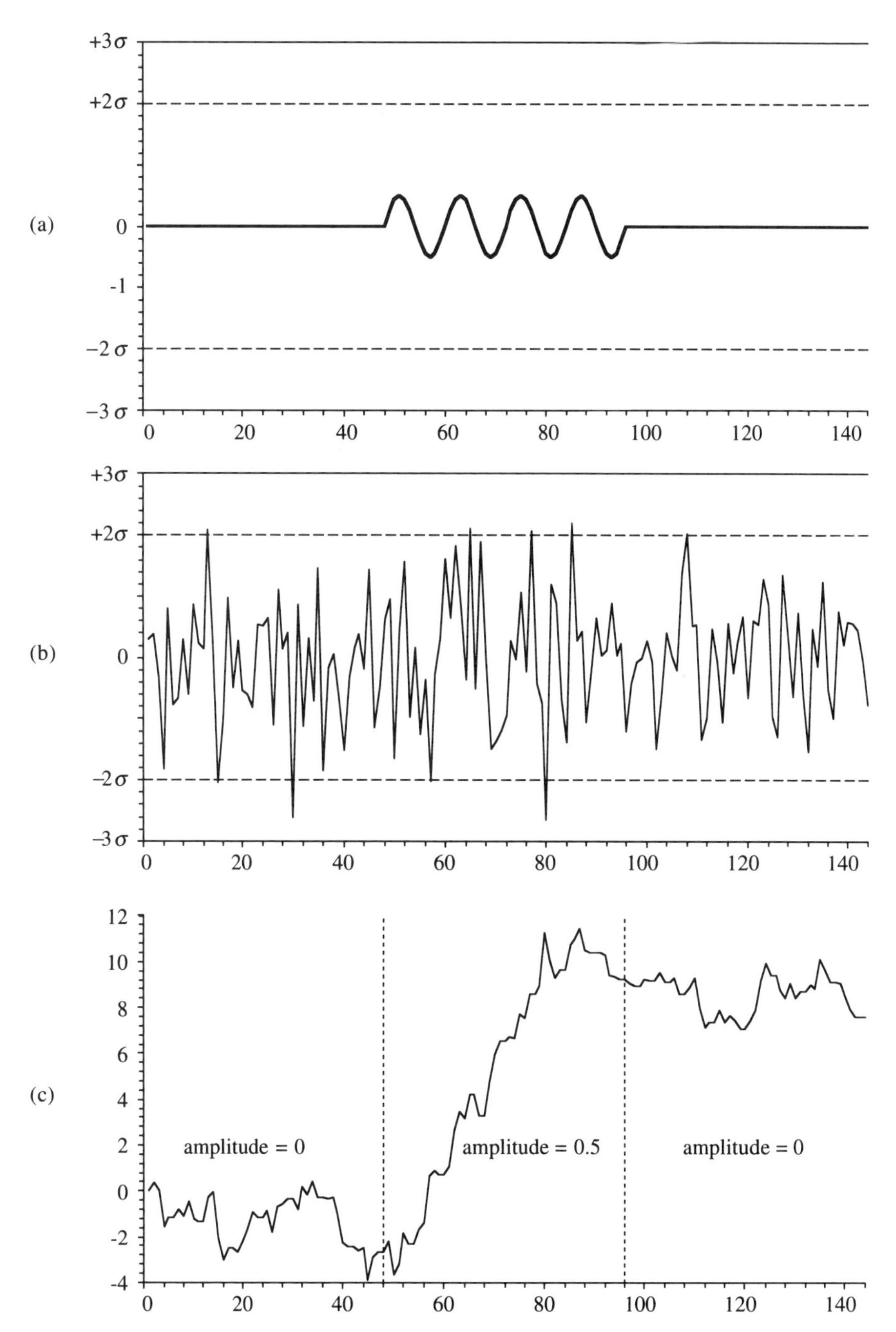

Figure 10.2 Detection of a sine wave: (a) an intermittent sine wave with amplitude 0.5σ; (b) the sum of the sine wave and white noise with $\sigma = 1$; and (c) the Cuscore statistic applied to the data of (b).

For this problem the appropriate Cuscore statistic is

$$Q = \sum (y_t - \beta t)t. \tag{10.2}$$

In this expression β is the normal rate of tool wear per unit interval, $t = 1, 2, 3, \ldots$ is the number of intervals of elapsed time, and $y_t - \beta t$ is the residual deviation from the line representing normal wear. The Cuscore statistic requires that each such residual is multiplied by t before summing.

For illustration, suppose the normal rate of tool wear is one unit per time interval (say, $\frac{1}{10{,}000}$ inch per unit interval), so $\beta = 1$; and also that the measured wear is subject to error with a standard deviation of $\sigma = 2$. In the data in Figure 10.3a it is supposed that after an initial period of ten units of time, during which the tool is wearing at the standard rate of 1 unit per time interval, the wear increases to 1.25 units per time interval. Although the change of slope is not particularly obvious in the original data, the Cuscore plot shown in Figure 10.3b reacts to this change very quickly.

For this problem it might have seemed appropriate simply to plot the Cusum of the residuals from the standard line. However, a more sensitive check will be provided by the Cuscore. Further examples will be presented after the following discussion.

10.2 HOW ARE CUSCORE STATISTICS OBTAINED?

It is evident that the Cuscore statistics can be of value to monitor process deviations of specific kinds and possibly to provide a more satisfactory basis for the choice of standard monitoring tools. The derivation of Cuscore statistics is discussed more fully in Appendix 10A. What follows provides a more intuitive understanding. As in Chapter 1 we regard an explanatory statistical model as a recipe for generating white noise from available data. For example, a model $y_t = T + a_t$ for a process randomly varying about its target may equally well be written $a_t = y_t - T$. This says that if you subtract the target value T from each of the observations y_t you will get white noise. Similarly, the null tool wear model that can be written $a_t = y_t - \beta t$ says that if you subtract βt from each observation y_t you will get white noise. Thought about in this way, these statistical models, *provided they are correct,* are recipes for generating informationless white noise.

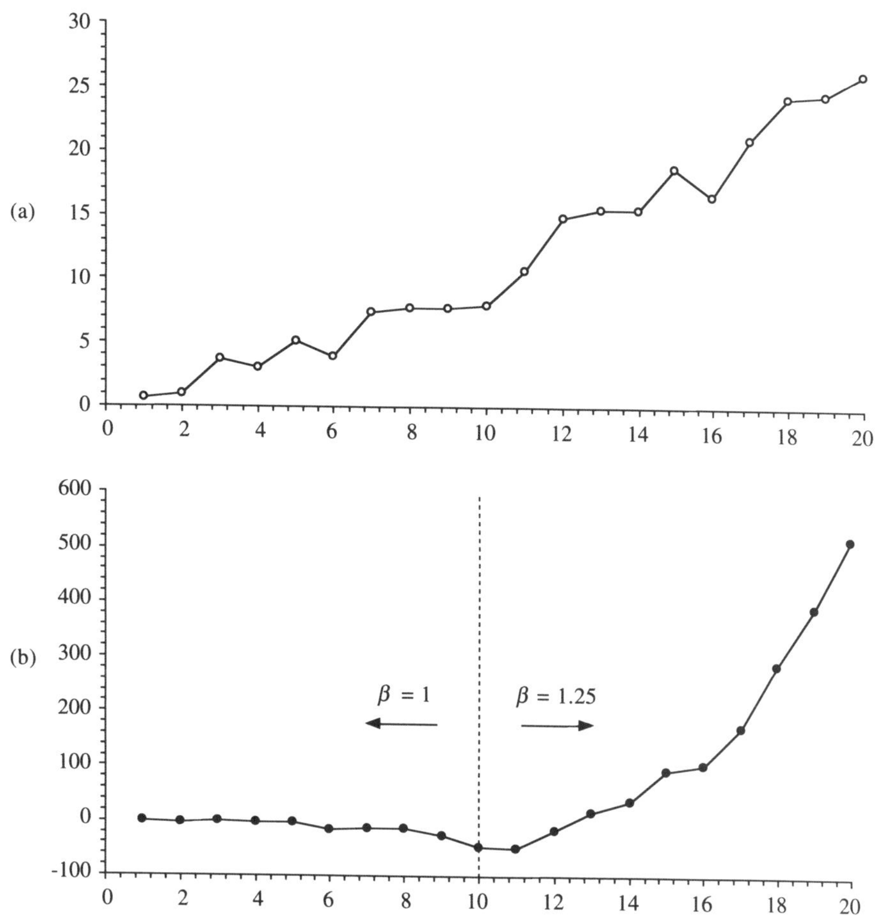

Figure 10.3 Tool wear example: (a) a linear trend, with gradient $\beta = 1.0$ for $n \leq 10$ and $\beta = 1.25$ for $n > 10$; with normal independent errors and $\sigma = 2$; and (b) a plot of the Cuscore.

To put it slightly differently, *if the model is correct* the quantities a_t on the left of each of the equations will contain no information about the model because the model on the right of the equation removes it all.[3] It is because the model may *not* be correct that careful inspection of residuals is so important in checking its adequacy.

More specifically, if, on the one hand, the residuals apparently consist of a random sequence[4] then you have no basis for disbelieving the

[3] These residual quantities do, of course, contain information about σ_a, which is, however, assumed known or in practice has been estimated from previous data.
[4] The fact that residuals are random does not, however, *prove* the adequacy of the model, because an omitted variable might itself be varying randomly.

model; if, on the other hand, they seem to contain a pattern, then this could point to model inadequacy and even allow you to make a guess at its source. It might be found, for instance, that the residuals were correlated with the particular ambient temperature occurring when each observation was made. Alternatively, a residual plot might raise the possibility that certain included variables had been modeled inadequately.

Unfortunately, however, while specific patterns in the residuals can bear the "signature" of the discrepancy, this signal may be buried in so much noise that you need very sensitive methods to detect it. In determining the appropriate form for the Cuscore statistic for a given signal, you are in fact determining the pattern of the residuals that this signal will produce and providing for optimal detection of that pattern. To understand how this is done, look at Table 10.1, which shows how the appropriate Cuscore may be derived for (a) a step change, (b) a sine wave, and (c) a slope change.

The first row of the table shows the null model: the model assuming that no discrepancy occurs of the feared kind. This is first shown in its usual form with y_t on the left of the equation. Immediately below is the same model rearranged with the residual quantity a_{t0} on the left. The zero in a_{t0} is added to the subscript of a_t to indicate that, in general, the a_{t0} values are just residuals; they are not a white noise sequence unless the null model happens to be true. The second row in the table shows the discrepancy model in both forms. The discrepancy model is supposed *true* when the correct value δ of the discrepancy parameter is employed; the residual a_t values then form a white noise sequence.

TABLE 10.1 Noise, Signals, Models, Detectors, and Cuscore Statistics for (a) a Step Change in the Mean, (b) an Intermittent Sine Wave in the Mean, and (c) a Change of Slope in the Mean

	(a)	(b)	(c)
Noise	White noise (a_t)	White noise (a_t)	White noise (a_t)
Signal	Step change	Sine wave	Slope change
Null model	$y_t = T + a_{t0}$	$y_t = T + a_{t0}$	$y_t = \beta t + a_{t0}$
	$a_{t0} = y_t - T$	$a_{t0} = y_t - T$	$a_{t0} = y_t - \beta t$
Discrepancy model	$y_t = T + \delta + a_t$	$y_t = T + \delta \sin x_t + a_t$	$y_t = \beta t + \delta t + a_t$
	$a_t = y_t - T - \delta$	$a_t = y_t - T - \delta \sin x_t$	$a_t = y_t - \beta t - \delta t$
Detector $r_t = (a_{t0} - a_t)/\delta$	1	$\sin x_t$	t
Cuscore $Q = \Sigma\, a_{t0} r_t$	$\Sigma\,(y_t - T)$	$\Sigma\,(y_t - T)\sin x_t$	$\Sigma\,(y_t - \beta t)t$
Value of Q when the discrepancy model is true	$\Sigma\, a_t + [\delta n]$	$\Sigma\, a_t \sin x_t + [\delta\, \Sigma \sin^2 x_t]$	$\Sigma\, a_t t + [\delta\, \Sigma\, t^2]$

As is explained more fully in Appendix 10A, the Cuscore statistic Q evaluated at $\delta = 0$ is

$$Q = \sum a_{t0} r_t.$$

This is the sequentially accumulated product of the "null" residuals a_{t0} with the appropriate "detector" r_t. The detector r_t measures the rate of change of a_t as the discrepancy parameter δ is changed from its zero value. For the examples so far considered, a change in δ produces a linear change in a_t and for such linear models the detector is

$$r_t = (a_{t0} - a_t)/\delta.$$

The r_t values for the various models are shown in the third row of the table and the resulting Cuscore statistics in the fourth row. For example, if you are looking for a step change buried in white noise, then from the first column of Table 10.1: $a_{t0} = y_t - T$, $a_t = y_t - T - \delta$, and $r_t = \delta/\delta = 1$. So finally, the Cuscore statistic is $\Sigma\, a_{t0} r_t = \Sigma(y_t - T)$. That is, it is the Cusum already discussed in Chapter 3.

The effectiveness of these efficient score statistics arises from the fact that, as soon as the feared discrepancy makes an appearance, the residuals a_{t0} will begin to contain a component r_t that resonates (correlates) with the multiplying detector. The fifth row of Table 10.1 shows the value that the Cuscore statistic takes when the discrepancy model is true. The first term in this expression is the value that the Cuscore would have had if the null model had been true (if δ had been equal to zero). When δ is not equal to zero, the quantities shown in square brackets, which have the same sign as δ, get bigger and bigger as the number of observations n increases.

Cuscore statistics work on the same principle as a radio tuner. The detectors—(1, 1, 1, 1, 1, . . .) for model (a), ($\sin x_1$, $\sin x_2$, $\sin x_3$, $\sin x_4$, $\sin x_5$, . . .) for model (b), and (1, 2, 3, 4, 5, . . .) for model (c)—are chosen to synchronize with any similar component pattern[5] existing in the residuals.

As will be illustrated by further examples, you can obtain Cuscore statistics for almost any kind of feared signal in almost any kind of noise. We illustrate with the problem of detecting a "rectangular bump" buried in white noise.

[5] If you are familiar with linear regression analysis or analysis of factorial designs, note that the sum Σxy, used to obtain the regression coefficients and to obtain the "effects" of the factors, works on the same principle. It is looking for a component in the data y that synchronizes with the "detector" x.

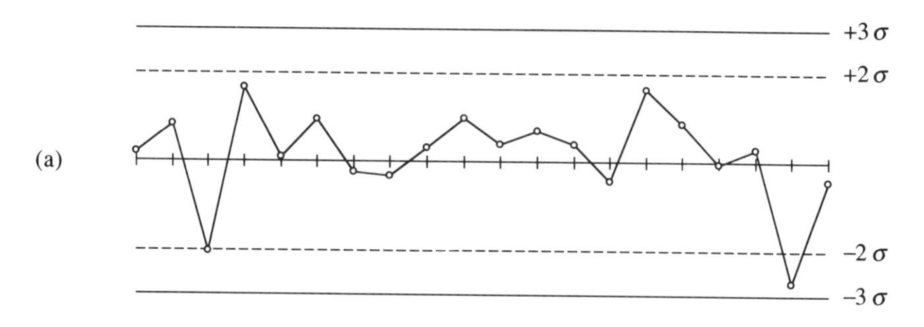

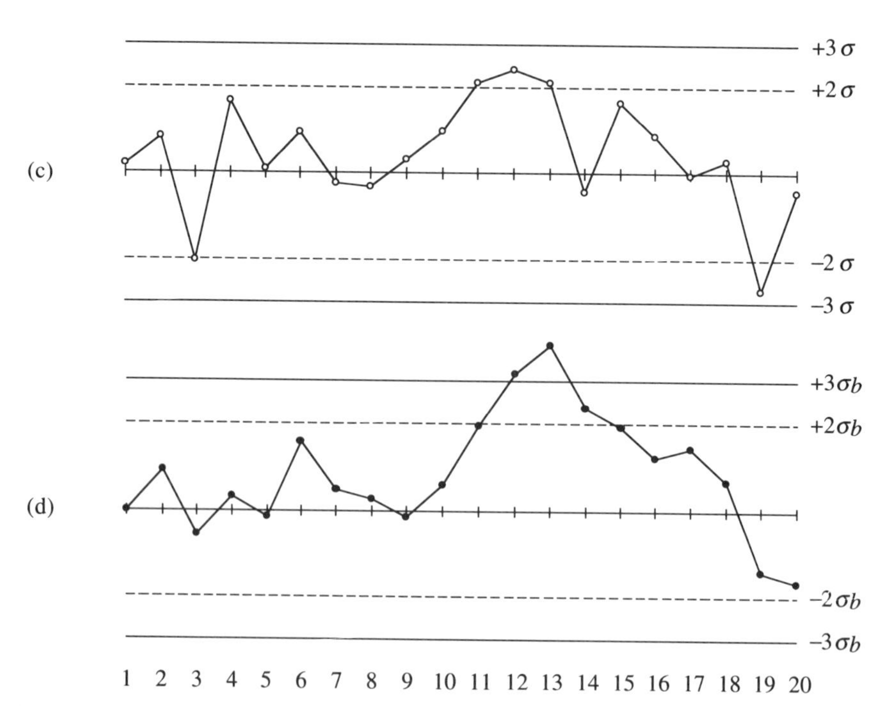

Figure 10.4 (a) A white noise sequence of 20 observations with $\sigma = 1$ representing a process in a state of control; (b) a rectangular bump of duration $b = 3$ and magnitude 1.7; (c) a Shewhart chart for the white noise sequence plus the bump; and (d) an AMA chart in which at a given time the last three observations are averaged with limit lines at $\pm 2\sigma_b$ and $\pm 3\sigma_b$.

Looking for a Rectangular Bump

It is shown in Sections 10.4 and 10.5 that, to detect a signal characterized by an individual "spike" in a background of white noise, the Cuscore reproduces the Shewhart chart: that is, it tells you to plot the individual deviations $y_t - T$. Now by a spike we mean a short-term deviation lasting only one unit interval. But suppose that for a particular process you thought that a deviation was likely to be somewhat more prolonged, lasting perhaps two or three unit intervals. Let's call a uniform change in level lasting for b unit intervals a *bump*. Then, as is more fully explained in Section 10.4, the appropriate Cuscore is the sum of the last b deviations,

$$(y_t - T) + (y_{t-1} - T) + (y_{t-2} - T) + \cdots + (y_{t-b+1} - T).$$

Equivalently, after dividing by b, it is the average of the last b observations, that is, an *arithmetic* moving average (AMA). Since we have supposed the original observations y_t have standard deviation σ and are independent, a moving average of b successive values has standard deviation $\sigma_b = \sigma/\sqrt{b}$. So an *arithmetic moving average chart* can be constructed for plotting the AMA on a Shewhart-like chart but with the control lines at $\pm 2\sigma_b$ and $\pm 3\sigma_b$, that is, at $\pm 2\sigma/\sqrt{b}$ and $\pm 3\sigma/\sqrt{b}$.

For illustration, suppose that for a process, which normally varies about its target value with standard deviation $\sigma = 1$, you want a chart that is particularly sensitive to a bump lasting for 3 unit intervals. Figure 10.4a shows 20 observations from a white noise series with standard deviation $\sigma = 1$ representing the process in "normal state." Figure 10.4b shows the bump signal lasting three intervals and terminating at $t = 13$. A Shewhart chart for the series plus the signal is shown in Figure 10.4c with $\pm 2\sigma$ and $\pm 3\sigma$ limit lines. The corresponding AMA chart with $b = 3$ is shown in Figure 10.4d with limit lines at $\pm 2\sigma_b$ and $\pm 3\sigma_b$, where the occurrence of the signal is clearly indicated by two points lying outside the $\pm 3\sigma_b$ limit lines. (Notice that a Shewhart chart augmented with the Western Electric rules also produces an action signal, but with greater risk of being wrong.)

10.3 EFFICIENT MONITORING CHARTS

The Cusum and other Cuscores, such as are given for illustration in Table 10.1, are examples of cumulated *efficient score* statistics. Roughly speaking, what Fisher meant by an "efficient" statistic was one that, given the model, made maximum use of the data[6].

As already argued, a system of process monitoring can be thought of as a search for a specific kind of signal in a specific kind of noise and the nature of the signal and the nature of the noise determine the appropriate Cuscore. A question of some interest is: For the detection of *what kind* of signals buried in *what kind* of noise are Shewhart charts, AMA charts, EWMA charts, and Cusum charts efficient? It turns out that, if the background disturbance is white noise, the Shewhart chart is efficient for looking for a spike, the AMA for looking for a rectangular bump, the EWMA for looking for an exponentially increasing signal (with discount factor the same as that for the EWMA statistic), and the Cusum for looking for a step change (Figure 10.5).

At first sight the exponential signal associated with the EWMA seems a trifle strange. However, an adaptation of an argument due to J. S. Hunter (1986) shows one way in which its usefulness might be explained. In discussing the Shewhart chart in Chapter 3, we mentioned the use by practitioners of a series of auxiliary rules (such as the Western Electric rules). Thus it might be decided that a process was not in a state of control if two out of three consecutive points were beyond the two-sigma limits or if four out of five points were beyond the one-sigma limits and so forth. If the discrepant points were *on the same side* of the target, then these rules are in effect looking for a change in a specific direction that might last more than one time interval. Figure 10.6 shows how, by adding together a series of such bumps, the aggregate can resemble an exponential signal. Technically, the EWMA is looking for fixed percentage change that does not end at any specific time. In practice, it may detect any temporary increase of unknown gradient.

10.4 A USEFUL METHOD FOR OBTAINING THE DETECTOR WHEN LOOKING FOR A SIGNAL IN NOISE

Figure 10.5 showed how a signal could be represented by a series of spikes. For illustration, suppose we are looking for a signal that could

[6] A particular parameter can be estimated by many different statistics; thus the mean of a normal distribution can be estimated by the sample average, median, or midpoint, for example. In general, suppose for a given set of data the *efficient* statistic (the sample average in the above illustration) has variance V_E. Then *for large samples,* every other statistic that provides an estimate will have larger variance. If the variance of some other statistic is V_0, then $E = V_E/V_0$ is called the *efficiency* of that statistic. For example, if $E = 0.5$, use of the inefficient statistic is equivalent to throwing away half the data. In practice, efficient statistics usually work well for small samples also.

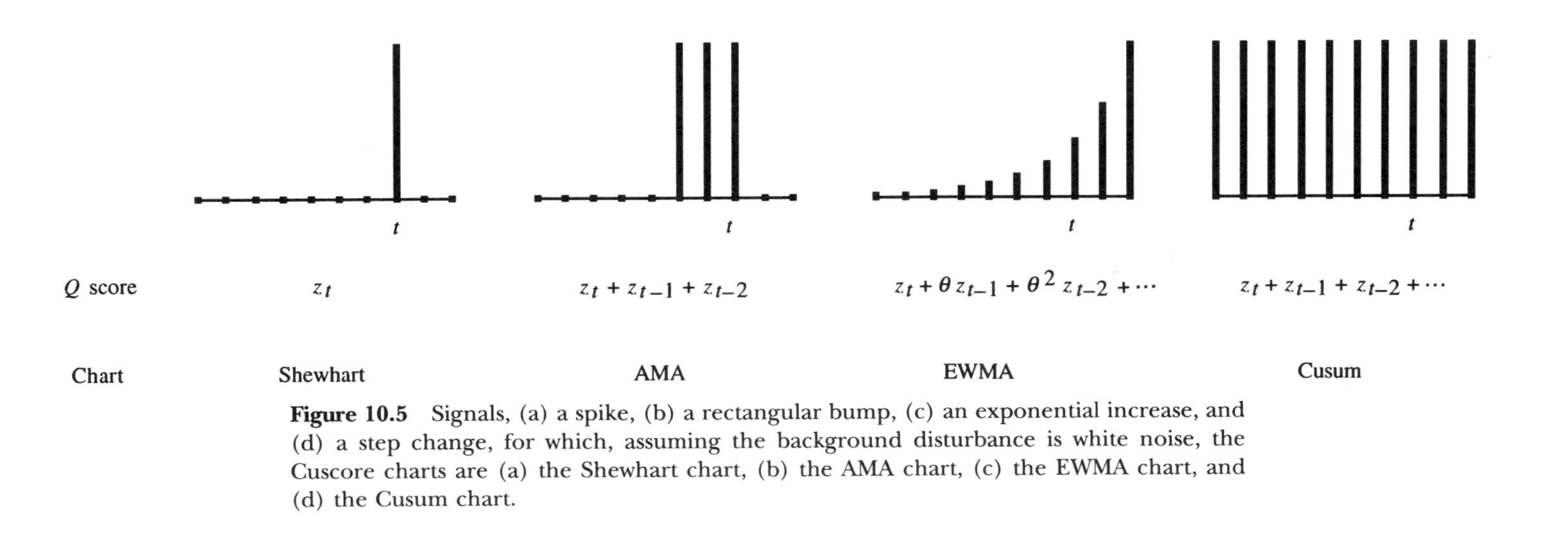

Figure 10.5 Signals, (a) a spike, (b) a rectangular bump, (c) an exponential increase, and (d) a step change, for which, assuming the background disturbance is white noise, the Cuscore charts are (a) the Shewhart chart, (b) the AMA chart, (c) the EWMA chart, and (d) the Cusum chart.

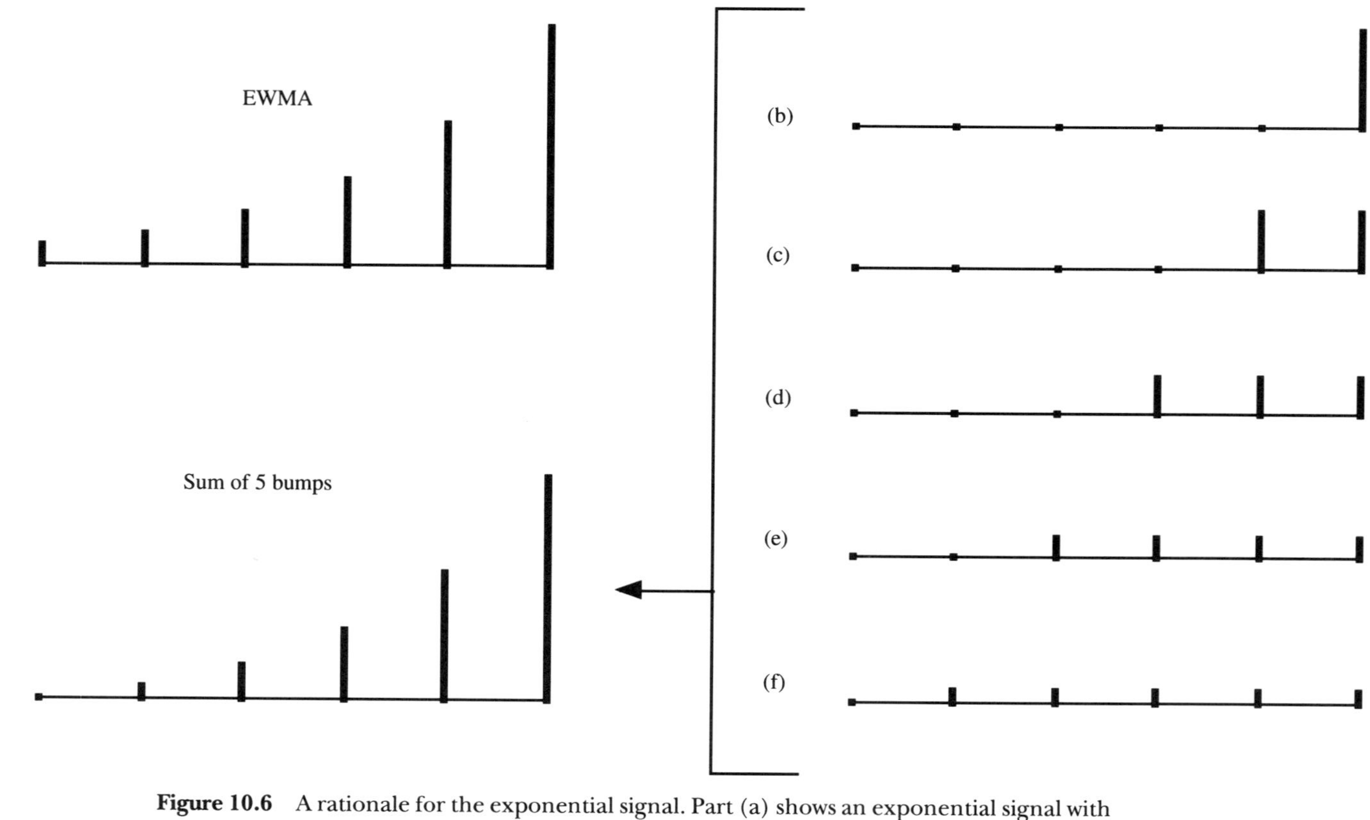

Figure 10.6 A rationale for the exponential signal. Part (a) shows an exponential signal with $\theta = 0.6$; (b–f) are bump signals of increasing length and decreasing magnitude, which sum to produce a signal (g) that mimics the exponential.

appear at some unknown time $t = i$ characterized by three successive spikes of magnitude

$$x_i = 3, \quad x_{i-1} = -2, \quad x_{i-2} = 2. \tag{10.3}$$

Then this signal could also be thought of as part of a time series $\{x_t\}$ that took the value zero except at times i, $i - 1$, and $i - 2$.

Now suppose that some multiple $-\delta$ of this signal is buried in a disturbance (noise) series $\{z_t\}$ represented by some known (or, in practice, estimated) stationary or nonstationary linear time series model, such as are discussed in Chapters 5 and 12. Then the observed series would be the sum of the disturbance series and the signal series $\{z_t - \delta x_t\}$.

Again, for illustration, let's suppose the disturbance $\{z_t\}$ is generated by a nonstationary IMA model with $\theta = 0.5$ so that at time t

$$z_t - z_{t-1} = a_t - 0.5a_{t-1}. \tag{10.4}$$

Now we have seen that any such model can be regarded as a recipe for converting the data $\{z_t\}$ to white noise $\{a_t\}$. In particular, the model (10.4) can be written

$$a_t = z_t - z_{t-1} + 0.5a_{t-1}. \tag{10.5}$$

It is shown in Appendix 10B that the detector r_t satisfies a relationship that exactly parallels Equation (10.5). Specifically, a detector series $\{r_t\}$ may be obtained from the signal series $\{x_t\}$ using the equation

$$r_t = x_t - x_{t-1} + 0.5r_{t-1}, \tag{10.6}$$

from which the Q statistic

$$Q = \sum a_{t0} r_t \tag{10.7}$$

may be obtained.

In this particular illustration the $\{x_t\}$ series, and hence the detector series$\{r_t\}$, is zero except at time $i - 2$, $i - 1$, and i and then

$$r_{i-2} = x_{i-2} - x_{i-3} + 0.5r_{i-3} = 2 - 0 + (0.5 \times 0) = 2,$$
$$r_{i-1} = x_{i-1} - x_{i-2} + 0.5r_{i-2} = -2 - 2 + (0.5 \times 2) = -3,$$
$$r_i = x_i - x_{i-1} + 0.5r_{i-1} = 3 + 2 + [0.5 \times (-3)] = 3.5.$$

The appropriate Cuscore at time t is thus

$$Q = 3.5a_{t0} - 3a_{t0-1} + 2a_{t0-2}.$$

Although we have described this method for obtaining a Cuscore for a special case, the procedure is perfectly general and is as follows. We can always write a linear time series model with a_t on the left and other terms on the right. Now suppose $\{x_t\}$ is a series with individual values all zero except where the signal occurs. To generate the detector series $\{r_t\}$, we replace the a_t terms by r_t terms and the z_t terms by x_t terms in the time series model formula as in Equations (10.5) and (10.6). Now (e.g., see Section 5.3) $a_t = z_t - \hat{z}_t$, where $\hat{z}_t$ is the minimum mean square error forecast of z_t made at time $t - 1$. Therefore we can write $r_t = x_t - \hat{x}_t$, where $\hat{x}_t$ is obtained by carrying out the same operations on the x's as were performed on the z's to find $\hat{z}_t$. So an alternative expression for the Cuscore is

$$Q = \sum (z_t - \hat{z}_t)(x_t - \hat{x}_t). \tag{10.8}$$

As an example, suppose you were using an EWMA with smoothing constant $\theta = 0.5$ to forecast the series assumed to be generated by Equation (10.4). Also, for some reason you were anxious to detect in the z's the signal $(3, -2, 2)$ of Equation (10.3). Since the x's before x_{i-2} are all zero, $\hat{x}_{i-2} = 0$, $\hat{x}_{i-1} = 1$, and $\hat{x}_i = -0.5$, and consequently $x_{i-2} - \hat{x}_{i-2} = 2$, $x_{i-1} - \hat{x}_{i-1} = -3$, and $x_i - \hat{x}_i = 3.5$. Finally, since $z_i - \hat{z}_i = a_{i0}$ by using Equation (10.8) we obtain the same expression for Q as before. Sensitive surveillance for this signal would therefore be provided by a run chart for Q which, if desired could show $\pm 2\sigma_Q$ and $\pm 3\sigma_Q$ limits, using

$$\sigma_Q = \sigma_a \sqrt{3.5^2 + 3^2 + 2^2} = 5.02\sigma_a.$$

Notice that depending on the nature of the noise the detector values are not necessarily proportional to those of the signal.

10.5 LOOKING FOR A SINGLE SPIKE

When a Shewhart chart is used to look for a single discordant deviation from target, it is equivalent to a Cuscore chart appropriate when the disturbance z_t is white noise and the signal x_t is a single spike occurring at some unknown time $t = i$. The signal series $\{x_t\}$ then consists of zeros with a single spike at $t = i$, and the null model is $a_{t0} = z_t$; so the Cuscore[7] is

$$Q = a_{t0} = z_t - \hat{z}_t = y_t - T.$$

More generally, from the argument of the last section it is clear that, for *any* linear time series model, the spike signal series $\{x_t\}$ and the detector series $\{r_t\}$ will be identical and the Cuscore statistic will always be

$$Q = a_{t0} = z_t - \hat{z}_t.$$

Thus to detect a single discrepant observation, you must always plot the residual series $\{a_{t0}\}$ and, if desired, this can be plotted in the form of a Shewhart chart with limit lines at $\pm 2\sigma_a$ and $\pm 3\sigma_a$. (See, for example, Berthouex, Hunter and Pallesen, 1978.) In particular for an AR(1) model, for example (see also Box and Ramírez, 1992; Ramírez, 1989), you should not plot z_t but

$$Q = a_{t0} = z_t - \hat{z}_t = z_t - \phi z_{t-1}.$$

This result can be useful for monitoring a system that is reasonably stationary but in which successive deviations are autocorrelated and may be modeled by an AR(1) process. (See also, e.g., Alwan and Roberts, 1988; Montgomery and Mastrangelo, 1991.)

As a further example consider the IMA model. The appropriate Cuscore to detect a single discrepant observation is

$$Q = a_{t0} = z_t - \hat{z}_t = z_t - \tilde{z}_{t-1} = z_t - \lambda(z_{t-1} + \theta z_{t-2} + \cdots).$$

This statistic makes very good sense because the nonstationary IMA series has no mean. So it uses $\hat{z}_t$ as a "local mean" and determines whether a deviation has occurred from that.

10.6 SOME TIME SERIES EXAMPLES

We now consider some examples of a somewhat different kind concerning time series models.

[7] Note the best forecast $\hat{z}_t$ of the white noise deviation that can be made at time $t-1$ is zero.

A Possible Change in the Parameter of an Autoregressive Time Series

Suppose you believe that variation about the process target T is represented by a first-order autoregressive time series model

$$z_t = \phi z_{t-1} + a_t \tag{10.9}$$

where z_t is the deviation $y_t - T$ from the target value. Suppose the parameter ϕ has been assumed to be equal to some value ϕ_0 but you wish for a sensitive check that looks for a possible change.

The null model is

$$a_{t0} = z_t - \phi_0 z_{t-1}$$

and the discrepancy model is

$$a_t = z_t - (\phi_0 + \delta) z_{t-1}.$$

Thus $r_t = z_{t-1}$ and the Cuscore statistic is

$$\begin{aligned} Q &= \sum_{i=1}^{t} (z_i - \phi_0 z_{i-1}) z_{i-1} \\ &= \sum_{i=1}^{t} z_i z_{i-1} - \phi_0 \sum_{i=1}^{t} z_{i-1}^2. \end{aligned} \tag{10.10}$$

The logic of the choice is that[8] $\hat{\phi}^{(t)} = \Sigma_{i=1}^{t} z_i z_{i-1} / \Sigma_{i=1}^{t} z_{i-1}^2$ is an efficient estimate of ϕ obtained from the data up to time t. Thus

$$Q = (\hat{\phi}^{(t)} - \phi_0) \sum_{i=1}^{t} z_{i-1}^2,$$

so Q is continuously looking for a discrepancy between $\hat{\phi}^{(t)}$ and ϕ_0.

A Cuscore Test for Lag 1 Autocorrelation

If in Equation (10.10) we set $\phi_0 = 0$, then the Cuscore becomes a test for lag 1 autocorrelation, that is, correlation between adjacent observations:

$$Q = \sum_{i=1}^{t} (y_i - T)(y_{i-1} - T) = \hat{\phi}^{(t)} \sum_{i=1}^{t} z_{i-1}^2, \tag{10.11}$$

[8] Also, $\hat{\phi}^{(t)}$ is an estimate at time t of the correlation coefficient between observations one step apart called the *lag 1 autocorrelation coefficient*.

where $\hat{\phi}^{(t)}$ is an estimate of the lag 1 autocorrelation made at time t.

Monitoring the Parameter of an IMA Time Series

The IMA is an important time series model first discussed in Chapter 5, where it appeared as the model that produced the EWMA as a best forecast and in later chapters it was frequently used as a model for a nonstationary disturbance.

Suppose we are monitoring the residual a_{t0} values from a time series that we believe can be represented by an IMA model

$$z_t - z_{t-1} = a_t - \theta a_{t-1} \tag{10.12}$$

and we want to be made aware of any major change in the smoothing constant θ from its assumed value θ_0. We might monitor these residuals to check either a forecasting system or a system of feedback control.

The null model would be

$$z_t - z_{t-1} = a_{t0} - \theta_0 a_{t0-1}$$

and the discrepancy model would be

$$z_t - z_{t-1} = a_t - (\theta_0 + \delta)a_{t-1},$$

where the change δ in the parameter θ could be in either direction.

It is shown in Appendix 10C that the detector r_t appropriate for this model is

$$r_t = -\hat{a}_{t0}/\lambda_0,$$

where $\hat{a}_t$ is an exponential average with smoothing constant θ of *past values of the* a_t terms

$$\hat{a}_t = \tilde{a}_{t-1} = \lambda(a_{t-1} + \theta a_{t-2} + \theta^2 a_{t-3} + \cdots),$$

which can continuously be updated like any other EWMA using the formula

$$\hat{a}_t = \lambda a_{t-1} + \theta \hat{a}_{t-1},$$

and a_{t0} is obtained by setting $\theta = \theta_0$.

The Cuscore statistic appropriate for the null model when $\theta = \theta_0$ is therefore

$$Q = -\frac{1}{\lambda_0} \sum a_{t0} \hat{a}_{t0}.$$

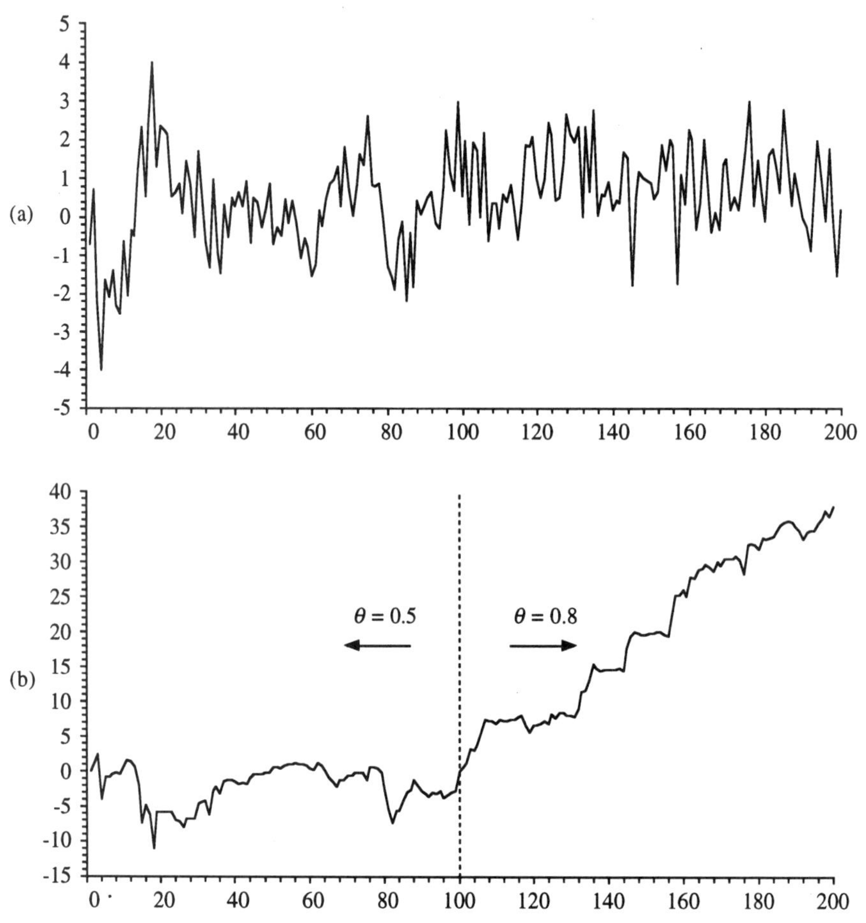

Figure 10.7 Monitoring the parameter of an IMA time series: (a) a time series of 200 observations generated with an IMA (0, 1, 1) model with $\theta = 0.5$ for the first 100 observations and $\theta = 0.8$ for the last 100 observations; and (b) a Cuscore chart, computed with $\theta_0 = 0.5$, for detecting this change.

with $\lambda_0 = 1 - \theta_0$.

This statistic makes excellent intuitive sense since $\hat{a}_{t0}$ can be regarded as a forecast of the residual a_{t0}. Then if θ_0 were the true value of θ, the residuals a_{t0} would be independent and it would not be possible at time $t - 1$ to make a useful "forecast" $\hat{a}_{t0}$ of the next residual a_{t0}. The Cuscore statistic is continuously checking that a_{t0} and its "forecast" $\hat{a}_{t0}$ made from past data are indeed uncorrelated.

For illustration we show in Figure 10.7a a generated IMA series of 200 values, with $\sigma_a = 1$ in which after 100 observations θ was changed from 0.5 to 0.8. Figure 10.7b shows the Cuscore for a null model with

$\theta_0 = 0.5$. While there is no particularly noticeable change in the series itself, a very marked change is visible in the Cuscore at about that time.

Is the Process in a State of Statistical Control? A Sensitive Check

We have seen that the IMA model, with $\lambda = 1 - \theta$ greater than zero, is of central importance in describing a process that is out of control.

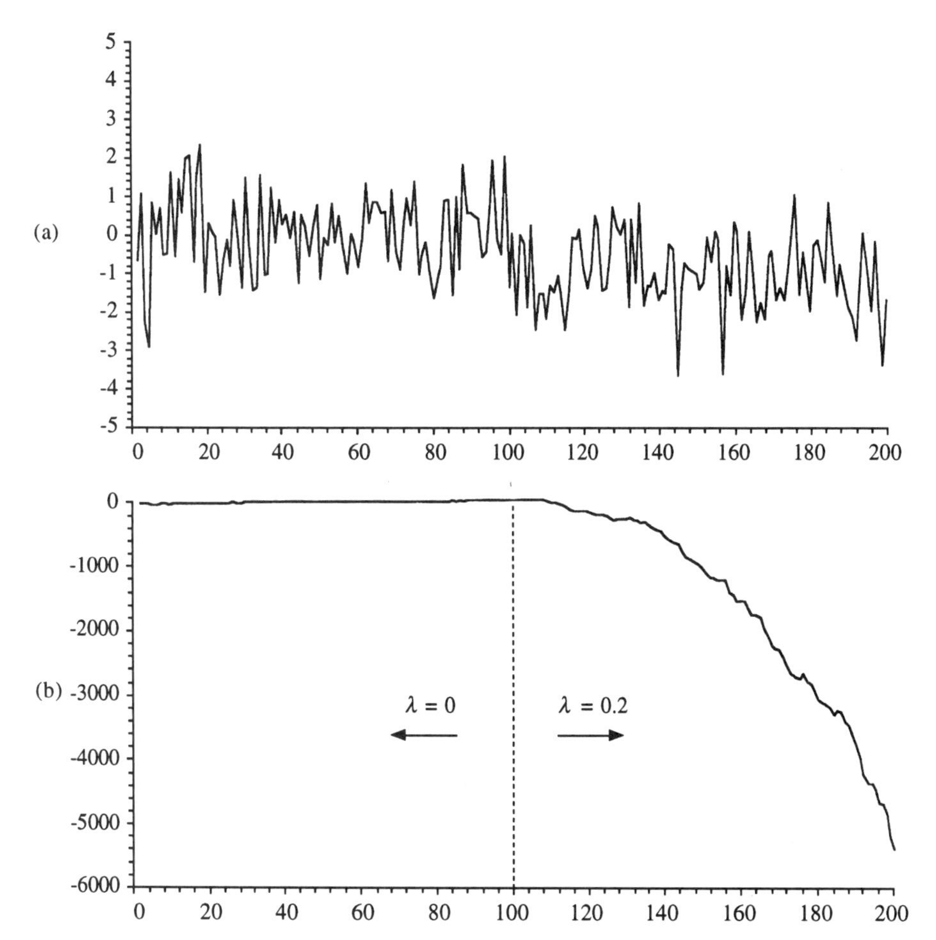

Figure 10.8 Monitoring the stability of a process: (a) a time series of 200 observations, the first 100 of which are white noise ($\lambda = 0$) and the second 100 generated by nonstationary IMA model with $\lambda = 0.2$; and (b) a Cuscore chart for detecting this change.

A sensitive monitor for checking on the state of control of a process is obtained using the analysis discussed above, with $\lambda_0 = 0$. The null model is then that the process is in a perfect state of control with $y_t - T = a_{t0}$ and the a_{t0} values white noise. The discrepancy model is an IMA process with $\delta = \lambda$, where λ has some small value representing a mild degree of nonstationarity.

For illustration, we have plotted in Figure 10.8 a time series of 200 observations with $\sigma_a = 1$. For the first 100 observations $\lambda_0 = 0$ and the deviations from target $y_t - T$ are thus white noise representing a process in a perfect state of statistical control. At time $t = 100$ the nonstationarity parameter in the IMA model was changed to $\lambda = 0.2$. The change from a stable to an unstable process is particularly evident in the Cuscore plot.

APPENDIX 10A THE LIKELIHOOD, FISHER'S EFFICIENT SCORE, AND CUSCORE STATISTICS

Cuscore statistics are particular examples of Fisher's efficient score statistics, which in turn are derived from the likelihood (Fisher, 1925). As was illustrated earlier, the statistical models we have considered in this chapter can all be written in the form

$$a_i = a_i(y_i, x_i, \theta) \qquad i = 1, 2, \ldots, t, \tag{10A.1}$$

where the y_i are observations, θ is some unknown parameter, and the x_i are known quantities and, in particular, the known levels of input variables. Thus, for example, the tool wear model could be written

$$a_i = y_i - i\theta \tag{10A.2}$$

with θ the rate of tool wear and $i = x_i$ the elapsed time.

The Likelihood

If the a_i values are supposed normal and independent, their joint probability density for any specific choice of θ can be written

$$p(\mathbf{a} \mid \theta) = p(a_1, a_2, \ldots, a_t \mid \theta) = \text{constant} \times \sigma^{-t} \exp\left[-\frac{1}{2}\sum\left(\frac{a_i^2}{\sigma^2}\right)\right]$$

and, unless otherwise stated, Σ will indicate summation from $i = 1, 2, \ldots, t$.

If σ is known, then, apart from a constant,

$$\ln[p(\mathbf{a} \mid \theta)] = -\frac{1}{2\sigma^2} \sum a_i^2. \tag{10A.3}$$

After the data have actually occurred, for each choice of θ, a set of a_i values can be calculated from Equation (10A.1). The logarithm of the likelihood associated with this choice of θ is then given by the same mathematical expression as Equation (10A.3), but with the understanding that it is θ that is now variable and the data values y_i that are fixed (since they have already happened). Thus, for each choice of θ, a different set of calculated a_i values will be obtained.

In particular, apart from a constant, the log likelihood for $\theta = \theta_0$ is

$$l_0 = -\frac{1}{2\sigma^2} \sum a_{i0}^2 \tag{10A.4}$$

where the a_{i0} values are obtained by setting $\theta = \theta_0$ in Equation (10A.1).

The Cuscore Statistic

Fisher's score statistic is obtained by differentiating the log likelihood with respect to the parameter θ. Thus

$$\left.\frac{\partial l}{\partial \theta}\right|_{\theta=\theta_0} = \frac{1}{\sigma^2} \sum a_{i0} r_i \quad \text{with} \quad r_i = -\left.\frac{\partial a_i}{\partial \theta}\right|_{\theta=\theta_0}. \tag{10A.5}$$

We refer to

$$Q = \sum a_{i0} r_i \tag{10A.6}$$

as the Cuscore associated with the parameter value $\theta = \theta_0$.

Now, exactly, if the model is linear in θ, and approximately, otherwise, by expanding a_i about a_{i0}, we have after rearrangement

$$a_{i0} = (\theta - \theta_0) r_i + a_i. \tag{10A.7}$$

Hence we see that, if the value of the parameter is not equal to θ_0, an increment of the pattern of r_i terms is added to that of the a_i terms. The Cuscore statistic (10A.6) sequentially correlates a_{i0} with r_i and is thus continuously searching for the presence of that particular pattern. Equivalently, since $(\hat{\theta} - \theta_0) = \Sigma\, a_{i0} r_i / \Sigma\, r_i^2$ is the least squares (and maximum likelihood) estimator of $\theta - \theta_0$,

$$Q = (\hat{\theta} - \theta_0) \sum_{i=1}^{t} r_i^2 . \tag{10A.8}$$

When plotted against t, the Cuscore can thus be expected to provide a very sensitive check for change in θ. Such changes will be indicated by a change in slope of the plot just as changes in the mean are indicated by a change in slope of the Cusum plot.

APPENDIX 10B A USEFUL PROCEDURE FOR OBTAINING AN APPROPRIATE CUSCORE STATISTIC

Denote by z_t the deviation of an observation from a target value (which may be its mean). Then any disturbance (noise) that can be represented by a linear time series model (which includes, in particular, the very wide class of ARIMA models; e.g., see Box and Jenkins, 1970) may be written

$$z_t = \hat{z}_t + a_t,$$

where, if a_t is a white noise error,

$$\hat{z}_t = w_1 z_{t-1} + w_2 z_{t-2} + \cdots$$

is a MMSE forecast of z_t made at time $t - 1$ whose weights $w_1, w_2, \ldots$ are determined by the particular parameters of the time series model and form a convergent series.

Then, as before, the null model is

$$a_{t0} = z_t - \hat{z}_t.$$

Now suppose the series may contain an added component due to a signal of arbitrary known form x_t, x_{t-1}, ... Then the discrepancy model is

$$a_t = (z_t - \delta x_t) - w_1(z_{t-1} - \delta x_{t-1}) - w_2(z_{t-2} - \delta x_{t-2}) - \cdots$$

and consequently the detector r_t is

$$r_t = x_t - w_1 x_{t-1} - w_2 x_{t-2} - \cdots;$$

that is,

$$r_t = x_t - \hat{x}_t,$$

where $x_t - \hat{x}_t$ is obtained from the x series in precisely the same way that $a_{t0} = z_t - \hat{z}_t$ is obtained from the z series.

So the Cuscore is

$$Q = \sum (z_t - \hat{z}_t)(x_t - \hat{x}_t) = \sum a_{t0} r_t.$$

APPENDIX 10C THE DETECTOR SERIES FOR AN IMA MODEL

The models discussed at the beginning of this chapter have the linear property that $r_i = -\partial a_i/\partial\theta$ does not depend on θ. This is not true, however, for the time series model of Equation (10.12) and in general whenever a_i is not a linear function of the parameter θ. The procedure we have outlined in Table 10.1 for obtaining r_i, however, may still be used as an approximation provided the nonlinearities are not too great.

In any case, the score statistic can be obtained by differentiation, as we illustrate below for the IMA time series model.

The IMA time series model can be written

$$a_i - \theta a_{i-1} = y_i - y_{i-1}, \qquad (10C.1)$$

where, from the point of view of likelihood calculations, the y's are fixed.

Differentiating with respect to θ, with $r_i = -\partial a_i/\partial\theta$, yields

$$-r_i + \theta r_{i-1} - a_{i-1} = 0. \qquad (10C.2)$$

Writing this equation for times $t, t-1, t-2, \ldots, t-p$ and multiplying the second equation by θ, the third by θ^2, and so on, we have

$$
\begin{aligned}
r_t - \theta r_{t-1} &= -a_{t-1}, \\
\theta r_{t-1} - \theta^2 r_{t-2} &= -\theta a_{t-2}, \\
\theta^2 r_{t-2} - \theta^3 r_{t-3} &= -\theta^2 a_{t-3}, \\
&\vdots \\
\theta^p r_{t-p} - \theta^{p+1} r_{t-p-1} &= -\theta^p a_{t-p-1}.
\end{aligned} \tag{10C.3}
$$

For p sufficiently large, the terms involving powers of θ higher than θ^{p+1} will be negligible and then after adding all the equations in (10C.3) we obtain

$$r_t = -(a_{t-1} + \theta a_{t-2} + \theta^2 a_{t-3} + \cdots).$$

After setting $\theta = \theta_0$ this is the detector signal $-\hat{a}_{t0}/\lambda_0$.

Problem 10.1. The data in Figure 10.2b are the sum of the noise and sine wave in the following tables:

Noise									
0.309	0.395	−0.310	−1.818	0.793	−0.772	−0.672	0.293	−0.589	0.876
0.233	0.136	2.095	0.271	−2.044	−1.033	0.975	−0.492	0.286	−0.526
−0.604	−0.808	0.534	0.521	0.653	−1.107	1.111	0.151	0.417	−2.618
0.874	−1.119	0.330	−0.696	1.465	−1.848	−0.159	0.063	−0.663	−1.520
−0.309	0.164	0.393	−0.187	1.441	−1.139	−0.489	0.626	0.710	−2.087
−0.039	1.125	−1.216	0.176	−1.013	0.074	−1.514	0.168	0.541	1.608
0.399	1.390	0.341	−0.798	1.869	−0.507	2.143	0.401	−0.993	−0.926
−0.941	−0.955	0.021	−0.455	0.555	−0.664	1.820	−0.429	−0.508	−2.229
1.686	1.316	−0.430	−1.375	1.951	−0.149	−0.059	−1.490	−0.482	0.656
0.285	0.554	1.380	0.472	0.486	−1.203	−0.426	−0.081	−0.008	0.289
−0.076	−1.501	−0.641	0.412	0.029	−0.200	1.382	2.025	0.523	0.534
−1.347	−0.993	0.466	−0.068	−1.063	0.571	−0.256	0.232	0.661	−0.659
0.604	0.538	1.290	0.864	−0.961	−1.305	1.339	0.393	−0.630	0.737
−0.651	−1.527	0.473	−0.144	1.251	−0.532	−0.989	0.756	0.222	0.584
0.557	0.449	−0.049	−0.773						

Sine Wave									
0	0	0	0	0	0	0	0	0	0
0	0	0	0	0	0	0	0	0	0
0	0	0	0	0	0	0	0	0	0
0	0	0	0	0	0	0	0	0	0
0	0	0	0	0	0	0	0.000	0.250	0.433
0.500	0.433	0.250	0.000	−0.250	−0.433	−0.500	−0.433	−0.250	0.000
0.250	0.433	0.500	0.433	0.250	0.000	−0.250	−0.433	−0.500	−0.433
−0.250	0.000	0.250	0.433	0.500	0.433	0.250	0.000	−0.250	−0.433
−0.500	−0.433	−0.250	0.000	0.250	0.433	0.500	0.433	0.250	0.000
−0.250	−0.433	−0.500	−0.433	−0.250	0.000	0	0	0	0
0	0	0	0	0	0	0	0	0	0
0	0	0	0	0	0	0	0	0	0
0	0	0	0	0	0	0	0	0	0
0	0	0	0	0	0	0	0	0	0
0	0	0	0						

Add the signal to the noise and apply the Cuscore statistic to confirm the result shown in Figure 10.2c (or, if you prefer, generate your own data). □

CHAPTER 11

Simultaneous Adjustment and Monitoring

"Come now, and let us reason together . . . ."
The Bible, ISAIAH

In this book the importance of a coordinated approach to control using *both* monitoring and adjustment has been emphasized. Concerns sometimes expressed about the employment of feedback control can help us better understand the need for proper coordination of these techniques. These concerns have two aspects: the first relates to short-term variation while the second relates to long-term variation.

1. It has been argued that the feedback process itself can smooth out and hence conceal spikes and other signals that might have been seen in the original uncontrolled data and could have pointed to removable causes.
2. Furthermore, since long-term trends in the original data will have been compensated, clues will have been removed that might have led to the discovery of long-term causes.

The above concerns may be met as follows:

1. The methods of Chapter 10 may be employed to develop appropriate Cuscore statistics that provide efficient detection of any signal of interest while the feedback system is in actual operation.
2. By summing the process adjustments that have been made, a smoothed estimate of the disturbance may be obtained (and from this, if desired and if the process adjustment dynamics are known, the disturbance itself).

Thus information about the trends that have been compensated is not lost.

The following example illustrates the second point. In a particular company, developments in chemical analysis had led to the discovery of a previously unsuspected impurity in the feedstock. When the level of this impurity was plotted against the reconstructed disturbance, a high correlation was found, suggesting the possible need to eliminate or, at least, control this impurity.[1]

Remember that you know a number of things when you use any system of feedback control. You know the adjustment equation being used, the series of adjustments that have been made, and the deviations from target (residual errors) that have occurred. Also, you probably have an approximate idea of the dynamics of the adjustment system. This makes it possible to recover any information that the original data contained.

11.1 MONITORING AN OPERATING FEEDBACK SYSTEM

In Chapter 6 we discussed discrete integral control of a responsive process where the adjustment x_t was made proportional to the last deviation from target e_t according to the adjustment equation

$$-gx_t = Ge_t. \tag{11.1}$$

In determining the performance of such a scheme, we supposed the disturbance z_t was adequately represented by the IMA model,

$$z_t - z_{t-1} = a_t - \theta a_{t-1} \tag{11.2}$$

with σ_a the standard deviation of the a_t values.

We found that if we set $G = \lambda$, then the scheme produced minimum mean square error with $\sigma_e = \sigma_a$, but by taking G less than λ at the cost of a slight increase in σ_e, the amount of needed manipulation measured by σ_x might be reduced considerably.

We now consider how such schemes may be monitored while in operation. Remember in what follows that e_t always represents the output error, that is the deviation from target.

[1] If it were not possible to control the level of such an impurity or to eliminate it, another device, which is not further discussed here, is to compensate for it by *feedforward control:* that is, to modify the process conditions depending on the level of the preanalyzed feedstock.

Looking for a Spike in a Disturbance z_t Subjected to Integral Control

We saw in Chapter 10 that the Shewhart chart can be regarded as a Cuscore statistic that is looking for an added spike in white noise. It is shown in Appendix 11A that the Cuscore statistic denoted by Q_S, which, in a similar way, is looking for an added spike in z_t during feedback adjustment by Equation (11.1), is

$$Q_S = De_t + (1 - D)(e_t - \tilde{e}_{t-1}), \tag{11.3}$$

where $D = G/\lambda$ and, as before, $\tilde{e}_{t-1}$ is an EWMA of the output errors $e_{t-1}, e_{t-2}, \ldots$ with smoothing constant θ.

To see the logic of this choice of monitoring statistic we consider the following special situations:

$G = \lambda$: In this case $D = 1$, Equation (11.1) defines the MMSE feedback control scheme, and Equation (11.3) gives the Cuscore statistic as

$$Q_S = e_t. \tag{11.4}$$

Thus, in the absence of a signal, $e_t = a_{t0}$ is a member of an uncorrelated (white noise) series. In this case then, as might be expected, you should run a chart on the residual output errors. Limit lines on the chart would use the standard deviation $\sigma_Q = \sigma_a$.

$G = 0$: In this case the disturbance is allowed to develop with no control so that $e_t = z_t$ is the deviation from the target value. Using Equation (11.3) with $D = 0$, the appropriate Cuscore statistic to look for a spike is

$$Q_S = e_t - \tilde{e}_{t-1} = z_t - \tilde{z}_{t-1} \tag{11.5}$$

or, with $\tilde{e}_{t-1} = \tilde{z}_{t-1}$ regarded as an estimate of the next value of the series,

$$Q_S = e_t - \hat{e}_t = z_t - \hat{z}_t = a_{t0}. \tag{11.6}$$

The time series $e_t = z_t$, being nonstationary, has no mean. To look for a discordant observation it makes sense, therefore, to compare e_t with the local exponential average $\tilde{e}_{t-1}$.

The above has immediate practical relevance for monitoring the bounded adjustment schemes discussed in Chapters 8 and 9. It will be recalled that, for such schemes, the disturbance is allowed to develop until $\hat{z}_t = \hat{e}_t$ falls outside the limit lines at $\pm L$. During each period of development the Cuscore of Equation (11.6) provides the means of detecting an added spike.

Again, Q_S may be plotted between Shewhart limit lines with $\sigma_Q = \sigma_a$.

$G < \lambda$: We saw in Chapter 6 that, by setting G less than λ, the needed manipulation measured by σ_x might be considerably reduced at the cost of a small increase in the output standard deviation σ_e.

The Cuscore that is looking for a spike is then given by Equation (11.3):

$$Q_S = De_t + (1 - D)(e_t - \tilde{e}_{t-1}).$$

This compromise is a linear interpolation between e_t and $e_t - \tilde{e}_{t-1}$. The rationale is that, with $G < \lambda$, the e_t series can exhibit trends (e.g., see Figure 7.7d). The statistic takes account of the possibility that, for example, e_t might deviate substantially from its forecast even though it did not show a large deviation from target.

It was shown in Chapter 10 that the Cuscore statistic to look for a spike buried in any kind of noise represented by a linear time series model was a_{t0}, the value of a_t calculated when the null model was true. It is shown in Appendix 11A that the Cuscore statistic of Equation (11.3) is simply an expression for calculating a_{t0} from the output errors $e_t, e_{t-1}, \ldots$ and can also be written

$$Q_S = a_{t0} = e_t - (1 - D)\tilde{e}_{t-1}. \qquad (11.7)$$

Looking for an Exponential Signal in a Disturbance Subject to Integral Control

Consider the EWMA monitoring chart of Chapter 3 with smoothing constant β. From the result in the last chapter, we know that this is equivalent to using a Cuscore statistic to seek for a corresponding exponential signal in a background of white noise.

Now suppose we are looking for such an exponential signal in the nonstationary disturbance of Equation (11.2) while the process is subjected to the integral control of Equation (11.1) and to avoid notational difficulties we will, in all of what follows, omit the zero subscript from the a_{t0}'s. The corresponding Cuscore turns out to be an EWMA[2] of the a_t values calculated from Equation (11.7) for the null IMA (0, 1, 1) model (see Section 10.4). Thus

$$Q_E = (1 - \beta)(a_t + \beta a_{t-1} + \beta^2 a_{t-2} + \cdots) \tag{11.8}$$

with standard deviation

$$\sigma_Q = \sigma_a \sqrt{\frac{1 - \beta}{1 + \beta}}. \tag{11.9}$$

So that the appropriate monitoring chart employs Q_E in Equation (11.8) which may be plotted between limit lines determined from σ_Q in Equation (11.9). The necessary computations consist of calculating the a_t's from Equation (11.7) and then continually updating their EWMA, Q_E, using

$$Q_t = (1 - \beta)a_t + \beta\, Q_{t-1}.$$

It is shown in Appendix 11B that an exponentially increasing rise of the disturbance level in small steps of sizes

$$\ldots, \xi\sigma_a\beta^3(1 - \beta),\ \xi\sigma_a\beta^2(1 - \beta),\ \xi\sigma_a\beta(1 - \beta),\ \text{and}\ \xi\sigma_a(1 - \beta),$$

corresponding to times

$$\ldots, t - 3,\ t - 2,\ t - 1,\ \text{and}\ t,$$

produces an eventual rise of the disturbance of size $\xi\sigma_a$ and an increase in Q_E of size

$$\sigma_a \frac{\xi}{1 - \beta\theta} \frac{1 - \beta}{1 + \beta}. \tag{11.10}$$

Then the standardized increase in Q_E, obtained as the quotient between Equations (11.10) and (11.9), is

$$\frac{\xi}{1 - \beta\theta} \sqrt{\frac{1 - \beta}{1 + \beta}}. \tag{11.11}$$

Values of the standardized increase in Equation (11.11) of about 3 or greater are likely to be detected by the Cuscore statistics Q_E.

[2] The concept of applying the signal weights to the a_t values, although correct for the two cases (spike and exponential) discussed here, is *not*, however, a general rule.

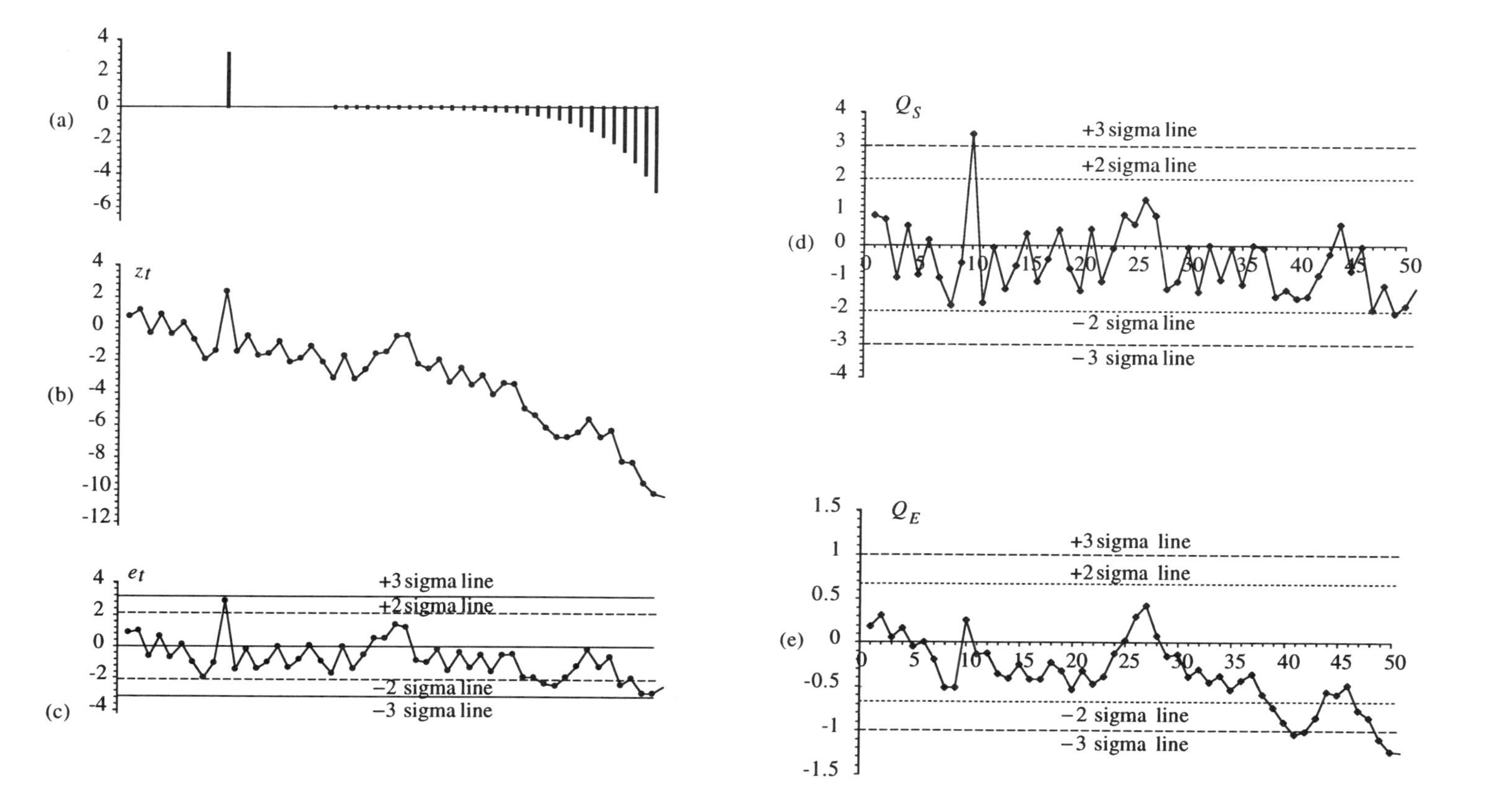

Figure 11.1 (a) Signal series with spike and exponential signals; (b) first 50 values of the disturbance series used earlier in Figures 7.3 and 7.7 with signals added; (c) chart for residual output errors resulting from integral control; (d) Q_S chart; and (e) Q_E chart.

Illustration of Monitoring During Feedback Adjustment

Figure 11.1 illustrates the monitoring of a constrained feedback scheme of the form of Equation (11.1) with $G = 0.2$ and $g = 0.08$ applied to a responsive process ($\delta = 0$). The disturbance, generated by an IMA model with parameters $\lambda = 0.4$ and $\sigma_a = 1$, was used earlier for illustration in Figures 7.3 and 7.7. In the present example, a spike signal of size 3.2 has been added to the 10th observation together with an exponential signal with $\beta = 0.8$ ending at the 50th observation with a spike of -5. The detection of signals buried in IMA noise may be very difficult irrespective of whether the process is being controlled or not. A direct plot of the output residuals e_t (from the controlled process) does suggest abnormalities near the 10th and 50th observations, but these are signaled much more clearly by the Q_S and Q_E charts. Such Cuscore charts respond to "concern 1" (expressed at the beginning of this chapter) that the feedback process can conceal signals in the data.

11.2 MONITORING A PROCESS WITH INERTIA REPRESENTED BY FIRST-ORDER DYNAMICS

When the control scheme of Equation (11.1) is employed for a process whose inertia is represented by first-order dynamics with parameter δ, the appropriate Cuscore statistics for detecting a spike and an exponential change are as before

$$Q_S = a_t \tag{11.12}$$

and

$$Q_E = (1 - \beta)(a_t + \beta a_{t-1} + \beta^2 a_{t-2} + \cdots), \tag{11.13}$$

where the a_t values are calculated assuming the null model true. The only thing that changes when $\delta \neq 0$ is the calculation of these a_t values. It turns out (see Appendix 11C) that

$$a_t = e_t - \tilde{e}_{t-1} + D\tilde{\tilde{e}}_{t-1}, \tag{11.14}$$

where as before $D = G/\lambda$ and $\tilde{e}_t$ means an EWMA with smoothing constant θ and $\tilde{\tilde{e}}_t$ is an EWMA of the $\tilde{e}_t$ values with smoothing constant δ, so that

$$\tilde{\tilde{e}}_{t-1} = (1 - \delta)(\tilde{e}_{t-1} + \delta\tilde{e}_{t-2} + \delta^2\tilde{e}_{t-3} + \cdots). \tag{11.15}$$

Substituting the expression for a_t in Equations (11.12) and (11.13) gives Q_S and Q_E.

11.3 RECONSTRUCTING THE DISTURBANCE PATTERN

An estimate ($\dot{z}_t$, say) of the uncontrolled disturbance that has affected the process can be obtained by calculating the total adjustment X_t that has been applied. This in turn is obtained by summing the individual adjustments[3] so that [see Equation (6.7)]

$$\dot{z}_{t+1} = -gX_t = -g(x_1 + x_2 + \cdots + x_t). \qquad (11.16)$$

For illustration, Figure 11.2a shows the individual adjustments x_t that were made in the control scheme presented in Figure 11.1. Figure 11.2b shows $-X_t$ obtained by summing these individual adjustments and reversing sign. It is seen to adequately reproduce the general trends of the uncontrolled disturbance. This ability to estimate the

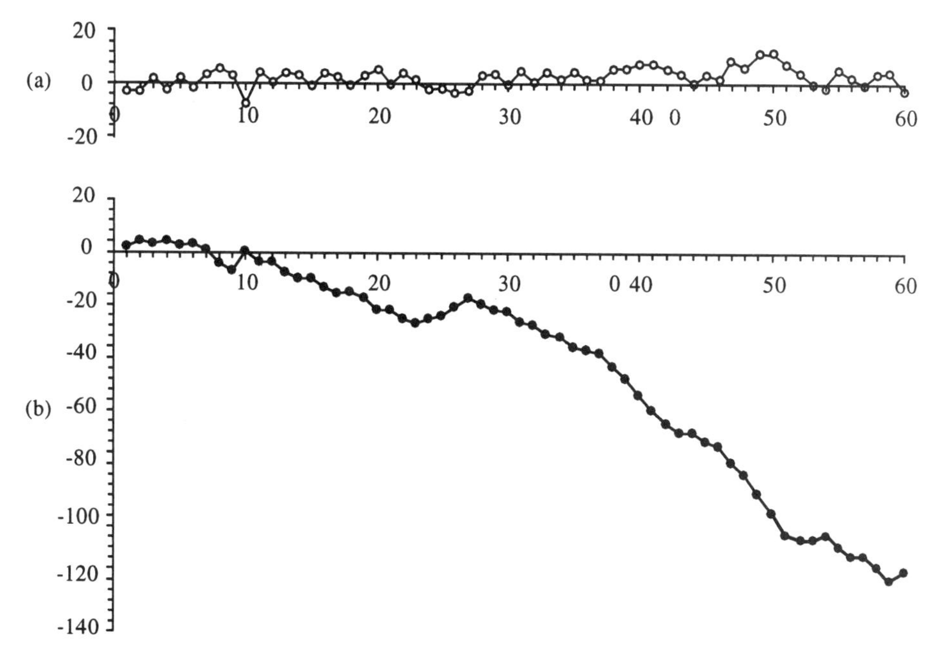

Figure 11.2 (a) Adjustments produced by the constrained control scheme of Figure 11.1. (b) Accumulated adjustments with sign reversed.

[3] This smoothed estimate of the disturbance series will usually be sufficient to aid any search for possibly informative trends. However, if desired, the signal disturbance itself can be reconstructed knowing the system models.

disturbance responds to "concern 2" (expressed at the beginning of this chapter) that possible clues to improvement supplied by the nature of long-term trends might be missed.

Calculation and Display

The calculations discussed above are readily made by a simple preprogrammed process computer or hand calculator. Familiarity with the process usually points to particular signals likely to be of special concern. These may be monitored by a *small battery* of Cuscores each of which can initiate an alarm pointing to the likely occurrence of a particular kind of signal. A reconstructed disturbance, as in Figure 11.2b, showing what the feedback control is actually doing should be shown simultaneously.

APPENDIX 11A DERIVATION OF EQUATION 11.3

The Cuscore statistic to look for a spike in any kind of noise adequately represented by a linear time series model is

$$Q = a_t, \tag{11A.1}$$

where the value of a_t is calculated as if the null model were true.

When the noise supposedly is generated by an IMA (0, 1, 1) model,

$$z_t - z_{t-1} = a_t - \theta\, a_{t-1}. \tag{11A.2}$$

Also, the control scheme of Equation (11.1) applied to a responsive system produces errors at the output satisfying

$$e_t = z_t + g\, X_{t-1},$$

so that, using Equation (11.1),

$$e_t - e_{t-1} = z_t - z_{t-1} + g\, x_{t-1} = z_t - z_{t-1} - Ge_{t-1};$$

that is,

$$e_t - (1 - G)e_{t-1} = z_t - z_{t-1}. \tag{11A.3}$$

Then combining Equations (11A.2) and (11A.3) gives

$$e_t - (1 - G)e_{t-1} = a_t - \theta\, a_{t-1},$$

from which

$$a_t = \frac{1}{\lambda}\tilde{e}_t - \left(\frac{1 - G}{\lambda}\right)\tilde{e}_{t-1} = e_t - (1 - D)\tilde{e}_{t-1},$$

according to Appendix 4A, where $D = G/\lambda$. Hence

$$Q_S = e_t - (1 - D)\tilde{e}_{t-1} = De_t + (1 - D)(e_t - \tilde{e}_{t-1}).$$

APPENDIX 11B DERIVATION OF EQUATION 11.10

Let the null model be given by Equation (11.2) and the alternative model by

$$(z_t - z_{t-1}) - \xi(w_t - w_{t-1}) = b_t - \theta b_{t-1}, \tag{11B.1}$$

where w_t is the signal. Suppose that ξ times the signal is composed of level increases of size

$$\ldots, \xi\sigma_a\beta^3(1 - \beta), \xi\sigma_a\beta^2(1 - \beta), \xi\sigma_a\beta(1 - \beta), \text{ and } \xi\sigma_a(1 - \beta), \tag{11B.2}$$

corresponding to times

$$\ldots, t - 3, t - 2, t - 1, \text{ and } t.$$

The sum of the step changes in (11B.2), which forms a geometric series, is $\xi\sigma_a$. Also, the computed shocks a_t in Equation (11.2) and b_t in Equation (11B.1) satisfy

$$(a_t - \theta a_{t-1}) - \xi(w_t - w_{t-1}) = b_t - \theta b_{t-1},$$

so that

$$a_t - b_t = \xi\sigma_a\frac{1 - \beta}{1 - \beta\theta},$$

$$a_{t-1} - b_{t-1} = \xi\sigma_a\beta\frac{1 - \beta}{1 - \beta\theta},$$

$$a_{t-2} - b_{t-2} = \xi\sigma_a\beta^2\frac{1 - \beta}{1 - \beta\theta}, \ldots. \tag{11B.3}$$

Then introducing Equations (11B.3) into Equation (11.8), where a_t is replaced with $a_t - b_t$, gives the increase in Q_E when the alternative model (11B.1) is true:

$$(1 - \beta)\xi\sigma_a \frac{1 - \beta}{1 - \beta\theta} (1 + \beta^2 + \beta^4 + \cdots) = \sigma_a \frac{\xi}{1 - \beta\theta} \frac{1 - \beta}{1 + \beta}.$$

APPENDIX 11C DERIVATION OF EQUATION 11.14.

When the noise is given by an IMA (0, 1, 1) model

$$z_t - z_{t-1} = a_t - \theta\, a_{t-1}, \tag{11C.1}$$

the control scheme of Equation (11.1) acting on a system with inertia δ produces errors at the output satisfying

$$e_t - \delta\, e_{t-1} = z_t - \delta z_{t-1} + g(1 - \delta)X_{t-1},$$

so that

$$(e_t - e_{t-1}) - \delta(e_{t-1} - e_{t-2}) = (z_t - z_{t-1}) - \delta(z_{t-1} - z_{t-2}) + g(1 - \delta)x_{t-1},$$

and using Equation (11.1)

$$\begin{aligned}(e_t - e_{t-1}) &- \delta(e_{t-1} - e_{t-2}) \\ &= (z_t - z_{t-1}) - \delta(z_{t-1} - z_{t-2}) - G(1 - \delta)e_{t-1}.\end{aligned} \tag{11C.2}$$

Then combining Equations (11C.1) and (11C.2) gives

$$\begin{aligned}(e_t - e_{t-1}) &- \delta(e_{t-1} - e_{t-2}) \\ &= (a_t - \theta a_{t-1}) - \delta(a_{t-1} - \theta a_{t-2}) - G(1 - \delta)e_{t-1},\end{aligned}$$

from which one can verify (after some algebra) that

$$a_t = e_t - \tilde{e}_{t-1} + D\tilde{\tilde{e}}_{t-1},$$

where $\tilde{e}_t$ is defined by Equation (11.15).

CHAPTER 12

A Brief Review of Time Series Analysis

"I wish he would explain his explanation."
Don Juan, LORD BYRON

This book is intended for use by the quality specialist as a practical guide for statistical control and certainly not as a text on the theory and practice of time series analysis. Nevertheless, various aspects of such analyses appear from time to time in these chapters and so, for completeness, it seems appropriate to append an elementary account here. Those wishing for a deeper discussion will find it, for example, in texts such as BJR.

In this review we first consider, somewhat more systematically than in the preceding chapters, some models appropriate for stationary and nonstationary time series; the former varying about a fixed mean in a stable manner and the latter exhibiting wandering behavior and having no fixed mean. We then illustrate how such models may be related to real data and may be used for forecasting.

It is supposed, as before, that the series to be modeled is observed at equal intervals of time and that y_t is the value of the series at time t; also, that $z_t = y_t - T$ represents the deviation of y_t from some reference value T. In addition, if the series is stationary, it is supposed that T is equal to the mean μ.

The simplest time series model is $z_t = a_t$, where $\{a_t\}$ is the white noise series defined earlier (see Section 5.1). An important characteristic of such a series is that its elements are serially independent, implying that the value z_t is not influenced in any way by the values of previous data $z_{t-1}, z_{t-2}, \ldots$. Frequently, however, we need to represent data where a probability model links z_t with previous values $z_{t-1}, z_{t-2} \ldots$, producing serial dependence. Before introducing such models, we consider how serial dependence may be measured.

12.1 SERIAL DEPENDENCE: THE AUTOCORRELATION FUNCTION AND THE VARIOGRAM

A measure of the dispersion of a stationary time series is its variance: that is, the long-term average of the *squares* of deviations from the mean. Thus

$$\sigma_y^2 = \sigma_z^2 = \frac{1}{n}\{z_t^2 + z_{t-1}^2 + z_{t-2}^2 + \cdots\},$$

where $z_t = y_t - \mu$ and the number of terms n in the sum is supposed to be very large (theoretically infinite).

In a similar way, a measure of the serial dependence between deviations one step apart is provided by the long-term average of the *products* of such deviations:

$$C_1 = \frac{1}{n}\{z_t z_{t-1} + z_{t-1} z_{t-2} + z_{t-2} z_{t-3} + \cdots\}.$$

The quantity C_1 is called the *lag 1 autocovariance.*[1] If each deviation is statistically independent of the one that preceded it, then C_1 will be zero. But if each deviation is related to the previous one, so that positive deviations tend to follow positive deviations and negative deviations tend to follow negative deviations, then C_1 will be positive. In the contrary case, where positive deviations tend to be followed by negative ones, C_1 will be negative. A measure of the serial dependence between observations two steps apart is the lag 2 autocovariance

$$C_2 = \frac{1}{n}\{z_t z_{t-2} + z_{t-1} z_{t-3} + z_{t-2} z_{t-4} + \cdots\}.$$

In a similar way, we can define a measure of serial dependence C_k for observations k steps apart. Note also that C_0 is an alternative notation for the variance σ_z^2.

Now these measures of serial dependence, $C_1, C_2, \ldots$, depend not only on the degree of similarity between adjacent deviations but also on the scale in which the data are measured. For example, for a series of temperatures the absolute values of $C_1, C_2, \ldots$ would be larger if the temperature was measured in degrees Fahrenheit than if it was measured in degrees Celsius. To obtain measures that do not depend on

[1] Also sometimes called the *serial* covariance.

the scale in which the measurements are made, the autocovariances are divided by the variance to obtain the *autocorrelation coefficients.* Such an autocorrelation is denoted by ρ (the Greek letter *rho*). Thus

$$\rho_1 = \frac{C_1}{C_0}, \quad \rho_2 = \frac{C_2}{C_0}, \quad \rho_3 = \frac{C_3}{C_0}, \quad \ldots$$

are the autocorrelations[2] at lags 1, 2, 3, ... The autocorrelation at lag zero, $\rho_0 = C_0/C_0 = 1$, can be thought of as the correlation of each deviation with itself. The series of autocorrelations, $1, \rho_1, \rho_2, \rho_3, \ldots$, serves to characterize a stationary time series model and is called the *autocorrelation function* or ACF.[3]

12.2 RELATION BETWEEN THE AUTOCORRELATION FUNCTION AND THE VARIOGRAM

In Section 5.4, we introduced the *variogram* to characterize certain time series discussed in Chapter 5. If $V_m = \text{var}(z_{t+m} - z_t)$ denotes the variance of the difference between observations taken m steps apart, then a corresponding scaleless quantity measuring variance inflation is $V_m/V_1 = \text{var}(z_{t+m} - z_t)/\text{var}(z_{t+1} - z_t)$. We called the series $\{1, V_2/V_1, V_3/V_1, \ldots\}$ the standardized variogram, or simply the variogram. For stationary time series, there is a direct relation between the variogram and the autocorrelation function, namely,

$$\frac{V_m}{V_1} = \frac{1 - \rho_m}{1 - \rho_1}. \tag{12.1}$$

Both the variogram and the ACF are useful, but, unlike the ACF, the variogram may be used to characterize series that are nonstationary.

12.3 SOME TIME SERIES MODELS

All the models we discuss for both stationary and nonstationary time series are expressions showing the relation between the series $\{z_t\}$, which is to be modeled, and a white noise generating series $\{a_t\}$. The elements $a_t, a_{t-1}, a_{t-2}, \ldots$ of this generating series are sometimes referred to as random *shocks* or *innovations.*

[2] Also sometimes called the *serial correlations* at lags, 1, 2, 3, ...
[3] The ACF is also sometimes called the *correlogram.*

Although this is often not their simplest representation, all these models are capable of being written so that $\{z_t\}$ is a weighted linear aggregate of past and present innovations $a_t, a_{t-1}, a_{t-2}, \ldots$. That is

$$z_t = w_0 a_t + w_1 a_{t-1} + w_2 a_{t-2} + \cdots, \tag{12.2}$$

where the w's are a series of constant weights related to the parameters of the time series model. Now Equation (12.2) defines a linear filtering operation such as was introduced in Chapter 4 in which $\{a_t\}$ is the input and $\{z_t\}$ is the output from some dynamic system defined by the weights $w_0, w_1, w_2, \ldots$. We begin by considering some *models* that can represent time series data that are stationary but serially dependent.

12.4 STATIONARY MODELS

Moving Average (MA) Models

The simplest of such models, called a *moving average* (MA) model, introduces dependence by using the first few terms of Equation (12.2) directly. For example, the first-order moving average model, MA(1), can be written

$$z_t = a_t - \theta a_{t-1}, \tag{12.3}$$

which is the same as Equation (12.2) with $w_0 = 1$, $w_1 = -\theta$, and all the other weights zero. To achieve stability, the constant θ must lie between -1 and $+1$. Now for time $t - 1$ this model gives

$$z_{t-1} = a_{t-1} - \theta a_{t-2}$$

and since z_t and z_{t-1} contain the same innovation a_{t-1}, they are correlated. However, at time $t - 2$,

$$z_{t-2} = a_{t-2} - \theta a_{t-3},$$

so z_t and z_{t-2} are not correlated because they do not contain a common innovation. So, for this MA(1) series, $\rho_2, \rho_3, \rho_4, \ldots$ are all zero and, as is illustrated in Figure 12.1b, the ACF *cuts off* after one lag.

In general, a moving average of order q, using $q + 1$ terms from Equation (12.2), is usually written

$$z_t = a_t - \theta_1 a_{t-1} - \theta_2 a_{t-2} - \cdots - \theta_q a_{t-q} \tag{12.4}$$

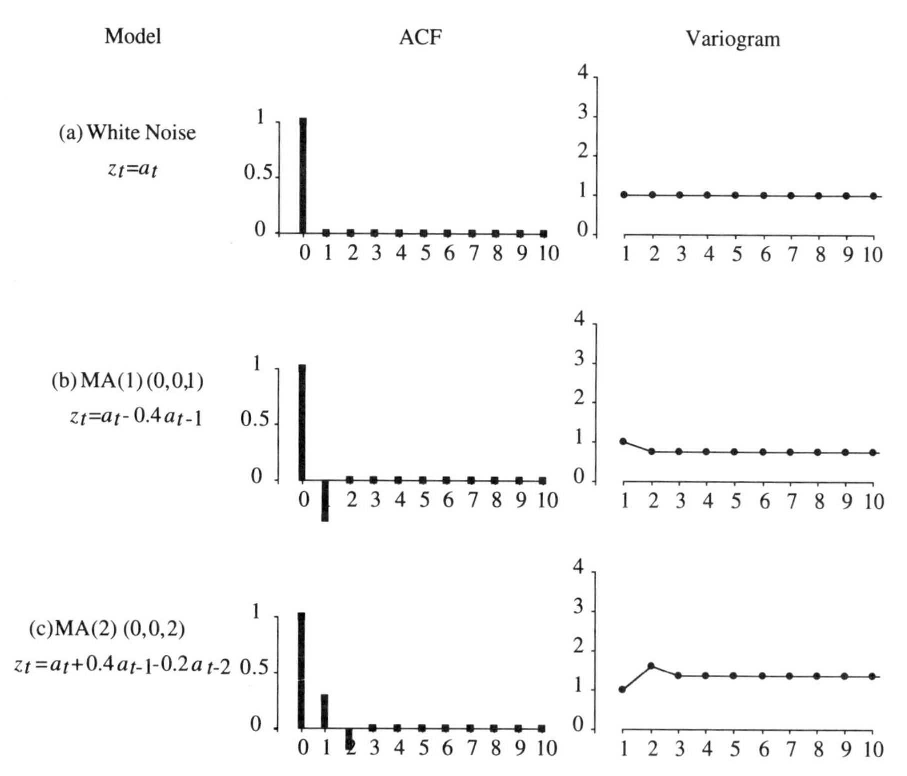

Figure 12.1 Examples of autocorrelation functions and associated variograms for some stationary models: (a) white noise, (b) MA(1), and (c) MA(2).

and, since for such a series no a's overlap after q lags, all the ρ's after ρ_q are zero. Thus, for a moving average model of order q, the ACF *cuts off* after q lags.

Times series that can be represented by stationary MA models are rather uncommon. However, the phenomenon of "carry-over" can produce a series characterized by a MA(1) model. Carry-over is sometimes evident in successive yields from a batch process. It can occur when a random amount of product remains in pipes and pumps after each batch, so that a part e_{t-1} of, say, yesterday's batch may be included in today's yield measurement and a part e_t missing from today's batch is included in tomorrow's and so on.

This phenomenon can be modeled by

$$z_t = -e_t + e_{t-1} + u_t,$$

$$z_{t-1} = -e_{t-1} + e_{t-2} + u_{t-1},$$

and so on.

In these expressions, u_t represents an additional white noise error, supposed independent of the e's and due, for example, to measurement. Because the same element e_{t-1} occurs both in the expression for z_t and in that for z_{t-1}, the model produces a lag 1 autocorrelation of

$$\rho_1 = \frac{-\sigma_e^2}{2\sigma_e^2 + \sigma_u^2},$$

but $\rho_2, \rho_3, \rho_4, \ldots$ are all zero. Consequently, since a stationary time series is characterized by its ACF, the carry-over series $\{z_t\}$ is modeled by a first-order moving average.

Figure 12.1 shows examples of ACFs and the corresponding variograms for white noise and for moving average models of orders $q = 1$ and $q = 2$. Note that for a MA(q) process, after the first q lags, the variogram does not increase and behaves like that for white noise. Equivalently, observations that are more than q steps apart will be uncorrelated.

Autoregressive Models

Of greater interest for representing stationary autocorrelated series are the autoregressive models. An autoregressive model of order p, referred to as an AR(p) model, supposes that the deviation z_t is a linear aggregate of p previous deviations, $z_{t-1}, z_{t-2}, \ldots, z_{t-p}$, and of a random shock a_t.

In particular, in Chapter 5 we introduced a first-order autoregressive AR(1) time series model

$$z_t = \phi z_{t-1} + a_t, \tag{12.5}$$

where the constant ϕ lies between -1 and $+1$. Now at time $t - 1$

$$z_{t-1} = \phi z_{t-2} + a_{t-1} \tag{12.6}$$

and, if we substitute this expression for z_{t-1} in Equation (12.5), we get

$$z_t = a_t + \phi a_{t-1} + \phi^2 z_{t-2}. \tag{12.7}$$

Also, substituting in Equation (12.7) for z_{t-2} we get

$$z_t = a_t + \phi a_{t-1} + \phi^2 a_{t-2} + \phi^3 z_{t-3}.$$

If we successively substitute i times in this way until $\phi^i z_{t-i}$ becomes negligible, we see that the AR(1) model (12.5) can be written as a MA model of infinite order,

$$z_t = a_t + \phi a_{t-1} + \phi^2 a_{t-2} + \phi^3 a_{t-3} + \cdots .$$

The weights for the linear filter of Equation (12.2) are now $w_0 = 1$, $w_1 = \phi$, $w_2 = \phi^2, \ldots$. They fall off exponentially and are all determined by the single parameter ϕ. In particular, the AR(1) model would describe the output from the first-order dynamic system of Section 4.3 with discount factor ϕ when the input was white noise. More generally, the AR(1) model can often describe approximately the output from some system in which variation has been smoothed by passage through the process (e.g., by mixing).

The autocorrelations of the first-order autoregressive model are $\rho_1 = \phi$, $\rho_2 = \phi^2$, $\rho_3 = \phi^3$, and so on. Thus this linkage of z_t with z_{t-1} induced by Equation (12.5) produces correlation not only at lag 1 but also at lags 2, 3, and so on. For instance, if $\phi = 0.9$, then $\rho_1 = 0.9$, $\rho_2 = 0.81$, $\rho_3 = 0.73, \ldots$, and—even at lag 12—the observations are still appreciably autocorrelated ($\rho_{12} = 0.28$). Thus the autocorrelation function "tails off" as a decaying exponential. Similar behavior is found with autoregressive models of higher order. In particular, it can be shown that the autocorrelation function of the second-order autoregressive process,

$$z_t = \phi_1 z_{t-1} + \phi_2 z_{t-2} + a_t, \tag{12.8}$$

is either a mixture of two decaying exponentials or a decaying sine wave. A useful general rule is that, while the autocorrelation function of a MA model *cuts off,* that for an AR model *tails off.*

12.5 AUTOREGRESSIVE MOVING AVERAGE (ARMA) MODELS

Time series can occur that are modeled by a mixture of autoregressive and moving average terms. In particular,

$$z_t - \phi z_{t-1} = a_t - \theta a_{t-1} \tag{12.9}$$

is a model with a first-order autoregressive term and a first-order moving average term. It is called an ARMA(1,1) model.

The autocorrelation function for this model is of the form

$$\rho_0 = 1, \quad \rho_1, \quad \rho_2 = \phi\rho_1, \quad \rho_3 = \phi^2\rho_1, \quad \rho_4 = \phi^3\rho_1, \ldots.$$

Note how this differs from the ACF of an AR(1) model. For this ARMA model, ρ_1 is not equal to ϕ but is determined by the values of both ϕ and θ, and is positive if ϕ is greater than θ, and negative otherwise (see Table 12.1 for details). Subsequent autocorrelations fall off exponentially with discount factor ϕ but *initiated from* ρ_1 rather than from $\rho_0 = 1$.

TABLE 12.1 Autocorrelations for Various Models Considered in This Chapter

AR(1) model	MA(1) model
$\rho_k = \phi^k$ (exponential decay)	$\rho_1 = \dfrac{-\theta}{1+\theta^2}, \quad \rho_2 = \rho_3 = \cdots = 0$
AR(2) model	**MA (2) model**
$\rho_0 = 1, \quad \rho_1 = \dfrac{\phi_1}{1-\phi_2},$ $\rho_k = \phi_1\rho_{k-1} + \phi_2\rho_{k-2}, \quad k > 0$ (mixture of damped sine waves or exponentials)	$\rho_1 = \dfrac{-\theta_1 + \theta_1\theta_2}{1+\theta_1^2+\theta_2^2}, \quad \rho_2 = \dfrac{-\theta_2}{1+\theta_1^2+\theta_2^2},$ $\rho_3 = \rho_4 = \cdots = 0$
ARMA(1, 1) model	
$\rho_1 = \dfrac{(1-\phi\theta)(\phi-\theta)}{1+\theta^2-2\phi\theta}$, (exponential decay after first lag) $\rho_k = \phi\rho_{k-1}, \quad k > 1.$	
ARMA(p, q) model	
$\rho_0 = 1, \rho_1, \rho_2, \ldots, \rho_q$ follow from a linear system of equations[a], then $\rho_k = \phi_1\rho_{k-1} + \phi_2\rho_{k-2} + \cdots + \phi_p\rho_{k-p}, \quad k > q$ (mixture of damped sine waves or exponentials after lag q)	

[a] The first p autocovariances $\gamma_0, \gamma_1, \ldots, \gamma_p$ can be obtained from the linear system of equations

$$\gamma_0 - \phi_1\gamma_1 - \cdots - \phi_p\gamma_p = \gamma_{ay}(0) - \theta_1\gamma_{ay}(-1) - \cdots - \theta_q\gamma_{ay}(-q),$$
$$\gamma_1 - \phi_1\gamma_0 - \cdots - \phi_p\gamma_{p-1} = -\theta_1\gamma_{ay}(0) - \cdots - \theta_q\gamma_{ay}(1-q),$$
$$\vdots$$
$$\gamma_p - \phi_1\gamma_{p-1} - \cdots - \phi_p\gamma_0 = -\theta_p\gamma_{ay}(0) - \cdots - \theta_q\gamma_{ay}(p-q),$$

where the $\gamma_{ay}(j)$ are obtained recursively using

$$\begin{cases} \gamma_{ay}(j) = 0, \quad j > 0, \\ \gamma_{ay}(j) = \phi_1\gamma_{ay}(j+1) + \cdots + \phi_p\gamma_{ay}(j+p) - \theta_{-j}\sigma_a^2, \quad j \le 0, \end{cases}$$

where $\theta_0 = -1$. If $q > p$, one can obtain $\gamma_{p+1}, \ldots, \gamma_q$ recursively from

$$\gamma_j - \phi_1\gamma_{j-1} - \cdots - \phi_p\gamma_{j-p} = \gamma_{ay}(j) - \theta_1\gamma_{ay}(j-1) - \cdots - \theta_q\gamma_{ay}(j-q), \quad j \ge 0.$$

Clearly, $\rho_j = \gamma_j/\gamma_0$.

Figure 12.2 shows examples of ACFs and corresponding variograms for the stationary autoregressive and mixed autoregressive–moving average models discussed above. Note that, for an AR(1) model, the autoregressive parameter ϕ determines how fast the correlation dies out. In particular, it determines the effect of changing the sampling rate. For example, if ϕ was equal to 0.7 and the sampling rate were halved, then the sampled series would follow an autoregressive process with parameter $\phi^2 = 0.7^2 = 0.49$. In general, if a series that followed a strictly stationary autoregressive model was sampled at sufficiently long intervals, the observations in the sampled series would be virtually independent. The point at which this would happen is where the ACF became negligible or, equivalently, where the variogram became constant. For example, by looking at Figure 12.2a you will see that the sampling interval required when $\phi = 0.7$ would be slightly larger than 10 unit intervals.

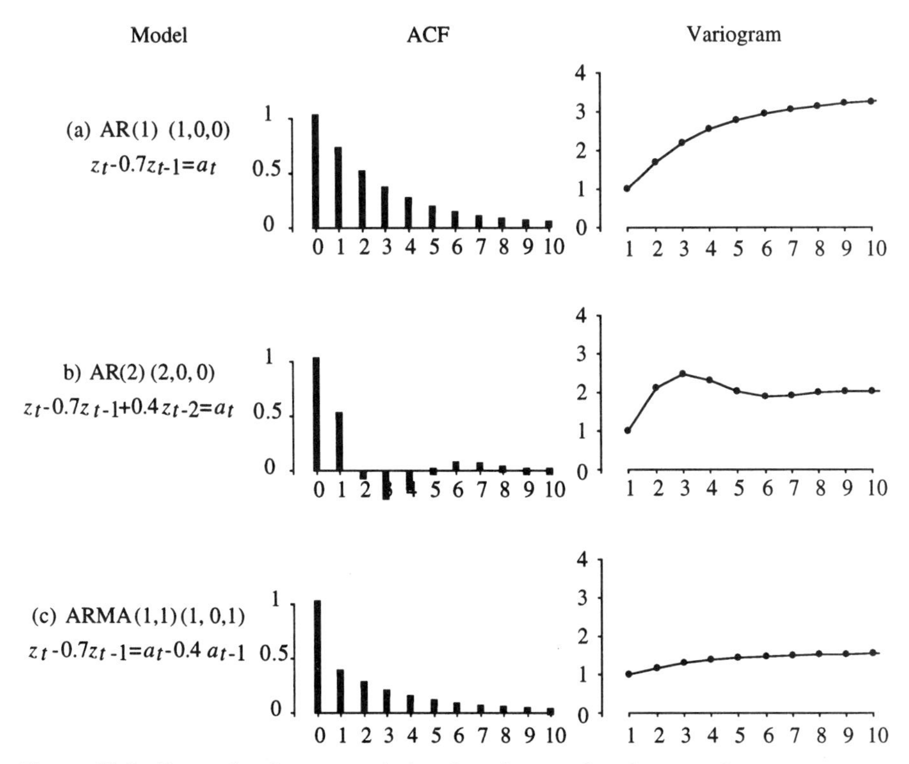

Figure 12.2 Example of autocorrelation functions and variograms for some autoregressive and mixed ARMA models: (a) AR(1), (b) AR(2), and (c) ARMA(1, 1).

Note that reducing the sampling rate "to produce statistical independence" in this way is not necessarily a good idea. In particular, as is evident from the discussion in Chapters 10 and 11, it could produce much less sensitive monitoring. Again, in practice, the possible reduction in cost due to less frequent sampling must be set against the increased chance of failing to detect a possible assignable cause. Finally, as we have frequently emphasized in this book, the assumption of strict stationarity is one that should not be made lightly.

By comparing Figures 12.2a and 12.2c, we see that the addition of a moving average term to the AR model does not greatly change the nature of the variogram. The important thing to remember is that, for any stationary model, the variogram eventually flattens out (in mathematical language, it approaches an asymptote[4]). For sensible values of the parameters of the model, this happens rather quickly. Thus stationary models cannot take into account the *continual* absorption of "sticky innovations" required for the characterization of a nonstationary disturbance allowed to develop without intervention.

12.6 NONSTATIONARY MODELS

Many naturally occurring series such as economic series, business series, as well as series representing uncontrolled process disturbances cannot be adequately represented by stationary models. However, although the *level* $\{z_t\}$ of such a series is nonstationary, it is frequently true that its rate *of change*, as measured by the first difference $\{z_t - z_{t-1}\}$, *is* stationary (at least to a reasonable approximation). Also, on very rare occasions, we might find a series that, although nonstationary in the first difference, is approximately stationary in the second difference; that is, in the difference of the differences. Thus $\{(z_t - z_{t-1}) - (z_{t-1} - z_{t-2})\}$, that is, $\{z_t - 2z_{t-1} + z_{t-2}\}$, might be modeled by a stationary series. This implies that the "acceleration" of the series is approximately stationary.

On this basis, a general class of models called ARIMA models (autoregressive integrated moving average models) has been widely used to represent time series (BJR). An ARIMA model of order (p, d, q) is such that its dth difference can be represented by a mixed autoregressive moving average (ARMA) model of order (p, q). That is, after

[4] It is easy to show that, when the model is stationary, as m increases V_m/V_1 approaches the value $1/(1 - \rho_1)$.

differencing d times, the model has an autoregressive part of order p and a moving average part of order q. In practice, d is rarely greater than one.

12.7 THE IMA—OR ARIMA (0, 1, 1)—MODEL

As we saw in Chapter 5, for the representation of an uncontrolled disturbance, reasonable theoretical arguments as well as practical experience pointed to the particular importance of the IMA model in which the first difference was a first-order moving average. In the more general (p, d, q) notation, this IMA model is an ARIMA model of order $(0, 1, 1)$.

At time t, the model is

$$z_t - z_{t-1} = a_t - \theta a_{t-1}. \tag{12.10}$$

By writing down this model for times $t, t - 1, t - 2, \ldots, 1$ and summing from 1 to t, we get

$$z_t = \text{constant} + a_t + \lambda(a_{t-1} + a_{t-2} + \cdots + a_1),$$

where $\lambda = 1 - \theta$. The model can thus be written[5]

$$z_t = \text{constant} + a_t + \lambda \sum_{i=1}^{t-1} a_i.$$

Thus z_t is a mixture of the current random shock a_t and the sum $\sum_{i=1}^{t-1} a_i$ of previous shocks. The sum of a white noise series is often called a *random walk*.

Now we have seen that another form for this model is

$$z_t = \tilde{z}_{t-1} + a_t,$$

where $\tilde{z}_{t-1}$ is the following exponentially weighted moving average of the z's:

$$\tilde{z}_{t-1} = \lambda(z_{t-1} + \theta z_{t-2} + \theta^2 z_{t-3} + \cdots).$$

[5] Equivalently, $z_t = \text{constant} + \theta a_t + \lambda \sum_{i=1}^{t} a_i$.

Thus if $\tilde{z}_{t-1}$ is used as a forecast $\hat{z}_t$ of z_t based on past data, $z_{t-1}, z_{t-2}, \ldots,$ then, equivalently,

$$z_t = \hat{z}_t + a_t$$

and a_t becomes the one-step-ahead forecast error. Alternatively, $\tilde{z}_{t-1} = \hat{z}_t$ can be thought of as the output from a first-order linear dynamic system in which the inputs are $z_{t-1}, z_{t-2}, z_{t-3}, \ldots$, and the "weights" are $\lambda, \lambda\theta, \lambda\theta^2, \ldots$.

Figure 12.3a shows the variogram for the model, $z_t - z_{t-1} = a_t - 0.4\, a_{t-1}$, that is, an IMA model with $\theta = 0.4$ ($\lambda = 0.6$). The *first difference* thus follows the model already used for illustration in Figure 12.1b. As we saw in Chapter 5, the variogram for the IMA model increases linearly with m. Figures 12.3b, (c) and (d) show variograms for (1, 1, 0), (2, 1, 0), and (1, 1, 1) models. Again we have chosen the parameters so that the models for the first differences are those used for illustration earlier. For all these models whose first difference is stationary, the variogram may show some curvature initially but eventually increases linearly without limit.[6]

For an ARIMA (p, 2, q) model, in which the first difference is not stationary but the second difference is a stationary ARMA(p, q) model, the variogram increases *as the square* of m irrespective of the values of the parameters. Thus, for these models, not only the variogram but also the *rate of increase* of the variogram increases without limit. Such models do not appear particularly useful for representing a process disturbance because they imply that the effect of the "sticky innovations" of Chapter 5 gets larger and larger as m increases. For the kinds of application discussed in this book, it is not easy to think of a situation where this could happen.

12.8 MODELING TIME SERIES DATA

Suppose you have a time series and are seeking a model that can approximately represent it. You can do this by following an "iterative"

[6] The eventual slope of the variogram for an ARIMA(p, 1, q) model is

$$\frac{\sigma_a^2}{\gamma_0}\left(\frac{1 - \theta_1 - \cdots - \theta_q}{1 - \phi_1 - \cdots - \phi_p}\right)^2,$$

where γ_0 is the variance of the first difference $z_t - z_{t-1}$ and the ϕ's and θ's are the parameters of the stationary ARMA (p, q) model for $z_t - z_{t-1}$. Also it was shown by MacGregor, 1976 (see also Tiao, 1972) that as the interval between observations increases the behavior of any (p, 1, q) model approaches that of the IMA (0, 1, 1).

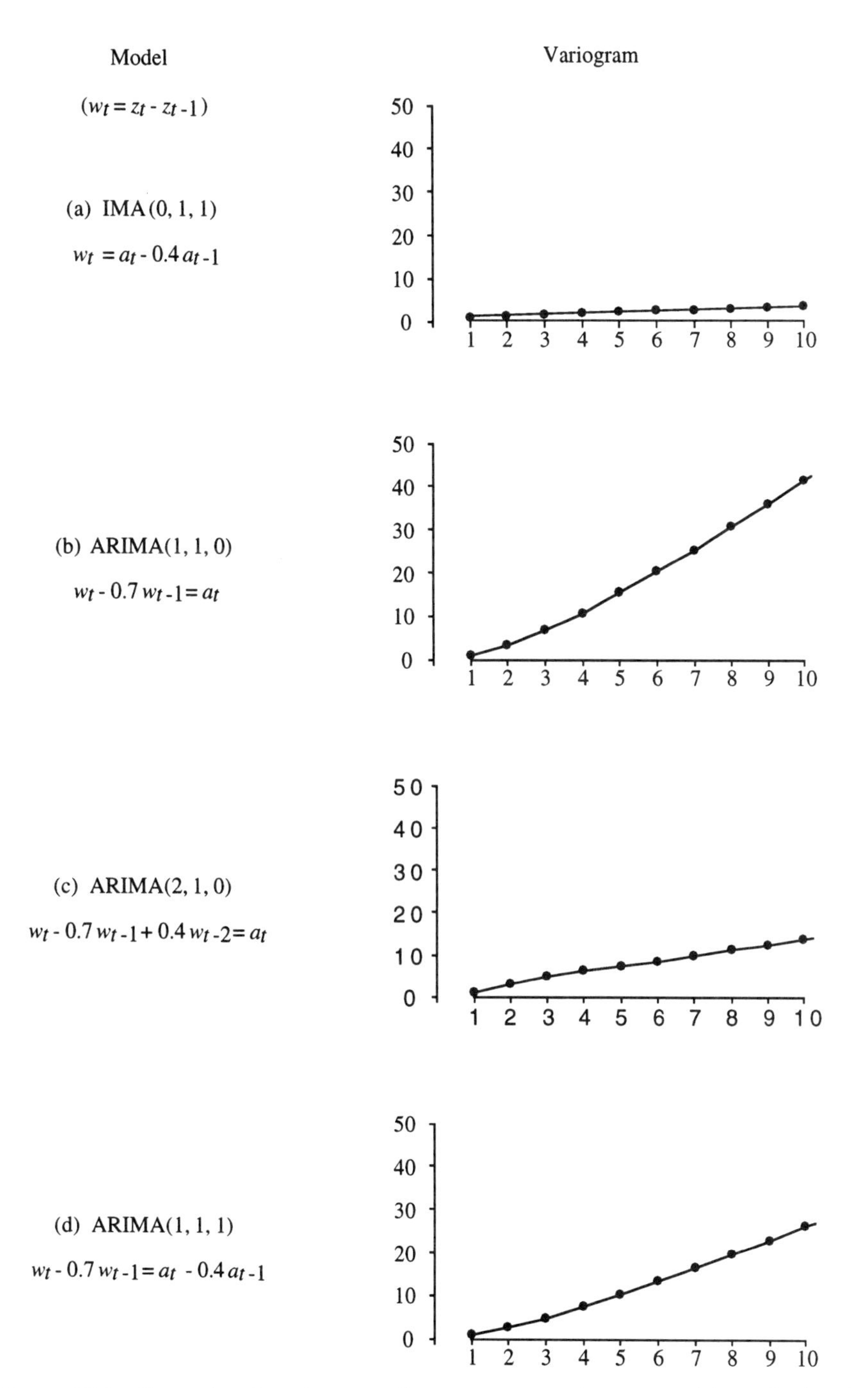

Figure 12.3 Example of variograms for some nonstationary models with $d = 1$: (a) IMA(0, 1, 1), (b) ARIMA(1, 1, 0), (c) ARIMA(2, 1, 0), and (d) ARIMA(1, 1, 1).

model building process suggested by Box and Jenkins (1970). This can be represented as follows:

$$\hookrightarrow \text{Identification} \rightarrow \text{Fitting} \rightarrow \text{Diagnostic Checking} \lrcorner$$

The *identification* stage can suggest a form of model that is worth entertaining. This form of model may then be *fitted* to the data (i.e., the best-fitting values of the parameters for the tentative model are determined). This is followed by a process of *diagnostic checking.* This uses the fact that, since a time series model is a recipe for transforming the data to white noise, we can check its adequacy by examining the series of residual quantities $\{\hat{a}_t\}$ obtained after fitting the model. If the model is adequate, these residuals will resemble a series of independent random errors.[7] The process is *diagnostic* in the sense that marked discrepancies or patterns in this series—or its ACF—can point to specific inadequacies of the model, leading to a fresh cycle of the model building process indicated by arrows in the diagram.

Computer Programs

Almost all major suppliers of computer programs now have time series packages available, which the reader can use to carry out the calculations we describe. Computer programs are also available which identify, fit, and check models for time series entirely automatically. In situations where thousands of series have to be fitted, these automatic programs have a place. However, for the building of models for individual series, we feel that the investigator must be in the loop looking at the plot of the empirical series, at its ACF, and at the residual plot and the autocorrelation of the residuals. For this purpose, it is necessary to have some idea of what the computer is actually doing—or failing to do—so we give here an elementary account of some of these model building tools even though, in practice, you will not need to actually perform the detailed calculations described.

12.9 MODEL IDENTIFICATION, MODEL FITTING, AND DIAGNOSTIC CHECKING

It is always important to first get a plot of the data. As we have seen (e.g., in Figure 1.1), the general appearance of the series can tell you a lot.

[7] The fact that we have *fitted* parameters to the series would introduce slight dependence between the $\hat{a}_t$ values even if the model was perfect, but unless the series is very short, this is not a serious problem.

For example, does it look stationary or nonstationary? Are there obvious discrepant observations?

The Sample Autocorrelation Function

A major tool for model identification is the *sample* autocorrelation function. Earlier in this chapter we discussed the *theoretical* autocorrelation function that characterized specific time series *models*. These theoretical autocorrelations used the concept of "long-term averages" (i.e., the theoretical means of products of deviations k steps apart corresponding to *infinitely long* series generated by the model).

Real series contain a limited number n of observations and are thought of as a *sample* of n consecutive observations from the series generated by the model. This model is initially unknown but clues to its possible nature are provided by the corresponding *sample* statistics.

In the sample statistics, the theoretical mean μ is replaced by the sample mean,

$$\bar{y} = \frac{1}{n} \sum_{t=1}^{n} y_t,$$

and the theoretical lag k autocovariance C_k is replaced by the corresponding sample autocovariance

$$c_k = \frac{1}{n} \sum_{t=1}^{n-k} (y_t - \bar{y})(y_{t+k} - \bar{y}).$$

Finally, the theoretical lag k autocorrelation ρ_k is replaced by the estimate $\hat{\rho}_k = r_k$, where

$$r_k = \frac{c_k}{c_0}.$$

Then $r_0 = 1$ and $r_0, r_1, r_2, \ldots$ is the *sample* autocorrelation function. By matching this sample autocorrelation function to a theoretical autocorrelation function, a preliminary identification of the model may be made.

The Sample Partial Autocorrelation Function

This is an important additional tool for identification of time series models and, in particular, of those containing autoregressive terms. To understand it, consider as an example the AR(1) model $z_t = \phi z_{t-1} + a_t$. As we have seen, for this model $\rho_1 = \phi$, $\rho_2 = \phi^2$, $\rho_3 = \phi^3$, and so on. So the single parameter $\phi = \rho_1$ induces all the subsequent correlations $\rho_2, \rho_3, \ldots$. The theoretical *partial autocorrelation coefficient* (PACF) at lag k, denoted by ϕ_{kk}, measures the *additional* correlation at lag k *not accounted for by previous correlations*. An estimate $\hat{\phi}_{kk}$ of ϕ_{kk} is obtained by fitting to the data an autoregressive model of order k:

$$z_t = \phi_1 z_{t-1} + \phi_2 z_{t-2} + \cdots + \phi_k z_{t-k} + a_t.$$

The estimate $\hat{\phi}_k$ of the last coefficient provides the estimated partial autocorrelation $\hat{\phi}_{kk}$. Thus the estimated partial autocorrelation function $\hat{\phi}_{11}, \hat{\phi}_{22}, \hat{\phi}_{33}, \ldots$ is obtained by successively fitting autoregressive models of orders 1, 2, 3, ... and recording for each model the *last* fitted coefficient.

The theoretical PACF has the property that, for an autoregressive model of order p, it *cuts* off after p lags, while, for a moving average model, it *tails* off. The behavior of the PACF is exactly opposite to that of the ACF. Thus:

For a MA(q) model, the ACF cuts off after q lags.
For an AR(p) model, the PACF cuts off after p lags.
For a mixed ARMA model, both the ACF and PACF tail off.

For a particular time series, the type of model worth considering may frequently be identified by jointly studying its sample ACF and PACF.

Appendix 12A contains references to some other identification tools.

Model Fitting

Time series models are best fitted using the *method of maximum likelihood* and many modern computer programs use this technique (see Appendix 12B). However, as we illustrated in Sections 4.8 and 6.7, a rough approximation to the maximum likelihood parameter estimates can be obtained using the method of least squares. In particular, for a single parameter, this can be done by graphing the sums of squares of the $\hat{a}_t$ terms for a range of values of the parameters and picking the minimum value. An autoregressive model such as $z_t = \phi_1 z_{t-1} + \phi_2 z_{t-2} + a_t$ is like an ordinary regression model except that the regressors, often denoted by x_1 and x_2, are replaced by the variables z_{t-1} and z_{t-2}. For such time series models you can obtain the least squares estimates by regressing

z_t on z_{t-1} and z_{t-2}. This regression calculation also yields approximate standard errors for the estimates of the parameters.

Diagnostic Checking

The process of diagnostic checking is accomplished by applying to the series of residuals from the fitted model very much the same treatment as that used for model identification of the original series—that is, careful visual inspection followed by study of the ACF, and possibly the PACF, for the residual series. It is helpful in this inspection to know that the standard deviations of the sample autocorrelations or partial autocorrelations calculated from a white noise series are approximately equal to $1/\sqrt{n}$.

Two Examples

We now illustrate the model building process for two series, which we will call the Temperature Series and the Metallic Film Series.

Temperature Series

This series will be found as Series C in the collection included at the back of the book. It consists of 80 consecutive values of temperature taken from the output of an industrial process to which feedback control was being applied. A plot of the series, shown in Figure 12.4a, suggests that it is stationary but autocorrelated. The sample ACF for the series is shown in Figure 12.4b. This ACF tends to tail off rather than cut off and, initially at least, it falls off roughly exponentially, suggesting that we might begin by fitting some kind of autoregressive model. The PACF in Figure 12.4c suggests that an autoregressive model of order 1 or possibly higher might represent the series. Figure 12.4d shows a plot of the residuals from the best fitting AR(1) model $z_t = 0.69\, z_{t-1} + a_t$. The ACF of the residual $\hat{a}_t$ values is shown in Figure 12.4e and indicates a reasonably good fit. (Remember that since $n = 80$ the standard deviation of these residuals is about $1/\sqrt{80} = 0.11$.) Figures 12.4f and 12.4g show the residual plot and its ACF after fitting an AR(2) model. The AR(2) does not seem to produce any appreciable improvement. It is concluded, therefore, that the series could be approximately represented by an AR(1) model with $\hat{\phi} = 0.69$.

Metallic Film Series

This series consists of the 100 measurements of the thickness of a metallic film previously plotted in Figure 6.1 and listed as Series A at the back of this book. The plot of the series, in Figure 12.5a, and of its

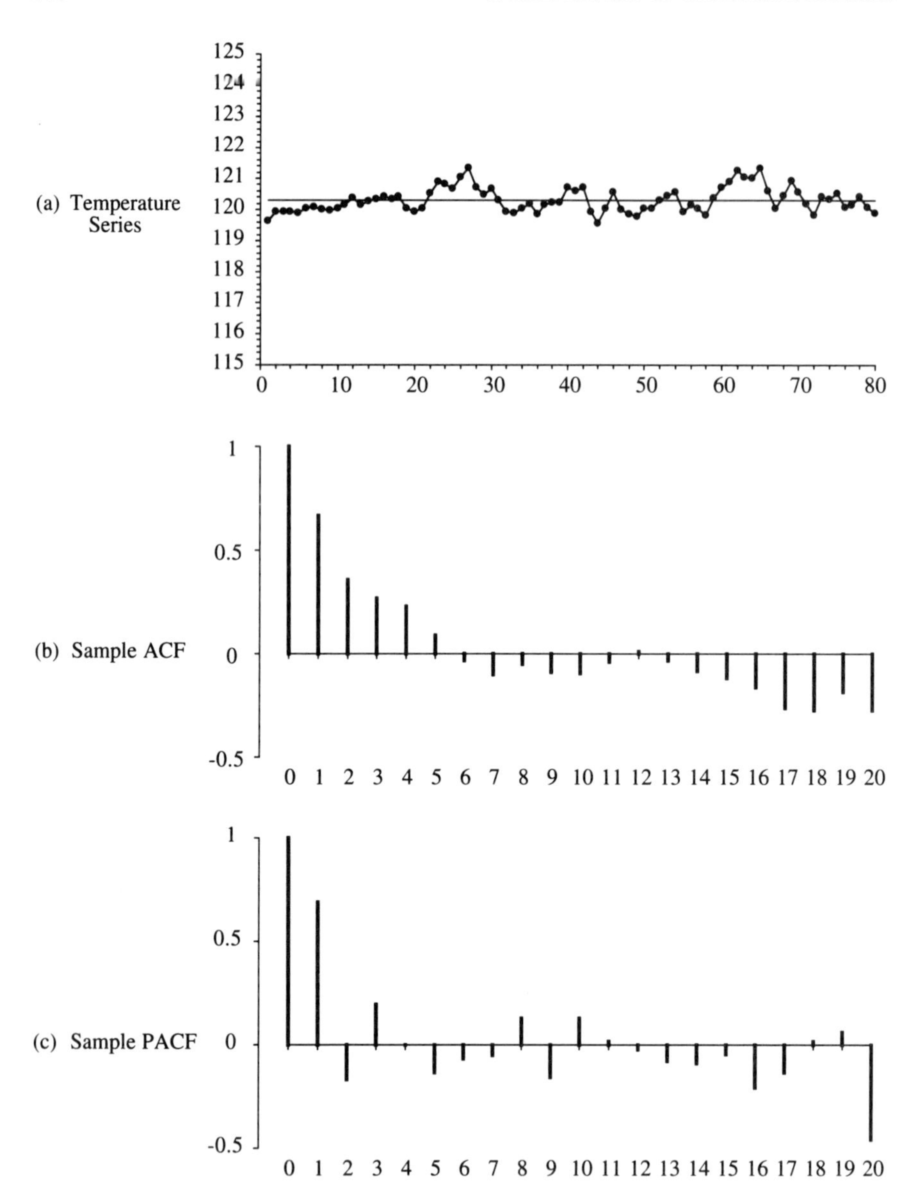

Figure 12.4 Graphical analysis of the Temperature Series: (a) the series, (b) the sample ACF, (c) the sample PACF, (d) the residuals after fitting an AR(1) model, (e) the sample ACF of the residuals in (d), (f) the residuals after fitting an AR(2) model, and (g) the sample ACF of the residuals in (f).

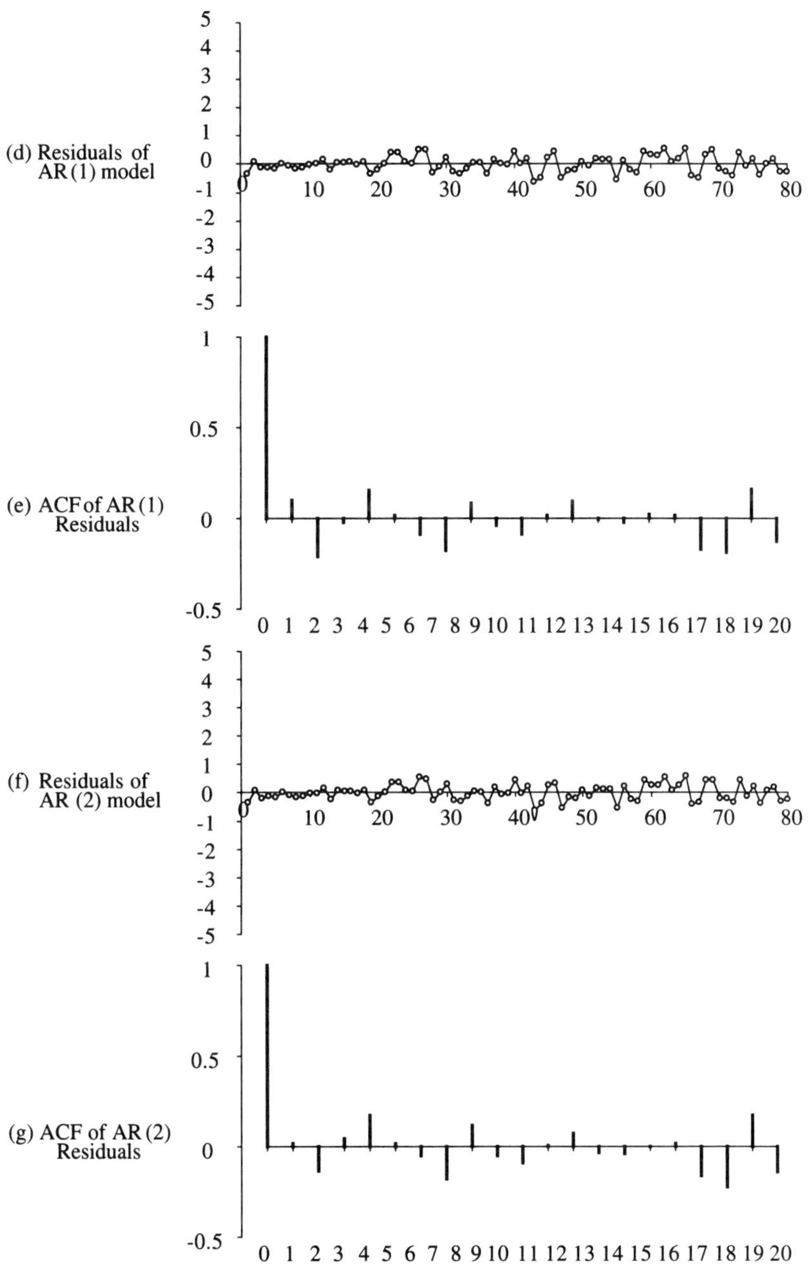

Figure 12.4 *(Continued)*

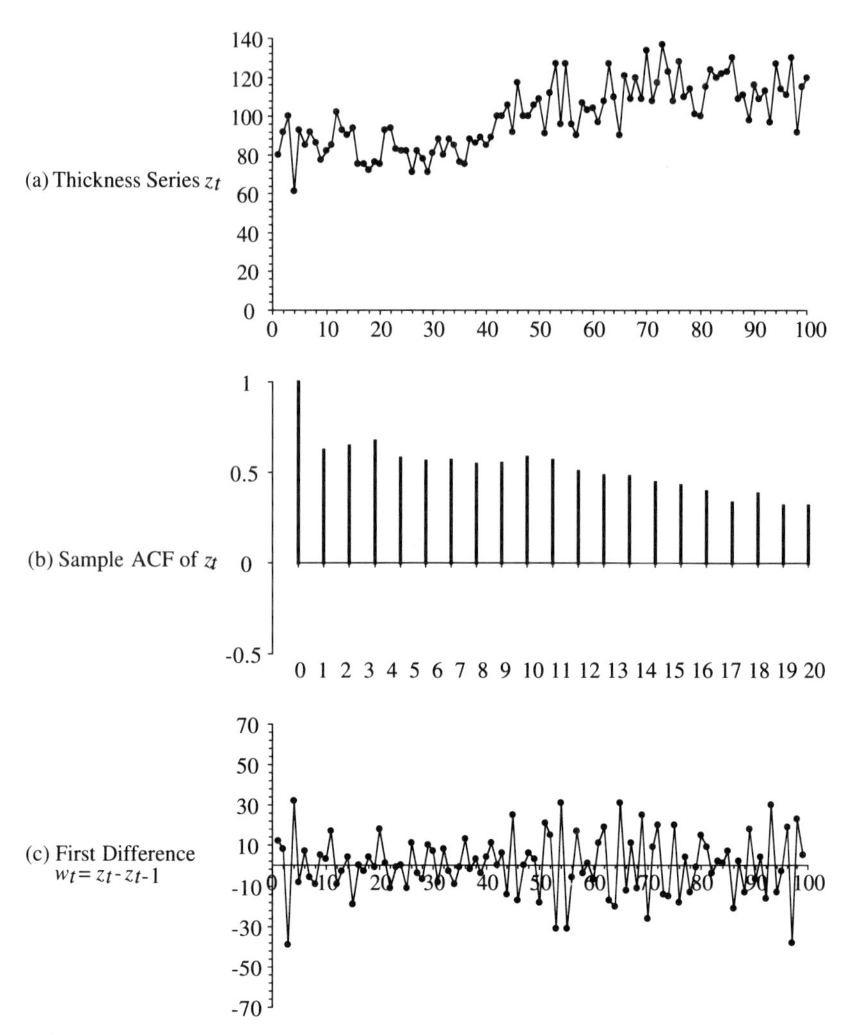

Figure 12.5 Graphical analysis of the Metallic Film Series: (a) the series, (b) the sample ACF, (c) the series of first differences, (d) the sample ACF for the first differences, (e) the sample PACF for the first differences, (f) the residuals for the first differences after fitting an IMA model, and (g) the sample ACF of the residuals in (f).

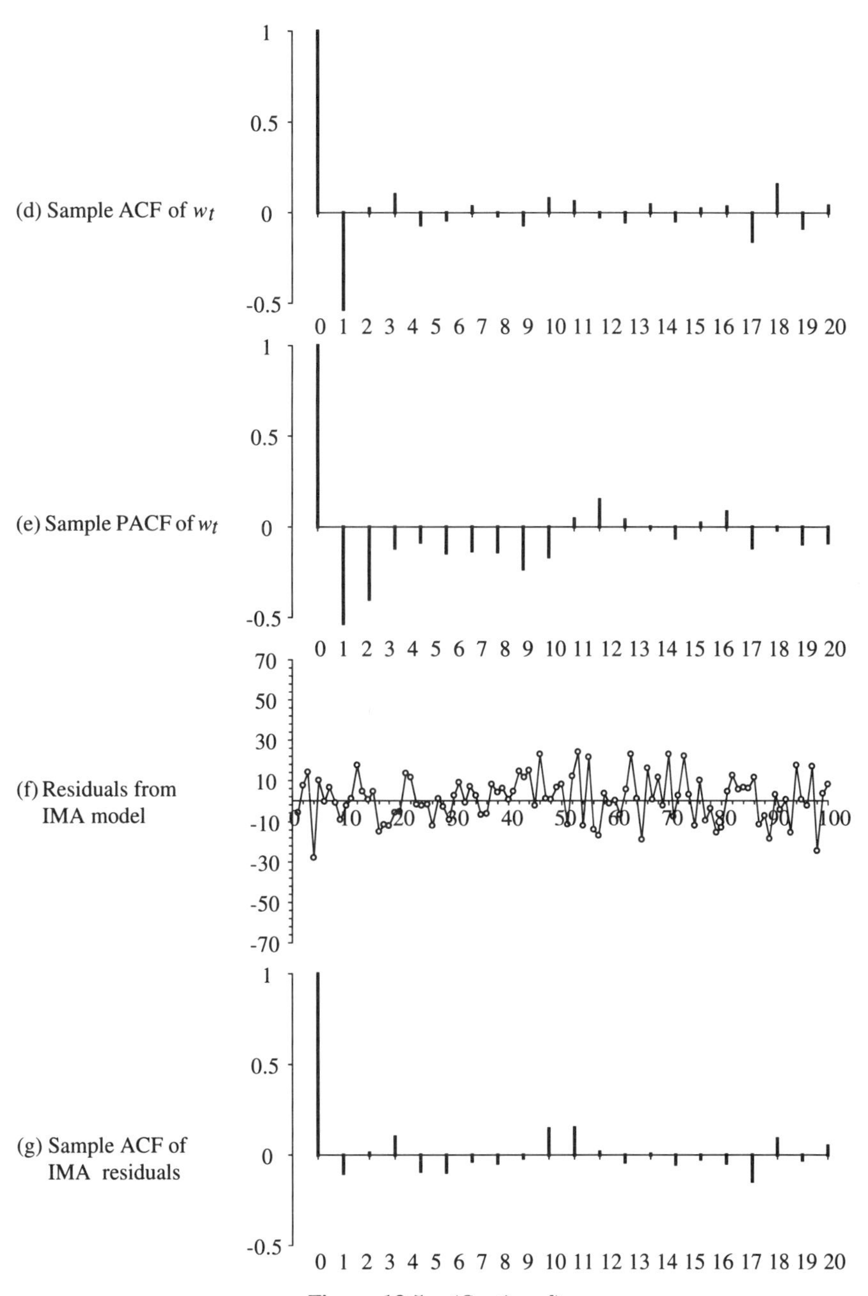

Figure 12.5 *(Continued)*

ACF, in Figure 12.5b, suggest that the series is nonstationary. However, the plot of the first difference, in Figure 12.5c, looks reasonably stationary. The ACF and PACF for the first difference are shown in Figures 12.5d and 12.5e. The ACF of the first difference cuts off after lag 1, suggesting that we should fit a moving average to the first difference. The PACF tends to tail off. Figure 12.5f shows a plot of residuals from the best-fitting IMA model with $\hat{\theta} = 0.78$. The ACF for the residuals is shown in Figure 12.5g and suggests a good fit. It was concluded, therefore, that the series could be approximately represented by an IMA(0, 1, 1) model with $\hat{\theta} = 0.78$ ($\hat{\lambda} = 0.22$).

12.10 FORECASTING

In Chapter 5 of this book and in later chapters, we frequently used the concept of forecasting a time series one step ahead. We consider now the problem of forecasting somewhat more generally.

If you knew the model for a time series, you could use it to make forecasts (predictions, estimates) of future values. In practice, you never *know* the model but let's suppose you have used the iterative model building process outlined above to obtain what you believe is a reasonably close approximation. Forecasting is then rather simple. Suppose you have data $z_t, z_{t-1}, z_{t-2}, \ldots$ up to the present time t and, using your model, you want to make a best[8] forecast of some future value z_{t+l}. We can then say that, from a *forecast origin* t, you want to forecast for *lead time* l.

A General Recipe for Forecasting z_{t+l} from a Forecast Origin t

You first write down the equation of the fitted model at time $t + l$, placing the value z_{t+l} on the left. Then, roughly speaking, you replace each term on the right of the model by the best estimate you can make at time t. More exactly, you substitute the "conditional mean" for each term on the right of the model equation. The *conditional mean* is the mean value, given that you have data only up to time t.

Like many things, forecasting is a lot easier to do than it sounds, so let's look at some examples.

[8] The forecasts we discuss can be justified as "best" in a number of different ways. In particular, assuming the model to be correct, they produce minimum mean square error.

Forecasting with an AR(1) Model

Suppose the model building process leads to an AR(1) model, say,

$$z_t = 0.7\, z_{t-1} + a_t,$$

and, given data up to the present time t, you want to forecast the series one step ahead. This is a problem we considered earlier in Chapter 5. According to the general recipe given above, you need first to write the model with z_{t+1} on the left. This is

$$z_{t+1} = 0.7 z_t + a_{t+1}. \tag{12.11}$$

Now look at the terms on the right. Conditional on the fact that you have data only up to time t, the value of z_t is already known—it is not a random quantity, it is a fact.[9] However, the quantity a_{t+1} has yet to happen. And you know (or can assume) that it is a member of a white noise series, which has mean *zero* and is *independent* of all the data up to time t. So its conditional mean is zero and, if the "hat" notation is used to denote a minimum mean square error forecast, then

$$\hat{z}_{t+1} = 0.7 z_t \tag{12.12}$$

or, in general, for any AR(1) model

$$\hat{z}_{t+1} = \phi z_t. \tag{12.13}$$

So for the AR(1) model, the best estimate made at time t of the deviation z_{t+1} from the process mean that will occur at time $t + 1$ is a proportion ϕ of the deviation z_t that occurred at time t.

The forecast error $z_{t+1} - \hat{z}_{t+1}$ is the difference between what actually occurred at time $t + 1$ and what you expected would occur. Subtracting Equation (12.12) from (12.11) you get

$$z_{t+1} - \hat{z}_{t+1} = a_{t+1}. \tag{12.14}$$

Thus the one-step-ahead forecast error is the white noise element a_{t+1}. As we mentioned earlier for *any* time series model of the kind discussed in this book and for a forecast made from origin t *one step ahead* (but *not* for a forecast more than one step ahead), the forecast error is always equal to the white noise innovation a_{t+1}.

[9] Omar Khayyám expresses the idea more eloquently: "The moving finger writes; and, having writ, moves on: nor all thy piety, nor wit shall lure it back to cancel half a line . . . ." (Fitzgerald translation).

Now let's try forecasting two steps ahead with an AR(1) model from origin t by applying the general recipe to the model

$$z_{t+2} = \phi z_{t+1} + a_{t+2}. \tag{12.15}$$

To do this you must substitute the best estimate you can make for each term on the right given data only up to time t. Now we saw in Equation (12.13) that the best estimate of z_{t+1} was $\hat{z}_{t+1} = \phi z_t$ and, arguing as before, the best estimate of a_{t+2} is zero. So if we use a double hat notation to mean an estimate made *two* steps ahead, we have from Equation (12.15)

$$\hat{\hat{z}}_{t+2} = \phi^2 z_t.$$

Thus, for example, if $\phi = 0.7$, $\hat{\hat{z}}_{t+2} = 0.49 z_t$. It is easy to see that, at origin t, the successive forecasts one, two, three steps ahead for any AR(1) model will be ϕz_t, $\phi^2 z_t$, $\phi^3 z_t$, . . . and, since ϕ must be less than 1 in absolute value, as the lead time gets longer the forecast exponentially decreases to the mean value zero. This corresponds to the fact that for an AR(1), and indeed for any stationary time series model, with sufficient separation between observations, the data will become essentially independent and the best forecast will become the mean of the series.

Now

$$\begin{aligned} z_{t+2} &= \phi z_{t+1} + a_{t+2} \\ &= \phi(\phi z_t + a_{t+1}) + a_{t+2} \\ &= \phi^2 z_t + \phi a_{t+1} + a_{t+2} \end{aligned}$$

and, since $\hat{\hat{z}}_{t+2} = \phi^2 z_t$, the forecast error two steps ahead is

$$z_{t+2} - \hat{\hat{z}}_{t+2} = \phi a_{t+1} + a_{t+2}.$$

Thus the *two*-steps-ahead forecast error is not given by a single white noise element. It depends on both a_{t+1} and a_{t+2} and will generally be larger than the one-step-ahead error. Also, successive errors will be correlated. In particular, for the AR(1) process the two-steps-ahead forecast error has variance $\phi^2\sigma_a^2 + \sigma_a^2$ or $(1 + \phi^2)\sigma_a^2$.

Exercise 12.1. Find the two-steps-ahead forecast $\hat{\hat{z}}_{t+2}$ for an AR(2), second-order autoregressive time series model, and the associated forecast error. □

Exercise 12.2. Write down the one-step-ahead forecast $\hat{z}_{t+1}$ for the ARMA(1,1) model $z_t - \phi z_{t-1} = a_t - \theta a_{t-1}$:

(a) In a recursive form suitable for updating the forecast.
(b) In terms of present and past values of z alone.
(c) What is the associated one-step-ahead forecast error? □

Forecasting with the IMA Model

For the IMA model,

$$z_t - z_{t-1} = a_t - \theta a_{t-1}.$$

Thus the model for z_{t+1} is

$$z_{t+1} = z_t + a_{t+1} - \theta a_t$$

and hence

$$\begin{aligned}\hat{z}_{t+1} &= z_t - \theta a_t \\ &= z_t - \theta(z_t - \hat{z}_t),\end{aligned}$$

which gives the well-known updating formula

$$\hat{z}_{t+1} = (1 - \theta)z_t + \theta \hat{z}_t. \tag{12.16}$$

As we saw earlier, successive substitution for $\hat{z}_t, \hat{z}_{t-1}, \ldots$ will then produce the result

$$\hat{z}_{t+1} = \tilde{z}_t,$$

where $\tilde{z}_t$ is an EWMA of $z_t, z_{t-1}, \ldots$.

For the forecast two steps ahead using

$$z_{t+2} = z_{t+1} + a_{t+2} - \theta a_{t+1},$$

we get

$$\hat{\hat{z}}_{t+2} = \hat{z}_{t+1} = \tilde{z}_t.$$

In the same way, it will be found that from forecast origin t, the EWMA $\tilde{z}_t$ is the forecast for all lead times. Although the forecast itself remains the same, the error of this forecast for increasing lead times increases. In fact, at lead time l, the variance of the forecast error is

$$\{1 + (l - 1)\lambda^2\}\sigma_a^2.$$

Exercise 12.3 Obtain the two-steps-ahead forecast error for the IMA model. □

12.11 ESTIMATION WITH CLOSED LOOP DATA

Suppose you are collecting data from an operating process and you wish to estimate or reestimate the parameters of the disturbance model. Occasionally, open-loop data $\{z_t\}$ are available—that is, you have observations made over a period of time when the process was allowed to run without adjustment. More likely, adjustments to the level X_t of the input variable will have been made for one reason or another—rational or otherwise. In that case, from a record of these adjustments and of the consequent output errors, you may be able to reconstruct $\{z_t\}$. For example, if the effect of an adjustment to X_t is realized at the output in one time interval, then

$$e_t = z_t + gX_{t-1} \tag{12.17}$$

and, in the commonly occurring case when—to an adequate approximation—the value of g is known, you can write Equation (12.17) in the form $z_t = e_t - gX_{t-1}$ and use it to reconstruct the series $\{z_t\}$.

If g is not known, one way to estimate it is as follows. A $\{z_t\}$ series is reconstructed for each of a number of values of g covering a suitable range. The value of g that results in a minimum value of the calculated residual sums of squares[10] will then provide an estimate of g and of the associated disturbance parameters.

For data collected from a feedback system subject to continual adjustment using some control equation, special care is needed. For example, suppose the disturbance can be represented by an IMA model and continual adjustments $x_t = X_t - X_{t-1}$ are made such that

$$gx_t = -Ge_t, \tag{12.18}$$

as in Chapter 6. Then, equivalently,

$$x_t = -ke_t \tag{12.19}$$

[10] Or essentially the maximum value for the calculated likelihood function.

with $k = G/g$, and, using Equations (12.10) and (12.17),

$$z_t - z_{t-1} = a_t - \theta a_{t-1} = e_t - e_{t-1} - gx_{t-1}. \qquad (12.20)$$

Substituting Equation (12.19) in (12.20) and rearranging, we obtain

$$e_t - (1 - gk)e_{t-1} = a_t - \theta a_{t-1}. \qquad (12.21)$$

Remember that k is a known quantity and so, if g is also known, $G = gk$ is known. Hence we can calculate $w_t = e_t - (1 - G)e_{t-1}$ for each value of t. Equation (12.21) then defines a MA(1) model

$$w_t = a_t - \theta a_{t-1}$$

from which θ may be estimated.

If g is not known, then writing $1 - gk = 1 - G = \phi$, we see that Equation (12.21) defines an ARMA(1, 1) times series model. The estimated autoregressive parameter $\hat{\phi}$ of model (12.21) provides an estimate $\hat{g} = (1 - \hat{\phi})/k$ for g along with the estimate $\hat{\theta}$ for θ.

A problem arises in the particular case where a minimum mean square error adjustment scheme is employed. In that case $G = \lambda = 1 - \theta$ and Equation (12.21) becomes

$$e_t - \theta e_{t-1} = a_t - \theta a_{t-1}. \qquad (12.22)$$

Then $\{e_t\} = \{a_t\}$, and the model (12.22) can provide no information about g and θ. Note, however, that if we were checking the *continued applicability* of a MMSE scheme assuming an IMA disturbance model and based on specific values of g and θ, then the fact that the residual e_t values appeared to be white noise innovations would provide no reason to recalibrate the scheme. Also, note that we are less likely to encounter the difficulty with a constrained adjustment scheme in which $G = kg$ is deliberately chosen to be less than λ. When k is not equal to, but is close to, λ/g, the parameter value $\phi = 1 - gk$ on the left-hand side of Equation (12.21) is close to θ on the right. In these circumstances, we are attempting to estimate the parameters of an ARMA(1, 1) when these parameters are nearly equal. Then the estimation of ϕ and θ, and hence of g and θ, becomes very unstable (e.g., see BJR, p. 266) and the estimates can have very large standard errors.

Several strategies have been suggested to solve problems of this kind. One is based on replacing Equation (12.18) by

$$gx_t = -Ge_t + \delta_t, \qquad (12.23)$$

where δ_t is a "dither" signal (e.g., see Box and MacGregor, 1974, 1976). Another strategy consists in using deterministic time-varying values G_t for G (e.g., see Ljung, Gustavsson, and Soderstrom, 1974; Luceño, 1997a), so that G is changed from time to time according to some deterministic rule—that is, a rule not depending on the random values of e_t or x_t. All such strategies are designed to break the relation x_t/e_t = constant, imposed by Equation (12.18).

The same kind of problem can occur with more complex models. Suppose, for example, that the process is controlled by some proportional integral scheme

$$gx_t = -G\{e_t + P(e_t - e_{t-1})\}$$

or, equivalently,

$$x_t = -k\{e_t + P(e_t - e_{t-1})\} \tag{12.24}$$

with $k = G/g$ and P chosen fixed constants. Also, suppose that the system inertia can be represented by a first-order dynamics so that

$$e_t = z_t + g\tilde{X}_{t-1},$$

where $\tilde{X}_{t-1} = (1 - \delta)(X_{t-1} + \delta X_{t-2} + \delta^2 X_{t-3} + \cdots)$.

After some algebraic manipulation, the error e_t at the output may be shown to follow the ARMA(2, 2) model

$$e_t - \{1 + \delta - gk(1 + P)(1 - \delta)\}e_{t-1} - \{-\delta + gkP(1 - \delta)\}e_{t-2}$$
$$= a_t - (\delta + \theta)a_{t-1} + \delta\theta a_{t-2}. \tag{12.25}$$

Estimates of g, θ, and δ may now be obtained by fitting this time series model to the e_t values.

Again, however, when the constants k and P of the control scheme are such as to produce minimum mean square error—that is, when $k = \lambda/g$ and $P = \delta/(1 - \delta)$—then the estimation process will become unstable and the earlier discussion again becomes relevant.

APPENDIX 12A OTHER TOOLS FOR IDENTIFICATION OF TIME SERIES MODELS

Although the sample autocorrelation and partial autocorrelation functions are extremely useful in model identification, they do not always provide unambiguous results, particularly in the case of mixed ARMA models. The problem is not as serious as might be thought, since identification of models is tentative and subject to diagnostic checking, often leading to appropriate modification. There has, however, been considerable investigation of additional methods of identification. The interested reader is referred to the R and S array approach proposed by Gray, Kelley, and McIntire (1978); the generalized partial autocorrelation function studied by Woodward and Gray (1981); the inverse autocorrelation function considered by Cleveland (1972) and Chatfield (1979); the extended sample autocorrelation function of Tsay and Tiao (1984); and the canonical correlation analysis of Akaike (1976), Cooper and Wood (1982), and Tsay and Tiao (1985).

APPENDIX 12B ESTIMATION OF TIME SERIES PARAMETERS

Earlier in this book, for illustration, we fitted IMA models in Chapters 4 and 6 using a least squares method whereby the sum of squares of the residuals is calculated for a series of values of the unknown parameter, the minimizing value of which provides the desired estimate. The method approximates more efficient methods of fitting times series models to data given in BJR and elsewhere. Such methods have been studied by many authors, in particular, Ali (1977), Ansley (1979), Brockwell and Davis (1987), Bruce and Martin (1989), Chang, Tiao, and Chen (1988), Dent (1977), Gardner, Harvey, and Phillips (1980), Gómez and Maravall (1994), Hall and Nicholls (1980), Harvey and Pierse (1984), Hillmer and Tiao (1979), Jones (1980), Kohn and Ansley (1986), Ljung (1982, 1989, 1993), Ljung and Box (1979), Luceño (1993, 1994a,b, 1996c, 1997b,c), Mélard (1984), Newbold (1974), Nicholls and Hall (1979), Pearlman (1980), Peña (1987), Peña and Tiao (1991), Shea (1987, 1989), Wincek and Reinsel (1986), and Wu, Hosking, and Ravishanker (1993).

Much of this work is concerned with the efficient calculation of maximum likelihood estimates possibly when there may be missing and/or outlying observations. Computer programs for calculating these more efficient estimates are available for general time series models and should be used in preference to the approximate procedures. However, as we have indicated earlier, because of the robustness of the control methods that we discuss, approximate estimates are usually adequate.

For the important case of estimating the parameter $\lambda = 1 - \theta$ in the IMA model, a useful computer program has been developed by Luceño (1995b). Some comparisons are made below using simulation of estimates obtained by exact maximum likelihood and by least squares for series of $n = 100$ values. As in this illustration, estimates obtained by minimizing the exact likelihood function are usually closer to the true value of λ than estimates obtained by least squares. The difference, however, tends to be smaller as the length n of the fitted series is increased.

True value of λ	0.500	0.200	0.050
Maximum likelihood estimate	0.498	0.230	0.080
Least squares estimate	0.495	0.245	0.125

Conclusion

"This is not the end . . . . But it is, perhaps, the end of the beginning."
WINSTON CHURCHILL, In a speech at the Mansion House, 1942

To augment the monitoring aspects of statistical process control with appropriate techniques for process adjustment has long been an evident need. Some 35 years ago, in response to a paper that attempted such enhancement, a discussant* remarked, "I welcome this flirtation between control engineering and statistics. I doubt, however, whether they can yet be said to be going steady." We believe that at long last, because of greater understanding, necessary accommodation, and certain serendipitous circumstances, a marriage can now be celebrated.

Investigations extending over a long period of time and especially recent research have shown:

(a) That appropriate adjustment techniques for SPC cannot be produced simply by borrowing from control engineering. This is principally because in the SPC environment we must often allow for the cost of *making* an adjustment and the cost of *getting* an observation as well as the cost of being off-target. Such considerations can totally change the form of the appropriate feedback methods.

(b) That simple techniques which have high efficiency and remarkable insensitivity (robustness) to assumption can meet most of the adjustment needs of SPC.

(c) That these methods are easily understood and, if desired, can be applied manually with simple charts.

* Professor J. H. Westcott in the discussion of "Some Statistical Aspects of Adaptive Optimization and Control" by Box and Jenkins, 1962.

We believe that the points made in Chapter 1 about model building and robustness have been amply demonstrated but that these principles have much wider applicability.

Of the two propositions:

- the study of mathematical models *alone* will lead to methods of practical use,
- the study of working empirical models will lead to fruitful mathematical models,

we believe that the latter is more likely to be true than the former.

More importantly, however, as was illustrated in Figure 1.2, once it is understood that investigation is an interactive rather than a "one shot" process, it is seen that empiricism and theory are not in conflict. It is the repeated interchange between them that results in the continuous never ending acquisition of knowledge and is the essence of scientific method.

Three Time Series

SERIES A

Data (in Rows) for Figure 6.1: Metallic Film Series

80	92	100	61	93	85	92	86	77	82	85	102	93	90	94
75	75	72	76	75	93	94	83	82	82	71	82	78	71	81
88	80	88	85	76	75	88	86	89	85	89	100	100	106	92
117	100	100	106	109	91	112	127	96	127	96	90	107	103	104
97	108	127	110	90	121	109	120	109	134	108	117	137	123	108
128	110	114	101	100	115	124	120	122	123	130	109	111	98	116
109	113	97	127	114	111	130	92	115	120					

SERIES B

Generated Disturbance Data (in Rows) for Figures 7.3 and 7.7: Only the *Last 100 Data* Are Used in the Figures

0.000	−1.239	−0.383	1.576	0.458	0.021	0.552	1.091	−1.166	−0.692
0.307	0.038	−0.155	−2.668	−0.745	−0.623	−1.684	0.828	−0.525	−0.323
−1.561	−1.453	−0.604	−0.598	−0.272	−2.374	−0.953	−0.710	−1.600	−0.872
−1.401	−1.379	0.114	0.643	0.545	−0.873	−0.543	−0.146	0.977	−2.055
−0.687	−0.177	−0.266	−0.760	−0.933	−1.039	−0.083	1.154	−1.307	1.484
−0.916	0.450	0.634	0.323	1.557	1.626	1.049	2.646	1.053	1.549
2.303	1.065	1.437	0.346	0.872	0.548	1.233	0.329	1.547	−0.199
0.961	−1.202	−0.724	−0.584	0.707	−0.102	0.543	−0.483	0.241	−1.663

SERIES B

Generated Disturbance Data (in Rows) for Figures 7.3 and 7.7: Only the *Last 100 Data* Are Used in the Figures (*Continued*)

−0.357	−0.510	−0.140	0.737	−0.136	0.341	−0.715	1.553	0.683	1.321
0.443	1.266	1.059	1.114	0.044	−0.472	−0.947	0.936	0.021	−0.550
0.772	1.177	−0.276	0.900	−0.334	0.378	−0.696	−1.925	−1.366	−0.885
−1.430	−0.435	−1.714	−1.552	−0.790	−2.117	−1.864	−1.117	−2.100	−3.059
−1.706	−3.109	−2.519	−1.540	−1.443	−0.444	−0.356	−2.196	−2.480	−1.863
−3.231	−2.353	−3.379	−2.789	−3.899	−3.143	−3.150	−4.591	−4.908	−5.588
−6.052	−5.850	−5.339	−4.279	−5.090	−4.261	−5.660	−5.068	−5.582	−5.177
−5.363	−4.518	−3.487	−2.864	−5.742	−5.020	−3.832	−5.456	−6.005	−3.241
−6.904	−4.685	−5.452	−5.966	−4.586	−6.198	−5.487	−4.920	−6.172	−2.891
−2.767	−2.848	−2.607	−2.768	−4.300	−3.945	−2.098	−3.039	−2.987	−4.126
−3.350	−4.117	−5.062	−5.607	−5.321	−6.708	−5.279	−5.709	−5.706	−5.722
−5.686	−6.100	−7.621	−4.354	−7.292	−6.463	−6.719	−6.965	−7.070	−5.575

SERIES C

Data (in Rows) for Figure 12.4a: Temperature Series

119.64	119.94	119.91	119.92	119.89	120.04	120.07	119.99
119.95	120.05	120.14	120.36	120.15	120.25	120.33	120.39
120.35	120.41	120.04	119.93	120.05	120.53	120.88	120.80
120.65	121.05	121.33	120.72	120.48	120.66	120.28	119.93
119.87	120.05	120.18	119.86	120.16	120.21	120.22	120.69
120.59	120.69	119.93	119.57	120.03	120.54	119.98	119.84
119.78	120.03	120.05	120.31	120.46	120.56	119.91	120.16
120.02	119.80	120.38	120.70	120.88	121.24	121.03	121.00
121.32	120.59	120.03	120.46	120.93	120.57	120.20	119.82
120.39	120.32	120.52	120.08	120.16	120.41	120.09	119.89

Solutions to Exercises and Problems

Exercise 2.1. (a) 0.1587; (b) 0.2752; (c) 0.0334.

Exercise 2.2. For this data set, $n = 5, \bar{y} = (6 + 4 + 3 + 8 + 4)/5 = 25/5 = 5$,

$$\sum (y - \bar{y})^2 = 1^2 + 1^2 + 2^2 + 3^2 + 1^2 = 16,$$

$$\hat{\sigma}^2 = \sum (y - \bar{y})^2/(5 - 1) = 16/4 = 4, \quad \text{and} \quad \hat{\sigma} = \sqrt{4} = 2.$$

Exercise 2.4. $\bar{y} = 19.46$, $\hat{\sigma}^2 = 82.26$. The process is clearly out of control because of increased variance.

Exercise 2.5. (a) $12(1 - p)/p = 108$ is smaller than $n = 120$, so the normal approximation may be used. (b) The warning limits are at 5.43 and 18.57. The action limits are at 2.14 and 21.86.

Exercise 2.6. (a) In this case there are two groups with variances $\hat{\sigma}_1^2 = 3.5$ and $\hat{\sigma}_2^2 = 6.21$. Then the overall estimate for the standard deviation is

$$\hat{\sigma}_G = \sqrt{\frac{(5 \times 3.5) + (7 \times 6.21)}{12}} = 2.25.$$

(b) The average of the 12 values of the moving range (five in the first group and seven in the second) is $\overline{MR} = 3.33$. Thus $\hat{\sigma}_M = 3.33/1.128 = 2.96$.

(c) The estimates agree reasonably well but, should be considered as tentative, since the data have been obtained during the start-up of a new machine.

(d) If the binomial distribution is adequate, the standard deviation should be smaller after adjustment, because the proportion failing the test has decreased from about 0.5 to about 0.35.

Exercise 3.1. $\tilde{y}_4 = 10.39$, $\tilde{y}_5 = 8.23$, $\tilde{y}_6 = 7.34$, $\tilde{y}_7 = 6.00$, $\tilde{y}_8 = 7.6$, $\tilde{y}_9 = 9.36$, $\tilde{y}_{10} = 11.62$.

Exercise 4.1. In this example $Y_{t+1} - Y_t = k\ X_t$, where k is some constant. Then setting $t = 1$ gives $k = 0.2$ and, since $X_t = 100 - Y_t$,

(a) $Y_{t+1} - Y_t = 0.2\ (100 - Y_t)$ or, equivalently, $Y_{t+1} = 20 + 0.8Y_t$.

(b) The recursion starts using $Y_1 = 0$. The values of $Y_2, Y_3, \ldots, Y_{10}$ are: 20.00, 36.00, 48.80, 59.04, 67.23, 73.79, 79.03, 83.22, and 86.58.

Exercise 4.2. Using $\theta = 0.6$, we have

	A			B		C	
t	z_t	$\hat{z}_t$	e_t	$\hat{z}_{t+1} - \hat{z}_t$	λe_t	$z_t - z_{t-1}$	$e_t - \theta e_{t-1}$
1	6	10.00	−4.00	−1.60	−1.60		
2	9	8.40	0.60	0.24	0.24	3	3
3	12	8.64	3.36	1.34	1.34	3	3
4	11	9.98	1.02	0.41	0.41	−1	−1
5	5	10.39	−5.39	−2.16	−2.16	−6	−6
6	6	8.23	−2.23	−0.89	−0.89	1	1
7	4	7.34	−3.34	−1.34	−1.34	−2	−2
8	10	6.00	4.00	1.60	1.60	6	6
9	12	7.60	4.40	1.76	1.76	2	2
10	15	9.36	5.64	2.26	2.26	3	3
11		11.62					

Exercise 5.1.

α_t	−1.2	−0.413	1.023	1.574	−0.157	1.751	0	−0.925	0.374	−0.59	−1.22	0.944
u_t	0	0.304	0.076	0.215	0.076	0.202	−0.278	−0.063	−0.013	−0.076	0.076	0.557
S_t	0	0.304	0.38	0.595	0.671	0.873	0.595	0.532	0.519	0.443	0.519	1.076
z_t	−1.2	−0.109	1.403	2.169	0.513	2.624	0.595	−0.393	0.893	−0.147	−0.701	2.02
$\hat{z}_t$	0	−0.24	−0.214	0.109	0.521	0.52	0.941	0.871	0.618	0.673	0.509	0.267
a_t	−1.2	0.131	1.617	2.059	−0.008	2.104	−0.346	−1.264	0.274	−0.821	−1.21	1.753

Problem 6.1. (a) and (b) According to Equation (6.3), the adjustment equation would be $1.2x_t = -0.08e_t$. Thus, for example, if the adjusted thickness was 110, then the error e_t would be $110 - 80 = 30$ and the adjustment equation would give $x_t = -2$. A plot of the points that would appear on the adjustment chart is shown in Figure E6.1a.

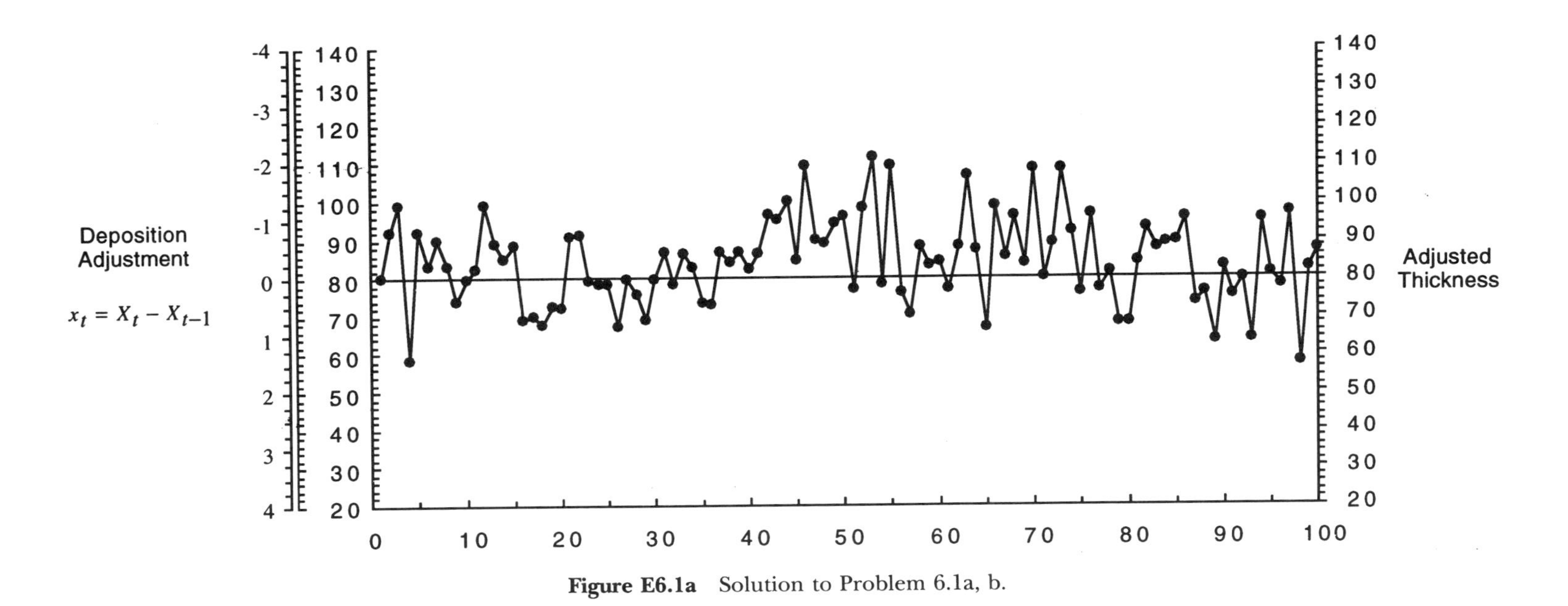

Figure E6.1a Solution to Problem 6.1a, b.

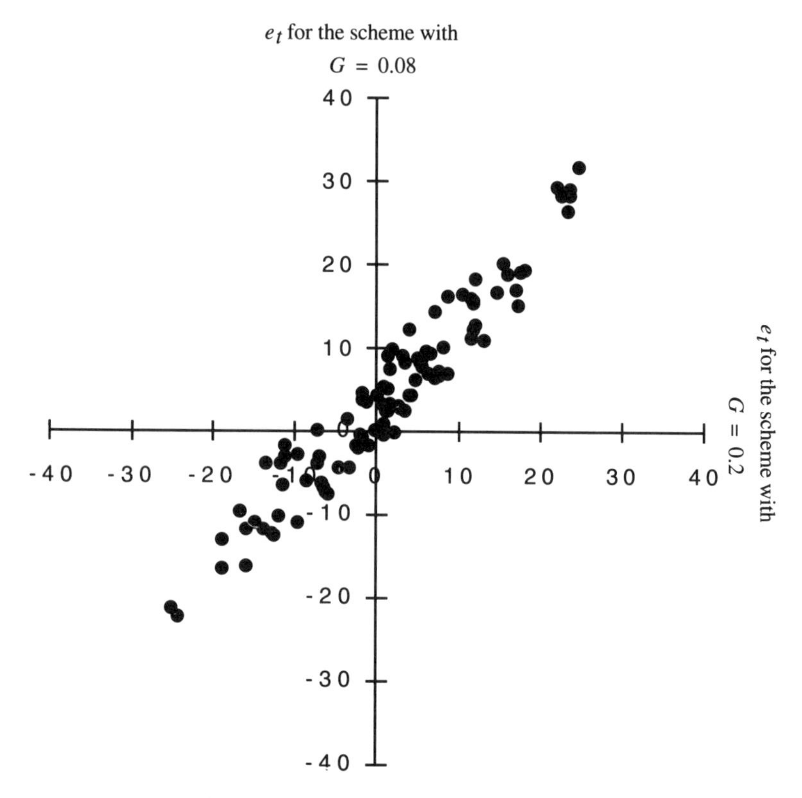

Figure E6.1b Solution to Problem 6.1c.

(c) The plot is shown in Figure E6.1b.

(d) Equation (6.12) with $\lambda_0 = 0.24$, $G = 0.08$, and $g = 1.2$ gives $\sigma_e/\sigma_a = 1.080$ and $\sigma_x/\sigma_a = (G/g)(\sigma_e/\sigma_a) = 0.072$. The values for the earlier scheme with $\lambda_0 = 0.24$, $G = 0.2$, and $g = 1.2$ are $\sigma_e/\sigma_a = 1.002$ and $\sigma_x/\sigma_a = 0.167$. Thus, using $G = 0.08$ rather than $G = 0.2$, produces an increase in σ_e/σ_a of 7.8% and a decrease in σ_x/σ_a of 56.9%.

(e) For the scheme with $G = 0.2$, $\hat{\sigma}_a = \hat{\sigma}_e = 11.13$ and $\hat{\sigma}_x = 1.85$. For the scheme with $G = 0.08$, $\hat{\sigma}_e = 11.38$ and $\hat{\sigma}_x = 0.76$ so $\hat{\sigma}_e/\hat{\sigma}_a = 1.022$ and $\hat{\sigma}_x/\hat{\sigma}_a = 0.068$.

Problem 7.1. (a) Using Figure 7.6, the chart should use $P = 0$ and $G = 0.46$. It would then produce an output standard deviation of $\sigma_e = 1.026\sigma_a$ and an adjustment standard deviation of $\sigma_x = 0.47\sigma_a/g$. (Concerning the assumptions made, see Sections 7.1 and 7.2.)

(b) Using again Figure 7.6, the chart should now use $P = 0$ and $G = 0.19$, so as to produce an output standard deviation of $\sigma_e = 1.10\sigma_a$ and an adjustment standard deviation of $\sigma_x = 0.21\sigma_a/g$.

(c) Using Figure 7.8, we have $P = -0.25$, $G = 0.42$, $\sigma_e = 1.046\sigma_a$, and $\sigma_x = 0.37\sigma_a/g$.

(d) Using again Figure 7.8, we get $P = -0.25$, $G = 0.20$, $\sigma_e = 1.10\sigma_a$, and $\sigma_x = 0.19\sigma_a/g$.

(e)

Scheme	Adjustment Equation	Output Standard Deviation	Input Standard Deviation
(a)	$2.8x_t = -0.46e_t$	0.62	0.10
(b)	$2.8x_t = -0.19e_t$	0.66	0.04
(c)	$2.8x_t = -0.42\{e_t - 0.25(e_t - e_{t-1})\}$	0.63	0.08
(d)	$2.8x_t = -0.20\{e_t - 0.25(e_t - e_{t-1})\}$	0.66	0.04

(f) The charts for schemes (a) and (b), using $P = 0$, would be like those in Figure 7.1 or 7.4, but no extrapolation or interpolation would be required. The target value would be at 70°C. The adjustment scales would be such that an error of 10°C would produce an adjustment of -1.64 in scheme(a), and an adjustment of -0.68 in scheme (b). Time lines would be plotted every hour.

The charts for schemes (c) and (d), using $P = -0.25$, would be like that in Figure 7.4, calling for interpolation at three-quarters of the way between each time line. Again, the target value would be at 70°C and the time lines would appear every hour. The adjustment scales would be such that an interpolated value, that is, a value of $e_t - 0.25(e_t - e_{t-1})$, of 10°C would produce an adjustment of -1.5 in scheme (c), and an adjustment of -0.71 in scheme (d).

Problem 8.1. (a) The target value is $T = (12.6 + 11.4)/2 = 12$. Then $L = 12.6 - T = 0.6$. Since the AAI is 9, interpolation in Table 8.5 gives $L/\lambda\sigma_a = 2.37$. Then $\sigma_a = (0.6/0.3)/2.37 = 0.84$.

Using $L/\lambda\sigma_a = 2.37$, a new interpolation in Table 8.5 gives $q = 1.19$, so that $\text{ISD} = 100 \times (\sqrt{1 + 0.3^2 \times 1.19} - 1) = 5.2\%$. Then $\sigma_e = \sigma_a \times (1 + \text{ISD}/100) = 0.89$.

(b) If ISD = 20%, then $q = (1.2^2 - 1)/0.3^2 = 4.89$, and interpolation in Table 8.5 gives $L/\lambda\sigma_a = 4.95$, so that $L = 4.95 \times 0.3 \times 0.84 = 1.25$. Further interpolation in Table 8.5 gives AAI = 31.1.

(c) The first scheme is for $R_A = 13.5$, so $C_A = 13.5 \times 0.3^2 \times 10 = \$12.2$. In the second scheme, $R_A = 161$, so $C_A = 161 \times 0.3^2 \times 10 = \$145$.

Note that this problem can also be solved using Table 8.3, but this would be less accurate than using Table 8.5.

Problem 9.1. For this scheme, $k_T = 1350/225 = 6$ and the off-target cost is $C_T = k_T\sigma_a^2 = 6 \times 9 = \$54$ (see Section 8.10), and since the cost of each adjustment to the process was $C_A = \$600$, we obtain

$$R_A = \frac{C_A/C_T}{\lambda^2} = \frac{600/54}{0.09} = 123.5 \quad \text{and} \quad R_S = \frac{C_S/C_T}{\lambda^2} = \frac{200/54}{0.09} = 41.2.$$

From Chart A of Figure 9.5a with $R_A = 123.5$ and $R_S = 41.2$ we see that ℓ is about 3.2, so that $L = \ell\lambda\sigma_a = 3.2 \times 0.3 \times 3 = 2.9$. Thus the *action limits* should be placed at 340 ± 2.9. Also, from Chart B for $\lambda = 0.3$ it will be seen that with $R_S = 41.2$, S is about 10 units; since the unit interval is one hour, the monitoring interval should be about ten hours. Using Figure 9.2 with $\lambda = 0.3$ and $S = 10$, we find that $\lambda_{10} = 0.66$, which is the value that must be used to compute the EWMAs when the process is monitored at intervals of $S = 10$. Figure 9.1c (for $\lambda = 0.3$) or the computer program in Luceño, González, and Puig-Pey (1996) shows that, for this scheme, the inflation of the standard deviation (ISD) would be 35.5% and the average interval between adjustments (AAI) would be 28.3 hours. Thus remembering that the sampling interval for the modified scheme is ten hours, you would need on average to adjust the process about every third sample.

Schemes of this kind are very robust. Thus, for this example, computer calculations show that for values of S of 9, 10, 11, and 12, the expected costs are \$140.63, \$140.32, \$140.39, and \$140.74 per unit interval with $L = 2.98$, 2.89, 2.80, and 2.71, and $\lambda_9 = 0.64$, $\lambda_{10} = 0.66$, $\lambda_{11} = 0.68$, and $\lambda_{12} = 0.69$, respectively. Thus there are several adjacent schemes that are very close to the absolute minimum, and in practice very precise results seem unnecessary.

Problem 10.1. The Cuscore statistic in Figure 10.2c takes the following values:

0.00	0.34	0.03	−1.54	−1.15	−1.15	−0.81	−1.06	−0.47	−1.23
−1.35	−1.35	−0.30	−0.07	−2.11	−3.00	−2.52	−2.52	−2.66	−2.20
−1.60	−0.90	−1.17	−1.17	−0.84	−1.80	−0.69	−0.56	−0.35	−0.35
−0.79	0.18	−0.15	0.45	−0.28	−0.28	−0.36	−0.30	−0.97	−2.28
−2.44	−2.44	−2.63	−2.47	−3.91	−2.93	−2.68	−2.68	−2.20	−3.63
−3.17	−1.82	−2.31	−2.31	−1.67	−1.36	0.65	0.88	0.73	0.73
1.06	2.64	3.48	3.16	4.22	4.22	3.28	3.30	4.80	5.97

6.57	6.57	6.70	6.69	7.74	7.54	8.58	8.58	8.95	11.26
10.07	9.31	9.65	9.65	10.75	11.00	11.44	10.52	10.40	10.40
10.39	10.28	9.40	9.37	9.25	9.25	9.04	8.97	8.96	9.21
9.17	9.17	9.49	9.14	9.11	9.28	8.59	8.59	8.85	9.31
7.97	7.11	7.34	7.34	7.87	7.38	7.63	7.43	7.10	7.10
7.40	7.87	9.16	9.91	9.43	9.43	8.76	8.42	9.05	8.41
8.73	8.73	8.97	8.84	10.10	9.63	9.14	9.14	9.03	8.52
7.97	7.58	7.60	7.60						

Exercise 12.1.

$$\hat{\hat{z}}_{t+2} = (\phi_1^2 + \phi_2)z_t + \phi_1\phi_2 z_{t-1},$$

$$z_{t+2} - \hat{\hat{z}}_{t+2} = \phi_1 a_{t+1} + a_{t+2}.$$

Exercise 12.2.

(a) $\hat{z}_{t+1} = \phi z_t - \theta a_t$ or $\hat{z}_{t+1} = (\phi - \theta)z_t + \theta\hat{z}_t$.

(b) $\hat{z}_{t+1} = \left(\dfrac{\phi - \theta}{1 - \theta}\right)\tilde{z}_t,$

where $\tilde{z}_t$ is the EWMA $(1 - \theta)(z_t + \theta z_{t-1} + \theta^2 z_{t-2} + \cdots)$.

(c) $z_{t+1} - \hat{z}_{t+1} = a_{t+1}$.

Exercise 12.3. $z_{t+2} - \hat{\hat{z}}_{t+2} = \lambda a_{t+1} + a_{t+2}$ with variance $(1 + \lambda^2)\sigma_a^2$. □

TABLE A

Normal Distribution Two-Sided Tail Probabilities

Normal Distribution Two-Sided Tail Probabilities

k	Two-Tail Probabilities	k	Two-Tail Probabilities	k	Two-Tail Probabilities	k	Two-Tail Probabilities	k	Two-Tail Probabilities	k	Two-Tail Probabilities
0.00	1.0000	0.50	0.6171	1.00	0.3173	1.50	0.1336	2.00	0.0455	2.50	0.0124
0.01	0.9920	0.51	0.6101	1.01	0.3125	1.51	0.1310	2.01	0.0444	2.51	0.0121
0.02	0.9840	0.52	0.6031	1.02	0.3077	1.52	0.1285	2.02	0.0434	2.52	0.0117
0.03	0.9761	0.53	0.5961	1.03	0.3030	1.53	0.1260	2.03	0.0424	2.53	0.0114
0.04	0.9681	0.54	0.5892	1.04	0.2983	1.54	0.1236	2.04	0.0414	2.54	0.0111
0.05	0.9601	0.55	0.5823	1.05	0.2937	1.55	0.1211	2.05	0.0404	2.55	0.0108
0.06	0.9522	0.56	0.5755	1.06	0.2891	1.56	0.1188	2.06	0.0394	2.56	0.0105
0.07	0.9442	0.57	0.5687	1.07	0.2846	1.57	0.1164	2.07	0.0385	2.57	0.0102
0.08	0.9362	0.58	0.5619	1.08	0.2801	1.58	0.1141	2.08	0.0375	2.58	0.0099
0.09	0.9283	0.59	0.5552	1.09	0.2757	1.59	0.1118	2.09	0.0366	2.59	0.0096
0.10	0.9203	0.60	0.5485	1.10	0.2713	1.60	0.1096	2.10	0.0357	2.60	0.0093
0.11	0.9124	0.61	0.5419	1.11	0.2670	1.61	0.1074	2.11	0.0349	2.61	0.0091
0.12	0.9045	0.62	0.5353	1.12	0.2627	1.62	0.1052	2.12	0.0340	2.62	0.0088
0.13	0.8966	0.63	0.5287	1.13	0.2585	1.63	0.1031	2.13	0.0332	2.63	0.0085
0.14	0.8887	0.64	0.5222	1.14	0.2543	1.64	0.1010	2.14	0.0324	2.64	0.0083
0.15	0.8808	0.65	0.5157	1.15	0.2501	1.65	0.0989	2.15	0.0316	2.65	0.0080
0.16	0.8729	0.66	0.5093	1.16	0.2460	1.66	0.0969	2.16	0.0308	2.66	0.0078
0.17	0.8650	0.67	0.5029	1.17	0.2420	1.67	0.0949	2.17	0.0300	2.67	0.0076
0.18	0.8572	0.68	0.4965	1.18	0.2380	1.68	0.0930	2.18	0.0293	2.68	0.0074
0.19	0.8493	0.69	0.4902	1.19	0.2340	1.69	0.0910	2.19	0.0285	2.69	0.0071
0.20	0.8415	0.70	0.4839	1.20	0.2301	1.70	0.0891	2.20	0.0278	2.70	0.0069
0.21	0.8337	0.71	0.4777	1.21	0.2263	1.71	0.0873	2.21	0.0271	2.71	0.0067
0.22	0.8259	0.72	0.4715	1.22	0.2225	1.72	0.0854	2.22	0.0264	2.72	0.0065
0.23	0.8181	0.73	0.4654	1.23	0.2187	1.73	0.0836	2.23	0.0258	2.73	0.0063
0.24	0.8103	0.74	0.4593	1.24	0.2150	1.74	0.0819	2.24	0.0251	2.74	0.0061
0.25	0.8026	0.75	0.4533	1.25	0.2113	1.75	0.0801	2.25	0.0245	2.75	0.0060

Table A *(Continued)*

k	Two-Tail Probabilities	k	Two-Tail Probabilities	k	Two-Tail Probabilities	k	Two-Tail Probabilities	k	Two-Tail Probabilities	k	Two-Tail Probabilities
0.26	0.7949	0.76	0.4473	1.26	0.2077	1.76	0.0784	2.26	0.0238	2.76	0.0058
0.27	0.7872	0.77	0.4413	1.27	0.2041	1.77	0.0767	2.27	0.0232	2.77	0.0056
0.28	0.7795	0.78	0.4354	1.28	0.2005	1.78	0.0751	2.28	0.0226	2.78	0.0054
0.29	0.7718	0.79	0.4295	1.29	0.1971	1.79	0.0735	2.29	0.0220	2.79	0.0053
0.30	0.7642	0.80	0.4237	1.30	0.1936	1.80	0.0719	2.30	0.0215	2.80	0.0051
0.31	0.7566	0.81	0.4179	1.31	0.1902	1.81	0.0703	2.31	0.0209	2.81	0.0050
0.32	0.7490	0.82	0.4122	1.32	0.1868	1.82	0.0688	2.32	0.0203	2.82	0.0048
0.33	0.7414	0.83	0.4065	1.33	0.1835	1.83	0.0673	2.33	0.0198	2.83	0.0047
0.34	0.7339	0.84	0.4009	1.34	0.1802	1.84	0.0658	2.34	0.0193	2.84	0.0045
0.35	0.7263	0.85	0.3953	1.35	0.1770	1.85	0.0643	2.35	0.0188	2.85	0.0044
0.36	0.7188	0.86	0.3898	1.36	0.1738	1.86	0.0629	2.36	0.0183	2.86	0.0042
0.37	0.7114	0.87	0.3843	1.37	0.1707	1.87	0.0615	2.37	0.0178	2.87	0.0041
0.38	0.7039	0.88	0.3789	1.38	0.1676	1.88	0.0601	2.38	0.0173	2.88	0.0040
0.39	0.6965	0.89	0.3735	1.39	0.1645	1.89	0.0588	2.39	0.0169	2.89	0.0039
0.40	0.6892	0.90	0.3681	1.40	0.1615	1.90	0.0574	2.40	0.0164	2.90	0.0037
0.41	0.6818	0.91	0.3628	1.41	0.1585	1.91	0.0561	2.41	0.0160	2.91	0.0036
0.42	0.6745	0.92	0.3576	1.42	0.1556	1.92	0.0549	2.42	0.0155	2.92	0.0035
0.43	0.6672	0.93	0.3524	1.43	0.1527	1.93	0.0536	2.43	0.0151	2.93	0.0034
0.44	0.6599	0.94	0.3472	1.44	0.1499	1.94	0.0524	2.44	0.0147	2.94	0.0033
0.45	0.6527	0.95	0.3421	1.45	0.1471	1.95	0.0512	2.45	0.0143	2.95	0.0032
0.46	0.6455	0.96	0.3371	1.46	0.1443	1.96	0.0500	2.46	0.0139	2.96	0.0031
0.47	0.6384	0.97	0.3320	1.47	0.1416	1.97	0.0488	2.47	0.0135	2.97	0.0030
0.48	0.6312	0.98	0.3271	1.48	0.1389	1.98	0.0477	2.48	0.0131	2.98	0.0029
0.49	0.6241	0.99	0.3222	1.49	0.1362	1.99	0.0466	2.49	0.0128	2.99	0.0028
0.50	0.6171	1.00	0.3173	1.50	0.1336	2.00	0.0455	2.50	0.0124	3.00	0.0027

Bibliography

Abraham, B., and Box, G. E. P. (1979). Sampling interval and feedback control. *Technometrics,* **21,** 1–7.

Adams, B. M. (1988). *Economically Optimal On-line Quality Control Procedures.* Unpublished Ph.D. thesis, University of Southwestern Louisiana.

Adams, B. M., Lowry, C., and Woodall, W. H. (1992). The use (and misuse) of false alarms probabilities in control chart design. *Frontiers in Statistical Quality Control 4,* H.-J. Lenz, G. B. Wetherill, and P.-Th. Wilrich (eds.). Physica-Verlag, Heidelberg, pp. 155–168.

Adams, B. M., and Woodall, W. H. (1989). An analysis of Taguchi's on-line process control procedure under a random walk model. *Technometrics,* **31,** 401–413.

Akaike, H. (1976). Canonical correlation analysis of time series and the use of an information criterion. *System Identification Advances and Case Studies,* R. K. Mehra and D. G. Lainiotis (eds.) Academic Press, New York, pp. 27–96.

Ali, M. M. (1977). Analysis of autoregressive–moving average models: estimation and prediction. *Biometrika,* **64,** 535–545.

Al-Osh, M. A., and Alzaid, A. A. (1987). First order integer valued autoregressive (INAR(1)) process. *Journal of Time Series Analysis,* **8,** 261–275.

Altham, P. M. E. (1978). Two generalizations of the binomial distribution. *Applied Statistics.* **27,** 162–167.

Alwan, L. C. (1992). Effects of autocorrelation on control chart performance. *Communications in Statistics—Theory and Methods,* **21,** 1025–1049.

Alwan, L. C., and Roberts, H. V. (1988). Time-series modeling for statistical process control. *Journal of Business and Economics Statistics,* **6,** 87–95.

Ansley, C. F. (1979). An algorithm for the exact likelihood of a mixed autoregressive–moving average process. *Biometrika,* **66,** 59–65.

Aström, K. J. (1970). Introduction to stochastic control. *Mathematics in Science and Engineering Series,* Vol. 70. Academic Press, New York.

Aström, K. J., and Wittenmark, B. (1984), *Computer Controlled Systems: Theory and Design.* Prentice Hall, Englewood Cliffs, NJ.

Aström. K. J., and Wittenmark, B. (1989). *Adaptive Control.* Addison-Wesley, Reading, MA.

Bachelier, L. (1900). Théorie de la spéculation. *Annales Scientifiques de l'École Normale Supérieure,* Series 3, **17,** 21–86.

Bagshaw, M., and Johnson, R. A. (1975). The effect of serial correlation on the performance of Cusum tests, II. *Technometrics,* **17,** 73–80.

Bagshaw, M., and Johnson, R. A. (1977). Sequential procedures for detecting parameter changes in a time-series model. *Journal of the American Statistical Association,* **72,** 593–597.

Bajaria, H. J., and Copp, R. P. (1991). *Statistical Problem Solving.* Multiface Publishing Company, Garden City, MI.

Barnard, G. A. (1959). Control charts and stochastic processes. *Journal of the Royal Statistical Society, Series B,* **21,** 239–271.

Barron, D. N. (1992). The analysis of count data: overdispersion and autocorrelation. *Sociological Methodology,* P. Marsden (ed.). Blackwell, Cambridge, MA, pp. 179–220.

Bather, J. A. (1963). Control charts and the minimization of costs. *Journal of the Royal Statistical Society, Series B,* **25,** 49–80.

Baxley, R. V., Jr. (1991). A simulation study of statistical process control algorithms for drifting processes. In *Statistical Process Control in Manufacturing,* J. B. Keats and D. C. Montgomery (eds.). Marcel Dekker, New York, pp. 247–297.

Baxley, R. V., Jr. (1994). Applications of the EWMA for algorithmic statistical process control. *Quality Engineering,* **7,** 397–418.

Bergh, L. G., and MacGregor, J. F. (1987). Constrained minimum variance controllers: internal model structure and robustness properties. *Industrial and Engineering Chemistry Research,* **26,** 1558–1564.

Bergman, B., and Klefsjö, B. (1994). *Quality, From Customer Needs to Customer Satisfaction.* Student Litteratur, Lund, Sweden.

Berthouex, P. M., Hunter, W. G., and Pallesen, L. (1978). Monitoring sewage treatment plants: Some quality control aspects. *Journal of Quality Technology,* **10,** 139–149.

Bissell, A. F. (1969). Cusum techniques for quality control. *Applied Statistics,* **18,** 1–30.

Bohlin, T. (1971). On the problem of ambiguities in maximum likelihood identification. *Automatica,* **7,** 199–210.

Box, G. E. P. (1980). Sampling and Bayes inference in scientific modeling and robustness. *Journal of the Royal Statistical Society, Series A,* **143,** 383–430.

Box, G. E. P. (1991a). Feedback control by manual adjustment. *Quality Engineering,* **4,** 143–151.

Box, G. E. P. (1991b). Bounded adjustment charts. *Quality Engineering,* **4,** 333–340.

Box, G. E. P. (1993). Process adjustment and quality control. *Total Quality Management,* **4,** 215–227.

Box, G. E. P., Hunter, W. G., and Hunter, J. S. (1978). *Statistics for Experimenters. An Introduction to Design, Data Analysis, and Model Building,* Wiley, New York.

Box, G. E. P., and Jenkins, G. M. (1962). Some statistical aspects of adaptive optimization and control. *Journal of the Royal Statistical Society, Series B,* **24,** 297–343.

Box, G. E. P., and Jenkins, G. M. (1963). Further contributions to adaptive quality control: simultaneous estimation of dynamics: non-zero costs. *ISI Bulletin,* 34th Session, Ottawa, Canada, pp. 943–974.

Box, G. E. P., and Jenkins, G. M. (1966). Models for prediction and control: VI diagnostic checking. *Technical Report No. 99,* Department of Statistics, University of Wisconsin–Madison.

Box, G. E. P., and Jenkins, G. M. (1968). Discrete models for feedback and feed-forward control. *The Future of Statistics,* D. G. Watts (ed.). Academic Press, New York, pp. 201–240.

Box, G. E. P., and Jenkins, G. M. (1970). *Time Series Analysis, Forecasting and Control.* Holden-Day, San Francisco.

Box, G. E. P., and Jenkins, G. M. (1976). *Time Series Analysis, Forecasting and Control, 2nd ed.* Holden-Day, San Francisco.

Box, G. E. P., Jenkins, G. M., and MacGregor, J. F. (1974). Some recent advances in forecasting and control, part II. *Applied Statistics,* **23,** 158–179.

Box, G. E. P., Jenkins, G. M., and Reinsel, G. C. (1994). *Time Series Analysis, Forecasting and Control,* 3rd ed. Prentice Hall, Englewood Cliffs, NJ.

Box, G. E. P., and Kramer, T. (1992). Statistical process monitoring and feedback adjustment. A discussion. *Technometrics,* **34,** 251–285.

Box, G. E. P., and Luceño, A. (1994). Selection of sampling interval and action limit for discrete feedback adjustment. *Technometrics,* **36,** 369–378.

Box, G. E. P., and Luceño, A. (1995). Discrete proportional-integral control with constrained adjustment. *Journal of the Royal Statistical Society, Series D—The Statistician,* **44,** 479–495.

Box, G. E. P., and Luceño, A. (1997). The anatomy and robustness of discrete proportional-integral adjustment and its application to statistical process control. *Journal of Quality Technology,* **29,** 248–260.

Box, G. E. P., and MacGregor, J. F. (1974). The analysis of closed loop dynamic-stochastic system. *Technometrics,* **16,** 391–398.

Box, G. E. P., and MacGregor, J. F. (1976). Parameter estimation with closed-loop operating data. *Technometrics,* **18,** 371–380.

Box, G. E. P., and Newbold, P. (1971). Some comments on a paper of Coen, Gomme, and Kendall. *Journal of the Royal Statistical Society, Series A,* **134,** 229.

Box, G. E. P. and Ramírez, J. (1992). Cumulative score charts. *Quality and Reliability Engineering International,* **8,** 17–27.

Breslow, N. E. (1984). Extra-Poisson variation in log-linear models. *Applied Statistics,* **33,** 33–48.

Brook, D., and Evans, D. A. (1972). An approach to the probability distribution of Cusum run lengths. *Biometrika,* **59,** 539–549.

Brockwell, P. J., and Davis, R. A. (1987). *Time Series: Theory and Methods.* Springer-Verlag, New York.

Bruce, A. G., and Martin, R. D. (1989). Leave-*k*-out diagnostics for time series. *Journal of the Royal Statistical Society, Series B,* **51,** 363–424.

Cameron, A. C., and Trivedi, P. K. (1986). Econometric models based on count data: comparisons and applications of some estimators and tests. *Journal of Applied Econometrics,* **1,** 29–53.

Camp, C. W., and Woodall, W. H. (1987). Exact results for Shewhart control charts with supplementary run rules. *Technometrics,* **29,** 393–399.

Chang, I., Tiao, G. C., and Chen, C. (1988). Estimation of time series parameters in the presence of outliers. *Technometrics,* **30,** 193–204.

Chatfield, C. (1979). Inverse autocorrelations. *Journal of the Royal Statistical Society, Series A,* **142,** 363–377.

Chatfield, C., and Goodhardt, G. J. (1970). The beta-binomial model for consumer purchasing behavior. *Applied Statistics,* **19,** 240–250.

Cleveland, W. S. (1972). The inverse autocorrelations of a time series and their applications. *Technometrics,* **14,** 277–298.

Collings, B. J., and Margolin, B. H. (1985). Testing goodness-of-fit for the Poisson assumption when observations are not identically distributed. *Journal of the American Statistical Association,* **80,** 411–418.

Cooper, D. M., and Wood, E. F. (1982). Identifying multivariate time series models. *Journal of Time Series Analysis,* **3,** 153–164.

Cox, D. R., (1983). Some remarks on overdispersion. *Biometrika,* **70,** 269–274.

Cox, D. R., and Solomon, P. J. (1988). On testing for serial correlation in large number of small samples. *Biometrika,* **75,** 145–148.

Cressie, N. (1988). A graphical procedure for determining nonstationarity in time series. *Journal of the American Statistical Association,* **83,** 1108–1116.

Crowder, M. J. (1978). Beta-binomial Anova for proportions. *Applied Statistics,* **27,** 34–37.

Crowder, M. J. (1985). Gaussian estimation for correlated binomial data. *Journal of the Royal Statistical Society, Series B,* **47,** 229–237.

Crowder, S. V. (1986). *Kalman Filtering and Statistical Process Control.* Unpublished Ph.D. thesis, Iowa State University, Department of Statistics.

Crowder, S. V. (1987). A simple method for studying run-length distributions of exponentially weighted moving average charts. *Technometrics,* **29,** 401–408.

Crowder, S. V. (1992). An SPC model for short production runs: minimizing expected cost. *Technometrics,* **34,** 64–73.

Crowder, S. V., Hawkins, D. M., Reynolds, M. R., and Yashkin, E. (1997). Process control and statistical inference. *A Discussion on Statistically-Based Process Monitoring and Control,* D. C. Montgomery and W. H. Woodall (eds.). *Journal of Quality Technology,* **29,** 134–139.

Dean, C. B. (1992). Testing for overdispersion in Poisson and binomial regression models. *Journal of the American Statistical Association,* **87,** 451–457.

Dean, C., and Lawless, J. F. (1989). Tests for detecting overdispersion in Poisson regression models. *Journal of the American Statistical Association,* **84,** 467–472.

Diels, H. (1924). *Antike Technik,* 3rd ed. Leipzig, pp. 203–207.

Deming, W. E. (1986). *Out of the Crisis.* Massachusetts Institute of Technology, Center for Advanced Engineering Studies, Cambridge, MA.

Dent, W. (1977). Computation of the exact likelihood function of an ARIMA process. *Journal of Statistical Computation and Simulation,* **5,** 193–206.

Duncan, A. J. (1974). *Quality Control and Industrial Statistics,* 4th ed. Richard D. Irwin, Homewood, IL.

Faltin, F. W., Hahn, G. J., Tucker, W. T., and Vander Wiel, S. A. (1993). Algorithmic statistical process control: some practical observations. *International Statistical Institute,* **61,** 67–80.

Fearn, T., and Maris, P. I. (1991). An application of Box–Jenkins Methodology to the control of gluten addition in a flour mill. *Applied Statistics,* **40,** 477–484.

Fisher, R. A. (1925). Theory of statistical estimation. *Proceedings of the Cambridge Philosophical Society,* **22,** 700–725.

Fisher, R. A. (1950). The significance of deviation from expectation in a Poisson series. *Biometrics,* **6,** 17–24.

Gan, F. F. (1991). An optimal design of Cusum quality control charts. *Journal of Quality Technology,* **23,** 279–286.

Gardner, G., Harvey, A. C., and Phillips, G. D. A. (1980). Algorithm AS 154. An algorithm for the exact maximum likelihood estimation of autoregressive–moving average models by means of Kalman filtering. *Applied Statistics,* **29,** 311–322.

Gómez, V., and Maravall, D. (1994). Estimation, prediction and interpolation for nonstationary series with the Kalman filter. *Journal of the American Statistical Association,* **89,** 611–624.

Gray H. L., Kelley, G. D., and McIntire D. D. (1978). A new approach to ARMA modeling. *Communications in Statistics,* **B7,** 1–77.

Grent, E. L., and Leavenworth, R. S. (1972). *Statistical Quality Control.* 4th ed. McGraw-Hill, New York.

Hahn, G. J., Faltin, F. W., Tucker, W. T., Richards, S., and Vander Wiel, S. A. (1988). ASPC: making SPC more proactive. Unpublished paper presented at the Fourth National Symposium on Statistics in Design and Process Control, Arizona State University.

Hall, A. D., and Nicholls, D. F. (1980). The evaluation of exact maximum likelihood estimates for VARMA models. *Journal of Statistical Computation and Simulation,* **10,** 251–262.

Harris, T. J. (1988). Interfaces between statistical process control and engineering process control. Unpublished manuscript.

Harris, T. J. (1992). Optimal controllers for nonsymmetric and nonquadratic loss functions. *Technometrics,* **34,** 298–306.

Harris, T. J., and MacGregor, J. F. (1987). Design of multivariate linear-quadratic controllers using transfer functions. *American Institute of Chemical Engineers Journal,* **33,** 1481–1495.

Harris, T. J., MacGregor, J. F., and Wright, J. D. (1982). An overview of discrete stochastic controllers: Generalized PID algorithms with dead-time compensation. *Canadian Journal of Chemical Engineering,* **68,** 425–432.

Harvey, A. C., and Pierse, R. G. (1984). Estimating missing observations in economic time series. *Journal of the American Statistical Association,* **79,** 125–131.

Haseman, J. K., and Kupper, L. L. (1979). Analysis of dichotomous response data from certain toxicological experiments. *Biometrics,* **35,** 281–293.

Hassemer, D. J., Wiebe, D. A., and T. Kramer (1989). The Wisconsin cholesterol study: An assessment of laboratory performance. *Clinical Chemistry Abstracts,* **35**(6), 1068.

Hillmer, S. C., and Tiao, G. C. (1979). Likelihood function of stationary multiple autoregressive–moving average models. *Journal of the American Statistical Association,* **74,** 652–660.

Holt, C. C. (1957). *Forecasting Seasonals and Trends by Exponentially Weighted Moving Averages.* Pittsburgh: Carnegie Institute of Technology.

Hromi, J. D. (ed.) (1996). *The Best on Quality.* ASQC Quality Press, Milwaukee, WI.

Hunter, J. S. (1986). The exponentially weighed moving average. *Journal of Quality Technology,* **18,** 203–210.

Hunter, J. S. (1989). A one-point plot equivalent to the Shewhart chart with Western Electric rules. *Quality Engineering,* **2,** 13–19.

Imai, M. (1986). *KAIZEN: The Key to Japan's Competitive Success.* Random House, New York.

Ishikawa, K. (1985). *What is Quality Control?* Prentice Hall, Englewood Cliffs, NJ.

Ishikawa, K. (1989). *Guide to Quality Control,* 2nd ed. (4th printing). Asian Productivity Organization, Tokyo 1982. Available from UNIPUB/Kraus, White Plains, New York, New York.

Ishikawa, K. (1990). *Introduction to Quality Control.* 3A Corporation, Tokyo, Japan.

Jensen, K. L., and Vardeman, S. B. (1993). Optimal adjustment in the presence of deterministic process drift and random adjustment error. *Technometrics,* **35,** 376–389.

Johnson, R. A., and Bagshaw, M. (1974). The effect of serial correlation on the performance of Cusum tests. *Technometrics,* **16,** 103–112.

Joiner, B. L. (1994). *Fourth Generation Management.* McGraw Hill, New York.

Jones, R. H. (1980). Maximum likelihood fitting of ARMA models to time series with missing observations. *Technometrics,* **22,** 389–395.

Jorgensen, B. (1987). Exponential dispersion models. *Journal of the Royal Statistical Society, Series B,* **49,** 127–162.

Jowett, G. H. (1952). The accuracy of systematic sampling from conveyor belts. *Applied Statistic,* **1,** 50–59.

Juran, J. (1988). *Quality Control Handbook,* 4th ed. McGraw-Hill, New York.

Kanji, G. K., and Asher, M. (1993). *Total Quality Management: A Systematic Approach.* Carfax Publishing Company, Oxfordshire, UK.

Kemp, K. W. (1962). The use of cumulative sums for sampling inspection schemes. *Applied Statistics,* **11,** 16–31.

Kohn, R., and Ansley, C. F. (1986). Estimation, prediction, and interpolation for ARIMA models with missing observations. *Journal of the American Statistical Association,* **81,** 751–761.

Kotz, S., and Johnson, N. L. (1993). *Process Capability Indices.* Chapman and Hall, London.

Kramer, T. (1989). *Process Control from an Economic Point of View.* Unpublished Ph.D. thesis, University of Wisconsin–Madison.

Kupper, L. L., and Haseman, J. K. (1978). The use of a correlated binomial model for the analysis of certain toxicological experiments. *Biometrics,* **34,** 67–76.

Lai, T. L. (1995). Sequential change point detection in quality control and dynamic systems. *Journal of the Royal Statistical Society, Series B,* **57,** 613–658.

Ljung, G. M. (1982). The likelihood function for a stationary Gaussian autoregressive–moving average process with missing observations. *Biometrika,* **69,** 265–268.

Ljung, G. M. (1989). A note on the estimation of missing values in time series. *Communications in Statistics, Simulation and Computation,* **18,** 459–465.

Ljung, G. M. (1993). On outlier detection in time series, *Journal of the Royal Statistical Society, Series B,* **55,** 559–567.

Ljung, G. M., and Box, G. E. P. (1979). The likelihood function of stationary autoregressive–moving average models. *Biometrika,* **66,** 265–270.

Ljung, L., Gustavsson, I., and Soderstrom, T. (1974), Identification of linear multivariable systems operating under linear feedback control. *IEEE Transaction on Automatic Control,* **19,** 836–840.

Lorenzen, T. J., and Vance, L. C. (1986). The economic design of control chart: a unified approach. *Technometrics,* **28,** 3–10.

Lucas, J. M. (1982). Combined Shewhart–Cusum quality control schemes. *Journal of Quality Technology,* **14,** 51–59.

Lucas, J. M., and Crosier, R. B. (1982). Fast initial response for Cusum quality-control schemes: give your Cusum a head start. *Technometrics,* **24,** 196–206.

Lucas, J. M., and Saccucci, M. S. (1990). Exponentially weighted moving average control schemes: properties and enhancements. *Technometrics,* **32,** 1–12.

Luceño, A. (1993). A fast algorithm for the repeated evaluation of the likelihood of a general linear process for long series. *Journal of the American Statistical Association,* **88,** 229–236.

Luceño, A. (1994a). Fast optimization of the exact likelihood of AR and ARMA processes. *Computational Statistics & Data Analysis,* **17,** 51–63.

Luceño, A. (1994b). A fast algorithm for the exact likelihood of stationary and partially nonstationary vector autoregressive–moving average processes. *Biometrika,* **81,** 555–565.

Luceño, A. (1995a). A family of partially correlated Poisson models for overdispersion. *Computational Statistics & Data Analysis,* **20,** 511–520.

Luceño, A. (1995b). Choosing the EWMA parameter in engineering process control. *Journal of Quality Technology,* **27,** 162–168.

Luceño, A. (1996a). A process capability index with reliable confidence intervals. *Communications in Statistics, Simulation and Computation,* **25,** 235–245.

Luceño, A.(1996b). A generalized Erlang distribution showing overdispersions. *Statistics and Probability Letters,* **28,** 375–386.

Luceño, A. (1996c). A fast likelihood approximation for vector general linear processes with long series: application to fractional differencing. *Biometrika,* **83,** 603–614.

Luceño, A. (1997a). Parameter estimation with closed-loop operating data under time varying discrete proportional-integral control. *Communications in Statistics, Simulation and Computation,* **26,** 215–232.

Luceño, A. (1997b). Estimation of missing values in possibly partially nonstationary vector time series, *Biometrika,* **84,** 495–499.

Luceño, A. (1997c). Detecting possibly nonconsecutive outliers in industrial time series. *Journal of the Royal Statistical Society, Series B,* **59.**

Luceño, A., and de Ceballos, F. (1995). Describing extra-binomial variation with partially correlated models. *Communications in Statistics, Theory and Methods,* **24,** 1637–1653.

Luceño, A., González, F. J., and Puig-Pey, J. (1996). Computing optimal adjustment schemes for the general tool-wear problem. *Journal of Statistical Computation and Simulation,* **54,** 87–113.

MacGregor, J. F. (1972). *Topics in the Control of Linear Processes Subject to Stochastic Disturbances.* Unpublished Ph.D. thesis, University of Wisconsin–Madison.

MacGregor, J. F. (1976). Optimal choice of the sampling interval for discrete process control. *Technometrics,* **18,** 151–160.

MacGregor, J. F. (1987). Interfaces between process control and on-line statistical process control. *Computing and Systems Technology Division Communications,* **10,** 9–20.

MacGregor, J. F. (1988). On-line statistical process control. *Chemical Engineering Progress,* **84,** 21–31.

MacGregor, J. F., and Wong, A. K. (1980). Multivariate model identification and stochastic control on a chemical reactor. *Technometrics,* **22,** 453–464.

Mayr, O. (1970). *The Origins of Feedback Control.* M.I.T. Press, Cambridge, MA.

McCullagh, P., and Nelder, J. (1989). *Generalized Linear Models.* Chapman and Hall, London.

McKenzie, E. (1988). Some ARMA models for dependent sequences for Poisson counts. *Advances in Applied Probability,* **20,** 822–835.

Mélard, G. (1984). Algorithm AS 197. A fast algorithm for the exact likelihood of autoregressive–moving average models. *Applied Statistics,* **33,** 104–114.

Montgomery, D. C. (1997). *Introduction to Statistical Quality Control,* 3rd ed. Wiley, New York.

Montgomery, D. C., Keats, B. J., Runger, G. C., and Messina, W. S. (1994). Integrating statistical process control and engineering process control. *Journal of Quality Technology,* **26,** 79–87.

Montgomery, D. C., and Mastrangelo, C. M. (1991). Some statistical process methods for autocorrelated data. *Journal of Quality Technology,* **23,** 179–193.

Montgomery, D. C., and Woodall, W. H. (1997). A discussion of statistically-based process monitoring and control. *Journal of Quality Technology,* **29,** 121–162.

Moore, D. F. (1987). Modeling the extraneous variance in the presence of extra-binomial variation. *Applied Statistics,* **36,** 8–14.

Muth, J. F. (1960). Optimal properties of exponentially weighted forecasts of time series with permanent and transitory components. *Journal of the American Statistical Association,* **55,** 299–306.

Newbold, P. (1974). The exact likelihood function for a mixed autoregressive–moving average process. *Biometrika,* **61,** 423–426.

Nelson, L. S. (1985). Interpreting Shewhart $\overline{X}$ control charts. *Journal of Quality Technology,* **17,** 114–116.

Nicholls, D. F., and Hall, A. D. (1979). The exact likelihood function of multivariate autoregressive–moving average models. *Biometrika,* **66,** 259–264.

Ott, E. R. (1975). *Process Quality Control.* McGraw Hill, New York.

Page, E. S. (1954). Continuous inspection schemes. *Biometrika,* **41,** 100–114.

Page, E. S. (1957). On problems in which a change in a parameter occurs at an unknown point. *Biometrika,* **44,** 248–252.

Page, E. S. (1961). Cumulative sum charts. *Technometrics,* **3,** 1–9.

Palm, A. C., Rodriguez, R. N., Spiring, F. A., and Wheeler, D. J. (1997). Some perspectives and challenges for control charts methods. *A Discussion on Statistically-Based Process Monitoring and Control.* D. C. Montgomery and W. H. Woodall (eds.). *Journal of Quality Technology,* **29,** 122–127.

Pearlman, J. G. (1980). An algorithm for the exact likelihood of a high-order autoregressive–moving average process. *Biometrika,* **67,** 232–233.

Peña, D. (1987). Measuring the importance of outliers in ARIMA models. *New Perspectives In Theoretical And Applied Statistics.* Puri et al. (ed.). pp. 109–118.

Peña D. and Tiao, G. C. (1991). A note on likelihood estimation of missing values in time series. *The American Statistician,* **45,** 212–214.

Pitt, H. (1994). *SPC for the Rest of Us.* Addison-Wesley, Reading, MA.

Prentice, R. L. (1986). Binary regression using an extended beta-binomial distribution, with discussion of correlation induced by covariate measurement error. *Journal of the American Statistical Association,* **81,** 321–327.

Prett, D. M., and Garcia, C. E. (1988). *Fundamental Process Control.* Boston, Butterworths.

Ramírez, J. G. (1989). *Sequential Methods in Statistical Process Monitoring.* Unpublished Ph.D. thesis, University of Wisconsin–Madison.

Ramírez, J. G., and Cantell, B. (1997). An analysis of a semiconductor experiment using yield and spatial information. *Quality and Reliability Engineering International,* **13,** 35–46.

Randall, S. C., Ramírez, J. G., and Taam W. (1996). Process monitoring in integrated circuit fabrication using both yield and spatial statistics. *Quality and Reliability Engineering International,* **12,** 195—202.

Roberts, S. W. (1959). Control chart tests based on geometric moving averages. *Technometrics,* **1,** 239–250.

Ryan, T. P. (1989). *Statistical Methods for Quality Improvement.* Wiley, New York.

Shea, B. L. (1987). Estimation of multivariate time series. *Journal of Time Series Analysis,* **8,** 95–109.

Shea, B. L. (1989). Algorithm, AS 242. The exact likelihood of a vector autoregressive–moving average model. *Applied Statistics,* **38,** 161–204.

Shewhart, W. A. (1931). Economic control of quality of manufacturing product. Van Nostrand Reinhold, Princeton, NJ.

Srivastava, M. S. (1995). Robustness of control procedures for integrated moving average process of order one. Technical Report 9501, Department of Statistics, University of Toronto.

Taguchi, G. (1981). *On-line Quality Control During Production.* Japanese Standard Association. Tokyo, Japan.

Taguchi, G. (1986). *Introduction to Quality Engineering: Designing Quality into Products and Processes.* Asian Productivity Organization. Kraus International Publications, White Plains, NY.

Taguchi, G. (1987). *System of Experimental Design,* Vol. 2. Unipub, Kraus International Publications, White Plains, NY.

Taguchi. G., Elsayed, E. A., and Hsiang, T. (1989). *Quality Engineering in Production Systems.* McGraw-Hill, New York.

Tarone, R. E. (1979). Testing the goodness of fit of the binomial distribution. *Biometrika,* **66,** 585–590.

Tiao, G. C. (1972). Asymptotic behavior of temporal aggregates of time series. *Biometrika,* **59,** 525–531.

Tsay, R. S., and Tiao, G. C. (1984). Consistent estimates of autoregressive parameters and extended sample autocorrelation function for stationary and non-stationary ARMA models. *Journal of the American Statistical Association,* **79,** 84–96.

Tsay, R. S., and Tiao, G. C. (1985). Use of canonical analysis in time series model identification. *Biometrika,* **72,** 299–315.

Tucker, W. T., Faltin, F. W., and Vander Wiel, S. A. (1993). Algorithmic statistical process control: an elaboration. *Technometrics,* **35,** 363–375.

Tunnicliffe-Wilson G. (1970). *Modeling Linear Systems for Multivariate Control.* Unpublished Ph.D. thesis, Department of System Engineering, University of Lancaster, England.

Vander Wiel, S. A., (1996). Monitoring processes that wonder using integrated moving average models. *Technometrics,* **38,** 139–151.

Vander Wiel, S. A., and Tucker, W. T. (1989). Algorithmic statistical process control: literature review, implementation, and research opportunities. Management Science and Statistical Program, Corporate Research and Development, General Electric Company.

Vander Wiel, S. A., Tucker, W. T., Faltin, F. W., and Doganaksoy, N. (1992). Algorithmic statistical process control: concepts and application. *Technometrics,* **34,** 286–297.

Wadsworth, H. M., Stephens, K. S., and Godfrey, A. B. (1986). *Modern Methods for Quality Control and Improvement.* Wiley, New York.

Wardell, D. G., Moskowitz, H., and Plante, R. D. (1994). Run length distributions of special-cause control charts for correlated processes. *Technometrics,* **36,** 3–17.

Western Electric (1956). *Statistical Quality Control Handbook.* Western Electric Corporation, Indianapolis, IN,

Wheeler, D. J. and Chambers, D. S. (1992). *Understanding Statistical Process Control.* SPC Press, Knoxville, TN.

Whittle, P. (1963). *Prediction and Regulation by Linear Least-Squares Methods.* English University Press, London.

Williams, D. A. (1975). The analysis of binary responses from toxicological experiments involving reproduction and teratogenicity. *Biometrics,* **31,** 949–952.

Williams, D. A. (1982). Extra-binomial variation in logistic linear models. *Applied Statistics,* **31,** 144–148

Wincek, M. A., and Reinsel, G. C. (1986). An exact maximum likelihood estimation procedure for regression-ARMA time series models with possibly nonconsecutive data. *Journal of the Royal Statistical Society, Series B,* **48,** 303–313.

Woodall, W. H. (1983). The distribution of the run length of one-sided Cusum procedures for continuous random variables. *Technometrics,* **25,** 295–301.

Woodall, W. H. (1984). On the Markov chain approach to the two-sided Cusum procedure. *Technometrics,* **26,** 41–46.

Woodall, W. H., and Adams, B. M. (1993). The statistical design of Cusum charts. *Quality Engineering,* **5,** 559–570.

Woodward, W. A. and Gray, H. L. (1981). On the relationship between the S array and the Box–Jenkins method of ARMA model identification. *Journal of the American Statistical Association,* **76,** 579–587.

Wu, L. S.-Y., Hosking, J. R. M., and Ravishanker, N. (1993). Reallocation outliers in time series. *Applied Statistics,* **42,** 301-313.

Yashchin, E. (1993). Performance of Cusum control schemes for serially correlated observations. *Technometrics,* **35,** 37–52.

Zeger, S. (1988). A regression model for time series of counts. *Biometrika,* **75,** 621–629.

INDEX

WILEY SERIES IN PROBABILITY AND STATISTICS

ESTABLISHED BY WALTER A. SHEWHART AND SAMUEL S. WILKS

Probability and Statistics

ANDERSON · An Introduction to Multivariate Statistical Analysis, *Second Edition*
*ANDERSON · The Statistical Analysis of Time Series
ARNOLD, BALAKRISHNAN, and NAGARAJA · A First Course in Order Statistics
BACCELLI, COHEN, OLSDER, and QUADRAT · Synchronization and Linearity: An Algebra for Discrete Event Systems
BARTOSZYNSKI and NIEWIADOMSKA-BUGAJ · Probability and Statistical Inference
BERNARDO and SMITH · Bayesian Statistical Concepts and Theory
BHATTACHARYYA and JOHNSON · Statistical Concepts and Methods
BILLINGSLEY · Convergence of Probability Measures
BILLINGSLEY · Probability and Measure, *Second Edition*
BOROVKOV · Asymptotic Methods in Queuing Theory
BRANDT, FRANKEN, and LISEK · Stationary Stochastic Models
CAINES · Linear Stochastic Systems
CAIROLI and DALANG · Sequential Stochastic Optimization
CHEN · Recursive Estimation and Control for Stochastic Systems
CONSTANTINE · Combinatorial Theory and Statistical Design
COOK and WEISBERG · An Introduction to Regression Graphics
COVER and THOMAS · Elements of Information Theory
CSÖRGŐ and HORVÁTH · Weighted Approximations in Probability Statistics
*DOOB · Stochastic Processes
DUDEWICZ and MISHRA · Modern Mathematical Statistics
DUPUIS and ELLIS · A Weak Convergence Approach to the Theory of Large Deviations
ETHIER and KURTZ · Markov Processes: Characterization and Convergence
FELLER · An Introduction to Probability Theory and Its Applications, Volume 1, *Third Edition,* Revised; Volume II, *Second Edition*
FREEMAN and SMITH · Aspects of Uncertainty: A Tribute to D. V. Lindley
FULLER · Introduction to Statistical Time Series, *Second Edition*
FULLER · Measurement Error Models
GHOSH, MUKHOPADHYAY, and SEN · Sequential Estimation
GIFI · Nonlinear Multivariate Analysis
GUTTORP · Statistical Inference for Branching Processes
HALD · A History of Probability and Statistics and Their Applications before 1750
HALL · Introduction to the Theory of Coverage Processes
HANNAN and DEISTLER · The Statistical Theory of Linear Systems
HEDAYAT and SINHA · Design and Inference in Finite Population Sampling
HOEL · Introduction to Mathematical Statistics, *Fifth Edition*
HUBER · Robust Statistics
IMAN and CONOVER · A Modern Approach to Statistics
JOHNSON and KOTZ · Leading Personalities in Statistical Sciences: From the Seventeenth Century to the Present

*Now available in a lower priced paperback edition in the Wiley Classics Library.

Probability and Statistics (Continued)

JUREK and MASON · Operator-Limit Distributions in Probability Theory
KASS and VOS · Geometrical Foundations of Asymptotic Inference
KAUFMAN and ROUSSEEUW · Finding Groups in Data: An Introduction to Cluster Analysis
LAMPERTI · Probability: A Survey of the Mathematical Theory, *Second Edition*
LARSON · Introduction to Probability Theory and Statistical Inference, *Third Edition*
LESSLER and KALSBEEK · Nonsampling Error in Surveys
LINDVALL · Lectures on the Coupling Method
McFADDEN · Management of Data in Clinical Trials
MANTON, WOODBURY, and TOLLEY · Statistical Applications Using Fuzzy Sets
MARDIA · The Art of Statistical Science: A Tribute to G. S. Watson
MORGENTHALER and TUKEY · Configural Polysampling: A Route to Practical Robustness
MUIRHEAD · Aspects of Multivariate Statistical Theory
OLIVER and SMITH · Influence Diagrams, Belief Nets and Decision Analysis
*PARZEN · Modern Probability Theory and Its Applications
PRESS · Bayesian Statistics: Principles, Models, and Applications
PUKELSHEIM · Optimal Experimental Design
PURI and SEN · Nonparametric Methods in General Linear Models
PURI, VILAPLANA, and WERTZ · New Perspectives in Theoretical and Applied Statistics
RAO · Asymptotic Theory of Statistical Inference
RAO · Linear Statistical Inference and Its Applications, *Second Edition*
*RAO and SHANBHAG · Choquet-Deny Type Functional Equations with Applications to Stochastic Models
RENCHER · Methods of Multivariate Analysis
ROBERTSON, WRIGHT, and DYKSTRA · Order Restricted Statistical Inference
ROGERS and WILLIAMS · Diffusions, Markov Processes, and Martingales, Volume I: Foundations, *Second Edition;* Volume II: Îto Calculus
ROHATGI · An Introduction to Probability Theory and Mathematical Statistics
ROSS · Stochastic Processes
RUBINSTEIN · Simulation and the Monte Carlo Method
RUBINSTEIN and SHAPIRO · Discrete Event Systems: Sensitivity Analysis and Stochastic Optimization by the Score Function Method
RUZSA and SZEKELY · Algebraic Probability Theory
SCHEFFE · The Analysis of Variance
SEBER · Linear Regression Analysis
SEBER · Multivariate Observations
SEBER and WILD · Nonlinear Regression
SERFLING · Approximation Theorems of Mathematical Statistics
SHORACK and WELLNER · Empirical Processes with Applications to Statistics
SMALL and McLEISH · Hilbert Space Methods in Probability and Statistical Inference
STAPLETON · Linear Statistical Models
STAUDTE and SHEATHER · Robust Estimation and Testing
STOYANOV · Counterexamples in Probability
STYAN · The Collected Papers of T. W. Anderson: 1943–1985
TANAKA · Time Series Analysis: Nonstationary and Noninvertible Distribution Theory
THOMPSON and SEBER · Adaptive Sampling
WELSH · Aspects of Statistical Inference
WHITTAKER · Graphical Models in Applied Multivariate Statistics
YANG · The Construction Theory of Denumerable Markov Processes

*Now available in a lower priced paperback edition in the Wiley Classics Library.

Applied Probability and Statistics

ABRAHAM and LEDOLTER · Statistical Methods for Forecasting
AGRESTI · Analysis of Ordinal Categorical Data
AGRESTI · Categorical Data Analysis
AGRESTI · An Introduction to Categorical Data Analysis
ANDERSON and LOYNES · The Teaching of Practical Statistics
ANDERSON, AUQUIER, HAUCK, OAKES, VANDAELE, and WEISBERG · Statistical Methods for Comparative Studies
ARMITAGE and DAVID (editors) · Advances in Biometry
*ARTHANARI and DODGE · Mathematical Programming in Statistics
ASMUSSEN · Applied Probability and Queues
*BAILEY · The Elements of Stochastic Processes with Applications to the Natural Sciences
BARNETT and LEWIS · Outliers in Statistical Data, *Second Edition*
BARTHOLOMEW, FORBES, and McLEAN · Statistical Techniques for Manpower Planning, *Second Edition*
BATES and WATTS · Nonlinear Regression Analysis and Its Applications
BECHHOFER, SANTNER, and GOLDSMAN · Design and Analysis of Experiments for Statistical Selection, Screening, and Multiple Comparisons
BELSLEY · Conditioning Diagnostics: Collinearity and Weak Data in Regression
BELSLEY, KUH, and WELSCH · Regression Diagnostics: Identifying Influential Data and Sources of Collinearity
BERRY · Bayesian Analysis in Statistics and Econometrics: Essays in Honor of Arnold Zellner
BERRY, CHALONER, and GEWEKE · Bayesian Analysis in Statistics and Econometrics: Essays in Honor of Arnold Zellner
BHAT · Elements of Applied Stochastic Processes, *Second Edition*
BHATTACHARYA and WAYMIRE · Stochastic Processes with Applications
BIEMER, GROVES, LYBERG, MATHIOWETZ, and SUDMAN · Measurement Errors in Surveys
BIRKES and DODGE · Alternative Methods of Regression
BLOOMFIELD · Fourier Analysis of Time Series: An Introduction
BOLLEN · Structural Equations with Latent Variables
BOULEAU · Numerical Methods for Stochastic Processes
BOX · R. A. Fisher, the Life of a Scientist
BOX and DRAPER · Empirical Model-Building and Response Surfaces
BOX and DRAPER · Evolutionary Operation: A Statistical Method for Process Improvement
BOX, HUNTER, and HUNTER · Statistics for Experimenters: An Introduction to Design, Data Analysis, and Model Building
BOX and LUCENO · Statistical Control by Monitoring and Feedback Adjustment
BROWN and HOLLANDER · Statistics: A Biomedical Introduction
BUCKLEW · Large Deviation Techniques in Decision, Simulation, and Estimation
BUNKE and BUNKE · Nonlinear Regression, Functional Relations and Robust Methods: Statistical Methods of Model Building
CHATTERJEE and HADI · Sensitivity Analysis in Linear Regression
CHATTERJEE and PRICE · Regression Analysis by Example, *Second Edition*
CLARKE and DISNEY · Probability and Random Processes: A First Course with Applications, *Second Edition*
COCHRAN · Sampling Techniques, *Third Edition*
*COCHRAN and COX · Experimental Designs, *Second Edition*
CONOVER · Practical Nonparametric Statistics, *Second Edition*
CONOVER and IMAN · Introduction to Modern Business Statistics

*Now available in a lower priced paperback edition in the Wiley Classics Library.

Applied Probability and Statistics (Continued)

CORNELL · Experiments with Mixtures, Designs, Models, and the Analysis of Mixture Data, *Second Edition*
COX · A Handbook of Introductory Statistical Methods
*COX · Planning of Experiments
COX, BINDER, CHINNAPPA, CHRISTIANSON, COLLEDGE, and KOTT · Business Survey Methods
CRESSIE · Statistics for Spatial Data, *Revised Edition*
DANIEL · Applications of Statistics to Industrial Experimentation
DANIEL · Biostatistics: A Foundation for Analysis in the Health Sciences, *Sixth Edition*
DAVID · Order Statistics, *Second Edition*
*DEGROOT, FIENBERG, and KADANE · Statistics and the Law
*DEMING · Sample Design in Business Research
DETTE and STUDDEN · The Theory of Canonical Moments with Applications in Statistics, Probability, and Analysis
DILLON and GOLDSTEIN · Multivariate Analysis: Methods and Applications
DODGE and ROMIG · Sampling Inspection Tables, *Second Edition*
DOWDY and WEARDEN · Statistics for Research, *Second Edition*
DRAPER and SMITH · Applied Regression Analysis, *Second Edition*
DUNN · Basic Statistics: A Primer for the Biomedical Sciences, *Second Edition*
DUNN and CLARK · Applied Statistics: Analysis of Variance and Regression, *Second Edition*
ELANDT-JOHNSON and JOHNSON · Survival Models and Data Analysis
EVANS, PEACOCK, and HASTINGS · Statistical Distributions, *Second Edition*
FISHER and VAN BELLE · Biostatistics: A Methodology for the Health Sciences
FLEISS · The Design and Analysis of Clinical Experiments
FLEISS · Statistical Methods for Rates and Proportions, *Second Edition*
FLEMING and HARRINGTON · Counting Processes and Survival Analysis
FLURY · Common Principal Components and Related Multivariate Models
GALLANT · Nonlinear Statistical Models
GLASSERMAN and YAO · Monotone Structure in Discrete-Event Systems
GNANADESIKAN · Methods for Statistical Data Analysis of Multivariate Observations, *Second Edition*
GREENWOOD and NIKULIN · A Guide to Chi-Squared Testing
GROSS and HARRIS · Fundamentals of Queueing Theory, *Second Edition*
GROVES · Survey Errors and Survey Costs
GROVES, BIEMER, LYBERG, MASSEY, NICHOLLS, and WAKSBERG · Telephone Survey Methodology
HAHN and MEEKER · Statistical Intervals: A Guide for Practitioners
HAND · Discrimination and Classification
*HANSEN, HURWITZ, and MADOW · Sample Survey Methods and Theory, Volume 1: Methods and Applications
*HANSEN, HURWITZ, and MADOW · Sample Survey Methods and Theory, Volume II: Theory
HEIBERGER · Computation for the Analysis of Designed Experiments
HELLER · MACSYMA for Statisticians
HINKELMAN and KEMPTHORNE: · Design and Analysis of Experiments, Volume 1: Introduction to Experimental Design
HOAGLIN, MOSTELLER, and TUKEY · Exploratory Approach to Analysis of Variance
HOAGLIN, MOSTELLER, and TUKEY · Exploring Data Tables, Trends and Shapes
HOAGLIN, MOSTELLER, and TUKEY · Understanding Robust and Exploratory Data Analysis
HOCHBERG and TAMHANE · Multiple Comparison Procedures
HOCKING · Methods and Applications of Linear Models: Regression and the Analysis of Variables

*Now available in a lower priced paperback edition in the Wiley Classics Library.

Applied Probability and Statistics (Continued)

HOEL · Elementary Statistics, *Fifth Edition*
HOGG and KLUGMAN · Loss Distributions
HOLLANDER and WOLFE · Nonparametric Statistical Methods
HOSMER and LEMESHOW · Applied Logistic Regression
HØYLAND and RAUSAND · System Reliability Theory: Models and Statistical Methods
HUBERTY · Applied Discriminant Analysis
IMAN and CONOVER · Modern Business Statistics
JACKSON · A User's Guide to Principle Components
JOHN · Statistical Methods in Engineering and Quality Assurance
JOHNSON · Multivariate Statistical Simulation
JOHNSON and BALAKRISHNAN · Advances in the Theory and Practice of Statistics: A Volume in Honor of Samuel Kotz
JOHNSON and KOTZ · Distributions in Statistics
Continuous Multivariate Distributions
JOHNSON, KOTZ, and BALAKRISHNAN · Continuous Univariate Distributions, Volume 1, *Second Edition*
JOHNSON, KOTZ, and BALAKRISHNAN · Continuous Univariate Distributions, Volume 2, *Second Edition*
JOHNSON, KOTZ, and BALAKRISHNAN · Discrete Multivariate Distributions
JOHNSON, KOTZ, and KEMP · Univariate Discrete Distributions, *Second Edition*
JUDGE, GRIFFITHS, HILL, LÜTKEPOHL, and LEE · The Theory and Practice of Econometrics, *Second Edition*
JUDGE, HILL, GRIFFITHS, LÜTKEPOHL, and LEE · Introduction to the Theory and Practice of Econometrics, *Second Edition*
JUREČKOVÁ and SEN · Robust Statistical Procedures: Aymptotics and Interrelations
KADANE · Bayesian Methods and Ethics in a Clinical Trial Design
KADANE AND SCHUM · A Probabilistic Analysis of the Sacco and Vanzetti Evidence
KALBFLEISCH and PRENTICE · The Statistical Analysis of Failure Time Data
KASPRZYK, DUNCAN, KALTON, and SINGH · Panel Surveys
KISH · Statistical Design for Research
*KISH · Survey Sampling
LAD · Operational Subjective Statistical Methods: A Mathematical, Philosophical, and Historical Introduction
LANGE, RYAN, BILLARD, BRILLINGER, CONQUEST, and GREENHOUSE · Case Studies in Biometry
LAWLESS · Statistical Models and Methods for Lifetime Data
LEBART, MORINEAU., and WARWICK · Multivariate Descriptive Statistical Analysis: Correspondence Analysis and Related Techniques for Large Matrices
LEE · Statistical Methods for Survival Data Analysis, *Second Edition*
LePAGE and BILLARD · Exploring the Limits of Bootstrap
LEVY and LEMESHOW · Sampling of Populations: Methods and Applications
LINHART and ZUCCHINI · Model Selection
LITTLE and RUBIN · Statistical Analysis with Missing Data
LYBERG · Survey Measurement
MAGNUS and NEUDECKER · Matrix Differential Calculus with Applications in Statistics and Econometrics
MAINDONALD · Statistical Computation
MALLOWS · Design, Data, and Analysis by Some Friends of Cuthbert Daniel
MANN, SCHAFER, and SINGPURWALLA · Methods for Statistical Analysis of Reliability and Life Data
MASON, GUNST, and HESS · Statistical Design and Analysis of Experiments with Applications to Engineering and Science
McLACHLAN and KRISHNAN · The EM Algorithm and Extensions

*Now available in a lower priced paperback edition in the Wiley Classics Library.

Applied Probability and Statistics (Continued)

McLACHLAN · Discriminant Analysis and Statistical Pattern Recognition
McNEIL · Epidemiological Research Methods
MILLER · Survival Analysis
MONTGOMERY and MYERS · Response Surface Methodology: Process and Product in Optimization Using Designed Experiments
MONTGOMERY and PECK · Introduction to Linear Regression Analysis, *Second Edition*
NELSON · Accelerated Testing, Statistical Models, Test Plans, and Data Analyses
NELSON · Applied Life Data Analysis
OCHI · Applied Probability and Stochastic Processes in Engineering and Physical Sciences
OKABE, BOOTS, and SUGIHARA · Spatial Tesselations: Concepts and Applications of Voronoi Diagrams
OSBORNE · Finite Algorithms in Optimization and Data Analysis
PANKRATZ · Forecasting with Dynamic Regression Models
PANKRATZ · Forecasting with Univariate Box-Jenkins Models: Concepts and Cases
PIANTADOSI · Clinical Trials: A Methodologic Perspective
PORT · Theoretical Probability for Applications
PUTERMAN · Markov Decision Processes: Discrete Stochastic Dynamic Programming
RACHEV · Probability Metrics and the Stability of Stochastic Models
RÉNYI · A Diary on Information Theory
RIPLEY · Spatial Statistics
RIPLEY · Stochastic Simulation
ROSS · Introduction to Probability and Statistics for Engineers and Scientists
ROUSSEEUW and LEROY · Robust Regression and Outlier Detection
RUBIN · Multiple Imputation for Nonresponse in Surveys
RYAN · Modern Regression Methods
RYAN · Statistical Methods for Quality Improvement
SCHOTT · Matrix Analysis for Statistics
SCHUSS · Theory and Applications of Stochastic Differential Equations
SCOTT · Multivariate Density Estimation: Theory, Practice, and Visualization
*SEARLE · Linear Models
SEARLE · Linear Models for Unbalanced Data
SEARLE · Matrix Algebra Useful for Statistics
SEARLE, CASELLA, and McCULLOCH · Variance Components
SKINNER, HOLT, and SMITH · Analysis of Complex Surveys
STOYAN, KENDALL, and MECKE · Stochastic Geometry and Its Applications, *Second Edition*
STOYAN and STOYAN · Fractals, Random Shapes and Point Fields: Methods of Geometrical Statistics
THOMPSON · Empirical Model Building
THOMPSON · Sampling
TIERNEY · LISP-STAT: An Object-Oriented Environment for Statistical Computing and Dynamic Graphics
TIJMS · Stochastic Modeling and Analysis: A Computational Approach
TITTERINGTON, SMITH, and MAKOV · Statistical Analysis of Finite Mixture Distributions
UPTON and FINGLETON · Spatial Data Analysis by Example, Volume 1: Point Pattern and Quantitative Data
UPTON and FINGLETON · Spatial Data Analysis by Example, Volume II: Categorical and Directional Data
VAN RIJCKEVORSEL and DE LEEUW · Component and Correspondence Analysis
WEISBERG · Applied Linear Regression, *Second Edition*
WESTFALL and YOUNG · Resampling-Based Multiple Testing: Examples and Methods for p-Value Adjustment

*Now available in a lower priced paperback edition in the Wiley Classics Library.

Applied Probability and Statistics (Continued)

WHITTLE · Optimization Over Time: Dynamic Programming and Stochastic Control, Volume I and Volume II
WHITTLE · Systems in Stochastic Equilibrium
WONNACOTT and WONNACOTT · Econometrics, *Second Edition*
WONNACOTT and WONNACOTT · Introductory Statistics, *Fifth Edition*
WONNACOTT and WONNACOTT · Introductory Statistics for Business and Economics, *Fourth Edition*
WOODING · Planning Pharmaceutical Clinical Trials: Basic Statistical Principles
WOOLSON · Statistical Methods for the Analysis of Biomedical Data
*ZELLNER · An Introduction to Bayesian Inference in Econometrics

Tracts on Probability and Statistics

BILLINGSLEY · Convergence of Probability Measures
KELLY · Reversability and Stochastic Networks
TOUTENBURG · Prior Information in Linear Models

*Now available in a lower priced paperback edition in the Wiley Classics Library.